Heaven and Earth in Ancient Greek Cosmology

Astrophysics and Space Science Library

Dirk L. Couprie

Heaven and Earth in Ancient Greek Cosmology

From Thales to Heraclides Ponticus

Springer

Dirk L. Couprie
St. Maartenslaan 51d
6221AW Maastricht
Netherlands

ISBN 978-1-4419-8115-8 e-ISBN 978-1-4419-8116-5
DOI 10.1007/978-1-4419-8116-5
Springer New York Dordrecht Heidelberg London

Library of Congress Control Number: 2011923655

Printed on acid-free paper

Springer is part of Springer Science+Business Media (www.springer.com)

Dedicated to the memory of my father
Leendert Couprie
September 20, 1896 – September 16, 1983
who showed me the panorama of the celestial vault

MOTHER
THERE'S SOMETHING
WRONG WITH THE SKY

RADIANT THE OCEAN SKY
SERENITY AND BLIGHT

graffiti by 

somewhere in Amsterdam

Contents

List of Figures

The picture on the front cover is a detail of the triptych *The Seven Wonders of the World* by Hans Exterkate. It is a pastiche of a well-known photograph of two Bornean tribesman with their gnomon. In the two figures one recognizes the artist and the author (left).

List of Tables

Foreword

Early Greek cosmology is a very difficult subject to deal with. First, the evidence we possess is fragmentary. Second, one cannot easily imagine what could be rational, but still mistaken, scientific theories – theories based on facts and reasoning, on the one hand, but quite different from modern views, on the other hand. Their reconstruction requires both sophisticated competence and compassion, even admiration for the early thinkers. But reconstruction is very well worth trying because all theoretical science has its beginning in early Greek cosmology. In this book, Dirk L. Couprie presents his efforts at clarifying the views of the pioneers of theoretical cosmology. It covers the crucial period from about the middle of the sixth until the middle of the fourth century B.C., with its focus on the magnificent figure of Anaximander. It is a sincere, modestly written, in no way pretentious, but clear book. Dirk Couprie invariably approaches opinions he criticizes with both due respect and straightforwardness. The author is particularly strong in visualizing the reconstructed views of the heaven and earth; he has no rivals in that among his predecessors. Despite various scholarly publications that appeared over the past century and a half, the history of early Greek cosmology is still a field in development. The book by Dirk Couprie constitutes an important and in several respects indispensable contribution to this field.

St. Petersburg State University — Dmitri Panchenko

Preface and Acknowledgements

This book is about the origin of our Western world-picture in Greek cosmology. The late Cornelis Verhoeven from the University of Amsterdam guided my first steps in the study of Anaximander, which in 1989 resulted in my doctoral dissertation on the translation and interpretation of the text of his only surviving fragment. At that time, I had the opportunity to discuss my earliest thoughts about Anaximander's cosmology with the also late Jan van Paradijs, who was, being a famous astronomer himself, interested in ancient astronomy as well. With Robert Hahn from Southern Illinois University Carbondale, I have had intensive discussions for many years, by email and during visits, on the understanding of Anaximander's cosmology. The fact that I disagree with a good deal of the tenor of his work does not diminish in the least the importance of this contact. I am especially grateful that he scrutinized the whole manuscript and made a lot of wise and helpful comments. This manuscript was already finished when his latest book, *Archaeology and the Origins of Philosophy*, appeared, on which I have been able, therefore, to insert only a few incidental remarks. For an extensive review, see Couprie (2010). Less frequent, but always stimulating, were the personal and email contacts with Gerard Naddaf from York University, Toronto. The book that the three of us wrote about Anaximander in 2003 has added much to my insights in ancient Greek cosmology. The exchange of thoughts by email with István Bodnár from Eötvös University, Budapest, Andrew Gregory from the University College London, and Carlo Rovelli from the Centre de Physique Théorique de Luminy, were very fruitful. The same holds especially for Dmitri Panchenko from the State University of St. Petersburg, who was so kind as to write a Foreword to my book. The contacts with Patricia O'Grady from Flinders University of South Australia at Adelaide were inspiring, although I fundamentally disagree with her interpretation of Thales. F. Richard Stephenson from the University of Durham kindly placed at my disposal his computer computations, based on the most recent insights, of some relevant solar eclipses. I was lucky to have had a couple of discussions with Teije de Jong from the University of Amsterdam on the prediction of eclipses as well as on the measuring of the size of the sun on a flat earth. I had the opportunity to discuss several items with other members of the Utrecht Study Group for the History of Astronomy (USHA) as well. W.H. von Soldt from the University of Leiden drew my attention to the text of a cuneiform tablet on the distances of celestial bodies. B.J. Mansvelt Beck, also from the University of Leiden, kindly answered my questions on Chinese astronomy. The same holds for J.J. de Jong as regards the measurements of columns and column drums. Olaf Kaper from Leiden University was so kind as to help me with the translation of a hieroglyphic text, although on a vital point I was so stubborn as not to follow his advice. With my wife Heleen Pott (Erasmus University of Rotterdam), together with whom I have written a couple of articles, I discussed continually every thought incorporated in this book. Ton Verschoor and Liesbeth van der Sluijs read earlier versions and offered many useful comments. My brothers Jan and Leendert Couprie helped me with the calculations in Chaps. 15 and 16. If not otherwise indicated, the pictures are drawn by myself. Hans Exterkate not only meticulously produced the illustrations for several figures, but also allowed me to include the beautiful picture of Anaximander teaching (detail of his triptych *The Seven Wonders of the World*) in the Introduction. The picture on the front cover shows another detail of the same triptych. Ton Lecluse was so kind as to make the figure of a dodecahedron inscribed in a sphere in Chap. 17 according to my directions. The acknowledgments for permission to print the other pictures are mentioned in the captions of figures. I am grateful to the editors and publishers of *Apeiron*, *Hyperboreus*, *Early Science and Medicine*, and *Greek Research in Australia*, who allowed me to make use of (parts of) formerly published articles. Where needed I have updated and (sometimes strongly) adapted the texts. I also took the opportunity to redress misunderstandings that some passages and drawings have given rise to.

Special thanks to my physical therapist Jos ("Jesus") Kremer, who cured me from the serious after effects of a whiplash. Without his healing hands, I would not have been able to finish this book.

Last but not least, I am very glad to mention Zdeněk Kratochvíl, Radim Kočandrle, and Vojtěch Hladký of the Charles University in Prague, who invited me to give a series of lectures on Anaximander and ancient Greek cosmology in June 2009. This visit offered me the opportunity to discuss at length the gist of this study as well as many details with them and with their students and colleagues.

Introduction

One of the dearest memories of my childhood is that my father took my brother and me with him to the Zeiss planetarium in the Hague (which unfortunately burned down in 1976). We were sitting under the white cupola, where after some time the light gradually dimmed. On the darkening vault, luminous points became visible. A few moments later a sigh of admiration passed through the crowd: from an inky black sky a magisterial legion of stars shone down upon us, a representation of a moon- and cloudless night, far from all city lights. It was only many years later that I began to realize that the planetarium is a perfect illustration of the fact that we cannot see how far away the stars are: whether they are at a distance of only a dozen meters or hundreds of light years, the image we experience is the same. This holds not only for all the celestial bodies taken together, but also for their mutual distances. As Upgren says: "The moon and a distant star look to be at the same distance, despite the ratio of many billions to one in their true distances" (2002: 121). I also realized that a planetarium is in fact a representation of the archaic world-picture, in which the flat earth is overarched by a hemispherical celestial vault that is cut off by the horizon, and on which the celestial bodies are painted, all at the same distance from us.

Immanuel Kant was one of the last philosophers who studied the stars. Since his time, this has become out of philosophical fashion. The tradition of stargazing, however, is as old as philosophy itself. Diogenes Laertius relates how Thales was looking at the stars while walking, and unfortunately fell into a well that he had not seen because his eyes were directed to the firmament. That event provoked a maidservant's laughter. Thus, the stage was set: whereas a normal person looks down and in front of him to find his way among the everyday obstacles, the philosopher has his eyes fastened upon the starry sky, trying to fathom the universe and man's place within it. For an astronomical interpretation of this anecdote, see Chap. 2. Tradition has it that Anaximander, a disciple of Thales, made a representation of the cosmos of sun, moon, and stars that unfortunately has been lost. Pythagoras could lose himself in the image of the literally unheard harmony of the celestial spheres. In Plato's dialog, the *Phaedo* that contains the conversation at Socrates' death bed, for the first time in history the sphericity of the earth is alluded to. Aristotle describes his astronomical convictions in his book *On the Heavens*, in which the question why the earth does not fall plays an important role, and in which proofs for the sphericity of the earth are put forward. Aristotle's opinion concerning the structure of the universe has for many centuries molded the picture of the world. In that conception, the earth floats in the center of a spherical universe that consists of a system of concentric spheres of sun, moon, and planets, within the surrounding sphere of the fixed stars.

Even much later, it continued to be a matter of course that a philosopher occupies himself with the structure of the universe and the place of the earth within it. Descartes, for instance, revived an old Greek idea when he imagined the universe as a system of gigantic vortices that produce the movements of the celestial bodies. But since the same Descartes made a clear distinction between thought and matter, the former became increasingly the dominion of philosophy and the latter that of science. It is typical that in the thin paper edition of his *Oeuvres et Lettres*, published in 1953, exactly those parts of *Les Principes de la Philosophie* were omitted that contain his ideas on the construction and origin of the universe and his theories about the universe of vortices.

Immanuel Kant is commonly known as the author of the *Kritik der reinen Vernunft*, a book that has had decisive influence on modern philosophy. Less known is that in the year 1755, when he was 31 years old, he wrote a fascinating anonymous book that bears the title *Allgemeine Naturgeschichte und Theorie des Himmels, oder Versuch von der Verfassung und dem mechanischen Ursprunge des ganzen Weltgebäudes nach Newtonischen Grundsätzen abgehandelt*. In those days, one did not bother about a word more or less in a book title. Unfortunately, the publisher went bankrupt and the book never appeared, but was published only posthumously. A short résumé of the main line of his argumentation appeared in *Der einzig möglichen Beweisgrund zu einer Demonstration des Daseyns Gottes* (1763). According to Kant's famous statement at the end of the *Kritik der praktischen Vernunft*, two things have always filled him with astonishment and awe: the starry sky above him and the moral law within him. The *Allgemeine Naturgeschichte und Theorie des Himmels* is the product of his astonishment and awe of the heavens. In this book, Kant not only speculates about the inhabitants of other planets, but also more in particular he unfolds his theory about the origin of the solar system which later became known as the theory of Kant–Laplace. Moreover, in this book, he is the first to give a correct explanation of the Milky Way and rightly explains the recently discovered nebulous spots as other galaxies.[1] He did all this, probably without ever having looked through a telescope, or perhaps only through the rather primitive instrument of Martin Knutzen, with whom he had studied philosophy and natural sciences.

However, it was Kant himself who eventually set philosophy on the track that led away from speculations about the universe. Since he, with the transcendental turn that was introduced in the *Kritik der reinen Vernunft*, looked for the conditions of the possibility of all knowledge and of metaphysical knowledge in particular, and

[1] For an extensive discussion of Kant's cosmological ideas, see Couprie (1996).

located them in the aprioristic categories that are hidden in the mind, philosophy has never been the same.[2] If philosophers after Kant dared to utter astronomical speculations at all, they went terribly wrong, like Georg Wilhelm Friedrich Hegel, who in 1801, in his *Dissertatio philosophica de orbitis planetarum*, solemnly declared that the relations of the distances between the planets, as settled by the Pythagoreans, rule out the existence of a celestial body between Mars and Jupiter. In the very same year, Giuseppe Piazzi discovered the asteroid Ceres, and in the next year Wilhelm Olbers discovered Pallas and Vesta. Yet 15 years later, in his *Vorlesungen über die Geschichte der Philosophie*, Hegel still stuck to his conviction, although it was already proven false.

Today, we are no longer inclined to associate philosophy, of whatever school of thought, with the study of the universe or with astronomical phenomena. Who thinks, when hearing the names of Husserl, Heidegger, Wittgenstein, Derrida or Rorty, of speculations about the nature and origin of the universe? Philosophers no longer fall into wells because they look up to the infinity of the heavens, but because they are absorbed by the infinity within themselves. Wonder, which has always been the source of philosophy, has for the last three centuries been less and less directed to the infinity outside us and seems to have lost sight of it almost completely. Present-day philosophers rather leave speculations about the universe to theoretical physicists like Stephen Hawking.

Although cosmology and philosophy have definitely separated, cosmological speculations are fundamental to the study of the history of Greek philosophy from Thales to Aristotle. The cosmology of the ancient Greeks initiated what we may call the Western world-picture. Usually, the discovery of the sphericity of the earth is seen as the start of the new world-picture and of modern astronomy at that. An example of this traditional opinion is Burkert: "From the point of view of the history of science, the most important points are the discovery of the spherical shape of the earth, the recognition of the five planets, and the explanation of the apparent irregularities of their courses" (1972: 302). I will argue, however, that the idea that the earth is a sphere, although it certainly was an epoch-making insight, is rather the coping-stone of the building of the new world-picture and was preceded, both temporally and logically, by other far-reaching insights that now perhaps look trivial to us. They can be enumerated thus:

1. The orbits of the celestial bodies do not stop at the horizon
2. The earth floats unsupported in the universe
3. The celestial bodies are at different distances from us

In all probability, we owe these three insights to Anaximander of Miletus (610–547 B.C.), which makes him the founding father of cosmology. The replacement of the archaic world-picture by the new one was not accomplished all at

---

[2] As regards the so-called schematism of categories, which is the rule of unity that expresses the category, Kant says: "Dieser Schematismus unseres Verstandes (...) ist eine verborgene Kunst in den Tiefen der menschlichen Seele" (*Kritik der reinen* Vernunft: A141, B180).

one stroke, and in particular the concept of a flat earth was still vindicated by most, if not all, Presocratic thinkers although they adopted these three insights of Anaximander. This means that a strictly diachronic exposition is not always appropriate, especially when important features of the archaic world-picture are discussed in Part I of this book.

We are so used to our modern world-picture that was introduced by the Greeks that we are hardly able to realize how unique in fact it is. Yet, our Western culture is the only one in which this picture has ever developed. Chapter 8 refers to an intriguing theory that might explain why this is so, as it throws a light on one of Anaximander's most strange-looking conceptions. All other cultures favored one or another variety of the picture of a flat earth, over which the heaven arches as a firmament, a big cupola, onto which the celestial bodies are attached, like in a planetarium. We may express the difference between the two world-pictures as a difference of opinion about space, with which I mean in this context the space of the universe. The Western concept of the universe is, so to speak, three-dimensional; it is a universe with *depth*, in which the celestial bodies are *behind* one another. The little story of the planetarium at the beginning of this Introduction already made clear that the depth of the universe is not something we can *see*. It is something we *know*. We do not *see* the celestial bodies standing behind each other, but we *know* it. The universe in our Western conception is, thus considered, to a high degree artificial, and it differs completely from the archaic conception of the universe as a firmament vaulted over a flat earth. Dreyer expresses the difference between what I call the archaic way of thinking and the new world-picture with these words: "Astronomy may be said to have sprung from Babylon, but cosmology (...) dates only from Greece." He does, however, not connect this with Anaximander, of whom he says: "Anaximander (...) did not advance much further in his ideas of the construction of the world." And elsewhere: "His cosmical ideas were as primitive as those of Homer" (1953: 1 and 11). The thesis defended in this book is that the opposite is the case. I try to reveal how the unique Western conception has originated, and therewith perhaps contribute to a better understanding of ourselves.

To achieve this, we have to place ourselves, both into the archaic way of thinking, and into the thought of those who developed the new world-picture, and especially into that of Anaximander. Accordingly, we have to suspend our own world-picture, as we have to learn to look "with Anaximander's eyes." This proves to be a difficult task. Time and again the danger of anachronism lies in wait. An anachronism, a pitfall into which many an author on early Greek cosmology has fallen, is nothing but a manifestation of our inability to put ourselves in the position of early thinkers. We encounter several instances of it.

The greatest hero of this book, Anaximander of Miletus, was a contemporary and co-citizen of Thales (Fig. 0.1). Thales is not only known as the most important of the Seven Sages, but also as the first Greek philosopher. Tradition has it that he was occupied with various questions of astronomy which still fall within the framework of the archaic world-picture. One may surmise that he awakened Anaximander's interest. As far as we know, Thales has not laid down his ideas in writing. Anaximander is said to have been the first to write a book in prose. Of this book,

Fig. 0.1 Artist's impression of Anaximander teaching his cosmological conceptions[3]

only a few lines and a few disconnected words have survived. These lines make up one of the most commented texts in the history of philosophy, the famous fragment of Anaximander. That Anaximander's book has been lost is not exceptional since no book of a Presocratic thinker has been preserved. All we know about Anaximander's cosmology or astronomy is handed down by the so-called doxography (a Greek neologism, meaning "the description of opinions"), that is to say from books of authors (the doxographers) who lived centuries after the Presocratics. These doxographers mainly borrowed from the work of Aristotle as well as from a book of his pupil and successor Theophrastus, called Φυσικῶν Δόξαι ("the opinions of the natural philosophers") that also has been lost.

[3] The picture is a detail of the triptych "The Seven Wonders of the World" by Hans Exterkate. As the artist is fond of anachronisms, the Greek letters on the plinth in the right under corner are written in characters that could not have been used in Anaximander's time. The person on the right shows admirable likeness with a well-known Dutch politician. The person left of him suggests with the gesture of his hands that he already thinks that the earth is spherical. The man with the staff points to Miletus on Anaximander's map which is drawn on a column drum with a diameter that is three times its height, as in Anaximander's conception of the earth. The numbers on the stone behind Anaximander indicate the dimensions of the universe according to Anaximander that is drawn on that stone. Anaximander himself makes theatrical gestures and wears pompous garments, as Diogenes Laertius tells (DK 12A8).

In his classic work *Doxographi Graeci*, published in 1879, Hermann Diels, a German scholar, has shown that the doxographers did not borrow immediately from Theophrastus, but made use, directly or indirectly, of a compilation by Aëtius (first century A.D.), who in his turn used a still older work that Diels named *Vetusta Placita* ("the oldest opinions"). Both books have been lost as well, but Aëtius' book was largely reconstructed by Diels. In 1903, the same Hermann Diels published a collection of classical texts from and about the Presocratic philosophers, called *Fragmente der Vorsokratiker*. Later imprints were edited by Walther Kranz, and usually the sixth impression (1952) in three volumes is quoted. The authors distinguish between A: the description of the opinions of the Presocratics, and B: *verbatim* quotations. An indication like "DK 12B1" means: the first literal quotation (=B1) of Anaximander (=no.12) in H. Diels and W. Kranz, *Die Fragmente der Vorsokratiker*. DK 12B1 is the abovementioned fragment that is handed down by Simplicius (sixth century A.D.) but goes back to Theophrastus. The material on Anaximander's astronomy belongs, with the exception of a few disconnected words and phrases, completely to the A-texts. For more elaborate expositions of the sources, see Mansfeld (1999: 22–44), and recently Runia (2008: 27–54).

The common consent is that when studying the Presocratics one has to start with the literal quotations (the B-texts), and draw upon the doxography (the A-texts) only with the utmost caution. This usual approach as regards the doxography has been expressed thus: "It is legitimate to feel complete confidence in our understanding of a Presocratic thinker only when the Aristotelian or Theophrastean interpretation, even if it can be accurately reconstructed, is confirmed by relevant and well-authenticated extracts from the philosopher himself" (Kirk et al. 2009: 6). As such, this is of course a reasonable rule to follow. We have to recognize, however, that the distinction between A- and B-texts is less sharp then one might perhaps expect. It has become an increasing matter of debate whether what Diels regarded as a literal quotation really is one. Gershenson and Greenberg, for instance, are very critical about the value of DK. Regarding the division between A- and B-texts, they say: "We find (...) the extent of the citations it brings as testimony arbitrary" (1964: xxiii). The ancient authors did not yet use our practice of putting quotations between quotation marks, and they did not always quote *verbatim*. The other way round, a paraphrase in an A-text may very well reproduce the thought of a Presocratic philosopher correctly. In the case of Anaximander, strict observing of Kirk's abovementioned rule would mean that we are definitely not able to put forward anything significant about his cosmology. Several scholars actually take this position. A typical representative of those who, more specifically, deny the importance of the doxography for the knowledge of Anaximander's astronomy is Detlev Fehling (1994: 67–70 and 149–157, 1985b: 222).[4] I, on the contrary, am of the opinion that in this case too much skepticism leads nowhere. The next section is meant to endorse a more positive attitude against the doxography on Anaximander.

---

[4] A critical confrontation with Fehling's ideas in Couprie (2004a: 127–143).

Diogenes Laertius reports that Apollodorus of Athens (second century B.C.) had laid hands on a copy of Anaximander's book, perhaps in the famous library of Alexandria, as Diels and Heidel suppose (DK 12A1(2); Diels 1879: 219, n. 3; Heidel 1921: 261). Recently, this report has won credibility by an interesting discovery in Taormina in Sicily. A fragment of the catalog of the gymnasium has been found there, on which the name of Anaximander can be read (see Fig. 7.1).[5] Since this discovery, we may take for granted "that in the second century B.C. Anaximander's text was still available at a Hellenistic gymnasium on Sicily", as the proud discoverer writes (Blanck 1997a: 507). However, this may be, as already said a bigger part of the doxography goes back to Theophrastus and Aristotle, who were still able to read the works of the Presocratics. One may suppose that Aristotle, Theophrastus, and maybe some others as well, who had before their very eyes the books of Anaximander and the other Presocratics, have rendered truthfully what they considered to be important, albeit that they read them through the spectacles of their own philosophical and astronomical conceptions, and sometimes with the intention to refute those opinions. Where we can check it with other Presocratics of whom relatively many original fragments have been preserved, we may conclude that the doxography, generally speaking and taking for granted the subjective and sometimes hostile coloration, provides a rather true rendering of the thoughts of these philosophers. This does not mean, of course, that any and every doxographical report can simply be trusted. We have to be continually on our guard to avoid the danger of an anachronistic interpretation, both from the doxographers and from ourselves. As concerns the doxography on Anaximander's cosmology in particular, I would like to add as a further condition that it has to allow a coherent and significant interpretation of the celestial phenomena for someone who thinks that the earth is flat. Unfortunately, the case of Thales is different because he has not left a written text. When no writings exist, the doxography has nothing serious to rely upon and has to be handled accordingly with utmost caution.

This book consists of three parts. In Part I, several aspects of the archaic world-picture are discussed. This part also has the purpose to familiarize the reader with looking at the heavens with archaic eyes, which is of help in appreciating the difference between the archaic and the new world-picture. The first chapter describes the main features of the archaic world-picture: the flat earth with the celestial vault. Special attention is paid to the bearers of the celestial vault, who or which prevent it from collapsing. In the second chapter, which is by far the largest, the archaic astronomical instruments are discussed, including new suggestions for how they might have been used. As far as I know, this subject never has been treated so extensively elsewhere. Special attention is paid to the gnomon, a very simple, but multifunctional instrument that it is argued to have stood at the cradle of science. In this context, also an attempt is made of a reconstruction of Anaximander's sundial. As a proponent of the archaic world-picture, Thales of Miletus has two chapters.

[5] Robert Hahn informed me that he looked for it at Taormina in June 2009 and that he was told that it is now in the museum of Giardini-Naxos, also on Sicily.

Chapter 3 explains by what kind of observations and by which mistake he could have "predicted" a solar eclipse, without taking refuge to alleged Mesopotamian wisdom. Chapter 4 is polemical against a recent author who claims that Thales already knew of the sphericity of the earth. Chapter 5 is about the much neglected and misinterpreted issue of the tilting of the celestial axis on the flat earth. It shows the kind of anachronistic misunderstandings which inhere in both ancient and modern expositions of the archaic world-picture. In Chap. 6, Anaximander's first map of the earth is reconstructed and is shown how it was linked with the astronomy of a flat earth.

Part II presents the transition from the archaic to the new world-picture by means of an exposition of Anaximander's opinions. In Chap. 7, a survey is presented of what we know about him. In Chaps. 8–10, Anaximander's "discovery of space" is extensively described and illustrated. These chapters cover the same field as my previous publications on Anaximander's cosmology, but with essential additions and improvements. Anaximander's three aforementioned main insights are exposed in Chap. 8. The issue of depth in the universe is elucidated by a discussion of Anaximander's numbers in Chap. 9. A three-dimensional visualization of his universe is offered in Chap. 10. In Chap. 11, the rather technical question of the interpretation of the translation of an image, used by Anaximander as an explanation of the light of the celestial bodies, is treated. It shows how a wrong translation, proposed by Diels, still haunts the contemporary reading of Anaximander. Chapter 12 is a polemic against a recent author who links Anaximander and the architects of his time.

The seven chapters that make up part III are about the reception of the new world-picture introduced by Anaximander, with special attention to the debate on the shape of the earth. Some of the additional insights that completed the new world-picture, such as the distinction between fixed stars and planets, the order and distances of the celestial bodies, and the explanation of eclipses are discussed in Chap. 13. Chapters 14–16 are devoted to Anaxagoras, who wholeheartedly accepted the new world-picture, but still believed, just like Anaximander himself, in a flat earth. Chapter 14 contains an exposition of his heretical conception of the celestial bodies as fiery stones. In Chap. 15, Anaxagoras' proof of the flatness of the earth is discussed, and Chap. 16 offers a reconstruction of what I think was his attempt to measure the distance and size of the sun, taking for granted that he believed in a flat earth. The development of the idea of the sphericity of the earth in Plato and Aristotle is the subject of the Chaps. 17 and 18. In the final chapter, it is shown how Anaximander's revolutionary destruction of the celestial vault, which was made undone by his successors, especially Aristotle, was only defended by the almost forgotten figure of Heraclides Ponticus. It is also described how he, as a consequence of his conception of the axial rotation of the earth, taught the infinity of the universe.

At this point of the Introduction, it is appropriate to make some remarks on the terminological distinction between astronomy and cosmology. To illustrate the terminological difficulty, let us look at the titles of two standard works that cover practically the same field. In 1970, *Early Greek Astronomy* by D.R. Dicks was published. The announced second volume of this book never appeared. In 1987, David Furley published his book *The Greek Cosmologists*. It was also intended as the first of two volumes but in this case, too, the second volume has never appeared. Although Furley's book is about cosmology and Dicks' book about astronomy, both works discuss practically the same issues.

Dicks reserves the word "astronomy" for what he calls "scientific astronomy.[6]" He does not define this any further, but it is clear that for Dicks astronomy is the careful and systematic observation of the celestial phenomena. This kind of astronomy reaches its summit in "mathematical astronomy," which is the mathematical description of the movements of the celestial bodies, resulting in predictions of when and where to find a celestial body, and of special phenomena as eclipses. So he may write, for instance: "The development of a mathematical theory to account for the planetary movements belongs to the most sophisticated stages of Greek astronomy" (1970: 26). Dicks contrasts Greek astronomy with cosmology and cosmogony, and particularly with the "cosmological fantasies of the Presocratics" (1970: 7). For him, the word "cosmology" has a pejorative flavor and is associated with wild speculations.

Furley describes the relation between the underlying world-picture and the science of astronomy that is based on it, although he never in so many words identifies world-picture with cosmology. He speaks, for instance, of the "picture of the world" of Homer and Hesiod as the starting-point of Greek astronomy, and of the "spherical picture" or the "spherical model" that marks the end of Greek astronomy (1987: 26–27). Such a world-picture is "a framework for explanations of the natural world"; it is a "cosmological system" (1987: 2 and 3). However, Furley is not always consistent, for instance, when he speaks of "the astronomy of the Pythagoreans," or the "astronomy of the atomists," obviously meaning their world-pictures (1987: 57 n.15, and 145). Dicks, too, describes time and again exactly the same relation between underlying world-picture and (scientific) astronomy, but without using the word "cosmology." For instance: "This quantity of empirical knowledge needed to be fitted into a *general picture of the universe as a whole, a celestial model*, in which the earth would occupy a special position as the standpoint of the observer, and the sun, moon, stars, and planets would follow their individual courses *inside the general framework*." And also: "*The general framework or model* that was eventually adopted was the celestial sphere with the spherical earth set immovably at the center" (1970: 60–61, both times my italics). On the other hand, Dicks speaks of the spherical shape of the earth as an *astronomical* discovery and of the spherical shape of the universe as an *astronomical* assumption. The same holds for the question whether the earth is or is not at the center of the universe, when he writes: "In *astronomy* the Pythagoreans also introduced a startling innovation. This was to displace the earth from the central position in the cosmos" (1970: 65, my italics). These are, I would say, ingredients of

[6] In the same sense O. Neugebauer: "We shall here call 'astronomy' only those parts of human interest in celestial phenomena which are amenable to mathematical treatment" (1983: 35).

what he calls the "general picture of the universe," and not results of scientific astronomy.

To make matters even more complicated, generally speaking the word "cosmology" is used by many authors to indicate the sum-total of opinions concerning what the Greeks called "nature" (φύσις), for instance, whether all things in the end consist of combinations of the four elements earth, water, air, and fire (and eventually a fifth, the aether), or of atoms. Characteristically, Furley's book bears the subtitle: *The Formation of the Atomic Theory an its Earlier Critics*. I would rather call this "natural philosophy" or "philosophy of nature." Cosmogony, which is about the origin of the cosmos, is also often treated under the head "cosmology." Recently, a thorough book on ancient Greek cosmogony has come available, in which there is hardly a word on cosmology in the sense intended here (Gregory 2007). With a certain amount of overstatement, one might say that my book starts where Gregory's stops. Important chapters of Charles Kahn's *Anaximander and the Origins of Greek Cosmology* are concerned with Anaximander's fragment, with the elements (philosophy of nature), and with the *apeiron*, out of which everything originates (cosmogony), as he acknowledges himself (1994: 7–8). Strictly speaking, only the pages 46–63 (the doxography) and 75–98 of chapter I cover the field of Anaximander's cosmology.

I prefer to use "astronomy" as the general term for the study of the cosmos and the celestial phenomena, and to distinguish within it between "speculative astronomy" or "cosmology," which is concerned with forming a general picture of the overall structure of the universe, "cosmogony," which is the study of the origin of the cosmos, and "descriptive astronomy" or "scientific astronomy" (or "astronomy in the narrower sense"), which is concerned with providing as accurate as possible a description (and thus prediction) of the movements of the celestial bodies. The relation between cosmology and astronomy in the narrower sense can be described more precisely. Cosmology results in a picture of the world, within which astronomy as the description of celestial phenomena becomes possible. Such a world-picture delivers the framework for the development of (scientific) astronomy.

This book is about cosmology in the sense meant above. This means that not only scientific astronomy, but also cosmogony and natural philosophy fall outside its scope. Yet, now and then some attention is paid to these three subjects, but only in so far as it is clarifying for the main theme. Thus, for instance, chapters are dedicated to archaic astronomy and the oldest astronomical instruments. By "archaic astronomy" is meant the observation and description of celestial phenomena as they appear to an observer who thinks that the earth is flat and that the celestial bodies are all at the same firmament. The study of archaic astronomy is important when we want to understand the kind of revolution Anaximander initiated in the way people imagine heaven and earth. The distinctions made above are visualized in Table 0.1.

The Presocratic Greeks, but Plato and Aristotle as well, were concerned with cosmological speculations, rather than with astronomical observations. As Von Fritz remarks: "Die Beobachtung der Himmelserscheinungen, die bei den Babyloniern mit solcher Konsequenz weiter getrieben worden ist, spielt in der frühen

Table 0.1 Terminological distinctions

Astronomy the study of the cosmos and the celestial bodies	**Cosmology** (speculative astronomy) resulting in a world-picture, an image of the overall structure of the universe
	Cosmogony about the origin of the universe
	(scientific) Astronomy (astronomy in the narrower sense) description and prediction of the movements of the celestial bodies
Natural Philosophy the study of the nature of things (elements)	

kosmologischen Astronomie der Griechen (...) eine untergeordnete Rolle" (1971: 151). Anaximander's cosmology was not descriptive astronomy, but *speculative* astronomy. Speculative astronomy or cosmology is the product of what I would call "creative imagination." Creative imagination is quite something other than fantasy or the "wilder flights of fancy" of which Dicks accuses the Presocratics (1966: 39). Fantasy creates things or images that do not help in understanding the celestial phenomena, but rather adapts them to a preconceived idea. Creative imagination, on the other hand, puts known empirical data into a new interpretative arrangement that helps us to understand the phenomena. Also the so-called discovery of the sphericity of the earth by Plato and Aristotle, as well as Anaxagoras' interpretation of the celestial bodies as fiery stones, belong to speculative astronomy or cosmology.

To illustrate the difference between fantasy and creative imagination one might say that the division of the celestial vault into starry constellations is a product of fantasy, whereas Kant's hypothesis of the band of light which is called the Milky Way as a gigantic disk of stars is the product of the power of creative imagination. Or with perhaps a rather tricky example, on which more in Chap. 13: the Pythagorean conception of the universe is a product of fantasy, based on the magic of numbers and other preconceived a priori or metaphysical ideas that as such have nothing to do with astronomy or cosmology. In their vision, for instance, the distances of the celestial bodies have to be derived from the musical harmonies. The earth had to be displaced from its central position because only the noble fire is worth to occupy that most honorable place. The counter-earth had to be invented because there must be ten celestial bodies, for ten is the sacred number. The same criticism was expressed by Aristotle when he said that the Pythagoreans are "not seeking accounts and reasons to explain the phenomena, but forcing the phenomena and trying to fit them into arguments and opinions of their own" (*On the Heavens* 293a25). Heraclitus already called Pythagoras a clever charlatan with fraudulent wisdom (DK 22B129, see also DK 22B40). I for one do not understand why Dicks, who treats the Ionian speculations with so much contempt, can talk about the Pythagoreans with great awe. A few examples, out of many: "The great novelty that the Pythagoreans introduced into Greek philosophical thought as a whole was their insistence on the concept of number"; and "In astronomy the Pythagoreans also introduced a startling innovation. This was to displace the earth from the central position in the cosmos" (1970: 64 and 65). The Pythagorean way of using numbers has nothing at all to do with the mathematical astronomy which Dicks so admires, and the position of the earth in their system was not an astronomical innovation but a metaphysical makeshift.

Anaximander's cosmological speculations, on the contrary, were not images of his fantasy, but they were meant to yield a better understanding of the celestial phenomena as they are observed. They created a completely new paradigm, a new world-picture within which empirical data got another meaning than they had before in the archaic world-picture. Sometimes, Dicks himself seems to understand this. See, for instance, his adequate formulation: "The attempt to look at the earth from the outside, as it were, and to consider its position in relation to the cosmos as a whole and in particular to the other celestial objects (...). This represents *a new departure in astronomical speculation*, though not in factual astronomical knowledge" (1970: 44, my italics).

Part I
Archaic Astronomy and the World-Picture of a Flat Earth

Chapter 1
The Archaic World Picture

The archaic world picture, the picture of a flat earth with the dome of the heaven vaulted above it, on which the celestial bodies are attached, is the basic world picture of many ancient cultures. Here "world picture" means the conception of the visible universe, not including all kinds of mythical or religious representations of what was imagined to be "under the earth." This archaic world picture (and also its penetration by a curious head) is beautifully rendered in a picture that is often thought to belong to the Renaissance period but was actually drawn in 1888 A.D. on the instructions of the famous French astronomer and popularizer Camille Flammarion (Fig. 1.1). The drawing refers to a story about Archytas (428–347 B.C.), who is supposed to have asked whether it would be possible to put a hand or a stick out of the heavens (DK 47A24). We will return to the implications of this question in the last chapter of this book.

According to the creation story of the Bible, the world originated when God made, as it were, a fissure in the primary waters: "Then God said, 'Let there be a firmament in the midst of the waters, and let it divide the waters from the waters'. And God called the firmament Heaven. Then, God said, 'Let the waters under the heavens be gathered into one place, and let the dry land appear'; and it was so" (Genesis 1: 6–9). Recently, it has been argued that in the opening text of the Torah, בראשית ברא אלהים את השמים ואת הארץ, the word ברא does not mean "create" but "separate," which means that this text is not about a *creatio ex nihilo* but that God's creative act consisted of making a separation between heaven and earth (Van Wolde 2009). Perhaps unconsciously, this is nicely pictured on a church window in Perugia, where God the Father appears in a light crack in the dark, creating with a majestic gesture the vaulted firmament above him and the flat earth under him (Fig. 1.2).

This image of the world as a kind of opening (an air bubble, as it were) in the surrounding chaos or primeval waters appears in various shapes in archaic myths. In the archaic world picture, the earth is the flat bottom of that gap in the primeval waters, and over the earth arches the firmament, on which the celestial bodies move along. As another example, we take the Egyptian picture of the world. The Egyptian word "heaven" (*pt*) is written with the signs for "p" and "t" over the ideogram for heaven. This ideogram is the simplified picture of a flat roof (Fig. 1.3).

D.L. Couprie, *Heaven and Earth in Ancient Greek Cosmology*,
Astrophysics and Space Science Library 1, DOI 10.1007/978-1-4419-8116-5_1,

Fig. 1.1 The archaic world picture of a flat earth with the celestial vault (Flammarion 1888: 163)

Fig. 1.2 God the Father on a church window in Perugia (author's photograph)

Fig. 1.3 The Egyptian hieroglyph for "heaven"

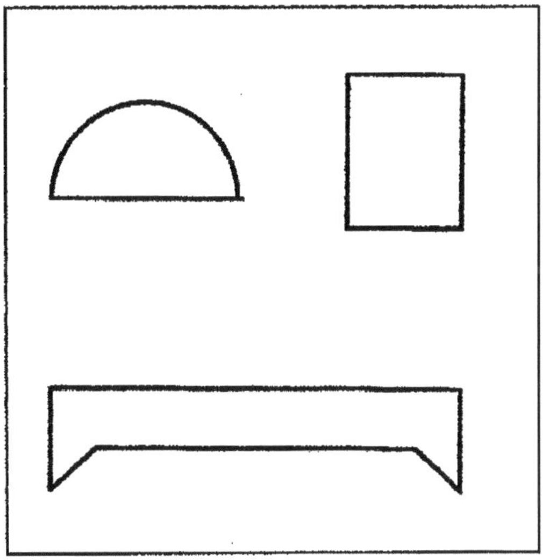

The same picture is used many times in Egyptian art as an indication of the heavens, as for instance on a painting in queen Nefertari's burial chamber, circa 1210 B.C. (Fig. 1.4).

Fig. 1.4 The hieroglyph for "heaven" painted in the tomb of Nefertari (drawing by Hans Exterkate)

Sometimes, however, the picture of the heaven is not flat but curved, as for example on the stele of Tanetperet, ± 850 B.C., shown in Fig. 1.5. Other examples of a curved *pt*-symbol are, for instance, the Semna-stele of Sesostris III (± 1280 B.C.), the Sphinx stele of Thutmoses IV (± 1400 B.C.), steles of Deniuënchons (± 800 B.C.), Nechtefmut (ca. 900 B.C.), Nastasen (320 B.C.), Djadamuniuanch (± 850 B.C.), Harsiese (± 664 B.C.), in the papyri of Ani (± 1420 B.C., the oldest one I know of) and Hunnefer (± 1300 B.C.), and on the boundary stones A and S from the time of Akhenaton (± 1350 B.C.). This kind of representation

Fig. 1.5 Curved celestial vault on the stele of Tanetperet (drawing by Hans Exterkate)

is remarkable, as the Egyptian temple architecture does not use the vaulted roof that enables builders to span bigger spaces (cf. Schäfer 1928: 98).[1]

Illustrative, however, is that the ceilings of the burial chambers in the tombs of Seti I (ca. 1270 B.C.), Ramesses II (ca. 1210 B.C.), Merenptah (ca. 1200 B.C.), Ramesses III (ca. 1150 B.C.), Ramesses VI (ca. 1130 B.C.), Ramesses IX (ca. 1100 B.C.), Ramesses X (ca. 1090 B.C.), Ramesses XI (ca. 1070 B.C.), and Tawosret/Setnachte (ca. B.C.) in the Valley of the Kings, as well as those of the high servants Sennedjem (ca. 1280 B.C.), Inherchau (ca. 1140 B.C.), and Pashedu (ca. 1175 B.C.) at Deir el-Medina, are vaulted. On some of these ceilings,

[1] The only exception I know of is the little sanctuary dedicated by Tuthmosis III to Hathor at Deir el-Bahari (ca. 1450 B.C.).

astronomical motifs are pictured, of which especially those of Seti I and Ramesses VI are famous (see Fig. 1.6; Weeks 2005: 338). Obviously, these vaulted ceilings were easier to realize as they were cut out in the rocks. Although they did not build vaulted temple roofs, the Egyptians apparently looked upon the celestial vault as a dome, and they adapted their symbolism accordingly.

Fig. 1.6 Curved ceiling of the burial chamber of Ramesses VI with astronomical painting (photograph by the courtesy of Travel Egypt)

Another widespread Egyptian representation of the heaven is that of the goddess Nut. Usually, she is pictured as a naked female, standing on all fours, her body spangled with stars. On the backside of the same stele of Tanetperet, she takes the same position as the curved roof in Fig. 1.5 (see Fig. 1.7). Mark that she swallows the sun in the evening, which travels through her body during the night, to be born again in the morning.

According to an obvious suggestion that has been made only recently, we can still see Nut every cloudless night as the belt on the firmament that we call the Milky Way. Figure 1.8 shows the Milky Way as it can be seen in Egypt, with Nut's legs to the left (East), her arms to the right (West), and her head at Gemini, where the ecliptic crosses the Milky Way (Wells 1997: 30, Wells 1992: 309).

Fig. 1.7 Arching Nut on the backside of the stele of Tanetperet (drawing by Hans Exterkate)

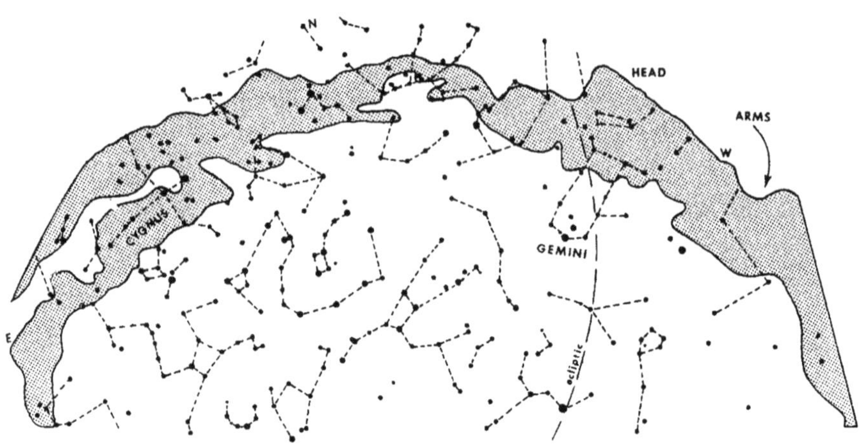

Fig. 1.8 Nut as the Milky Way (picture by the courtesy of British Museum)

On many pictures, Nut is supported by the sky god Shu, who stands upon the earth god Geb. His arms make the *ka*-sign as a symbol of the cosmic power that upholds the heaven. Apparently, the Egyptians were afraid that the heavens would fall down, unless they were supported. According to the myth, Nut and Geb originally were entangled in an incestuous embrace (as they were sister and brother). They were divided by Shu, who had to prevent them from coming together again. Shu, bearing the heavens, is the personification of the archaic fear that the celestial vault will collapse and fall upon the earth. An example is Fig. 1.9.[2]

Fig. 1.9 Shu, standing upon the earth god Geb, supports Nut, the goddess of the heaven (author's drawing, after Piankoff and Rabova 1957, vol. II, no. 8: papyrus of Nisti-ta-Nebet-Tawi, scene 11)

In a dramatic way, this motif appears in the Bible. At the time of the Deluge, the heaven actually collapsed, and the primeval waters took possession of the earth again. "In the six hundredth year of Noah's life, the seventeenth day of the month, that same day were all the fountains of the great deep broken up, and the windows of the heaven were opened" (Genesis 7: 11). After the Deluge, God gives Noah the assurance that he will never again destroy the earth by means of water, and as a token of his promise, he makes the rainbow (Genesis 9: 12–17). The rainbow is, as it were, a temporary glimpse of the invisible support that prevents the heaven from collapsing again, just like Shu in the Egyptian representation (Fig. 1.10).

In other cultures, the celestial vault is represented by other symbols, such as the mantle or the tent of the heaven. An example is an Assyrian seal cylinder, on which

[2] After Piankoff and Rabova (1957, vol. II, no. 8: papyrus of Nisti-ta-Nebet-Tawi, scene 11). This representation is dated in the New Kingdom (1570–1070 B.C.). The authors' suggestion that the god with the head of a baboon is not Shu but Hapi (vol. I: 101 n.) makes no sense.

Fig. 1.10 The rainbow supports the celestial vault (photograph by Kees Floor)

the holy tree is covered with the mantle of the heaven (Fig. 1.11). Similar pictures are used in the Bible as well: "Who coverest thyself with light as with a garment: who stretchest out the heavens like a curtain," and: "who stretches out the heavens like a curtain, and spreads them out like a tent to dwell in" (Psalm 104: 2; Isaiah 40: 22).

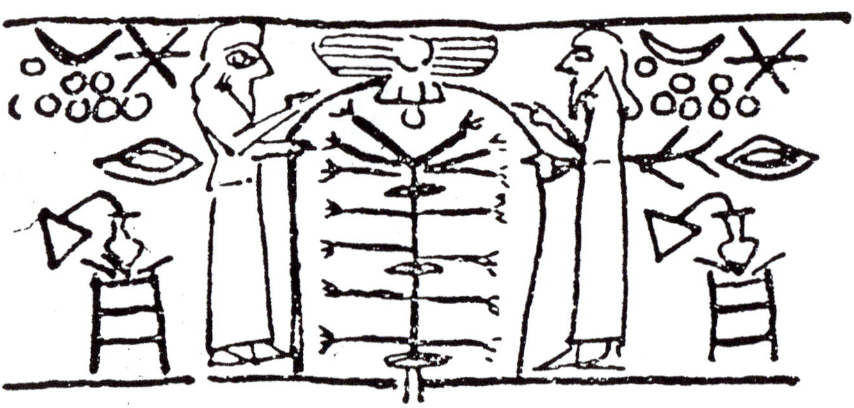

Fig. 1.11 The mantle of the heaven on an Assyrian seal cylinder (Andrae 1933: 15, Fig. 11a)

The oldest Greek conception of the heaven was another modification of the archaic world picture. It can be found in Homer, Theognis, and Pindar, where the heaven is considered as a brazen or iron vault that is supported by pillars.[3]

[3] Homer mentions "the brazen heaven" in *Iliad* XVII, 425 and in *Odyssey* III, 2. In *Odyssey* XV, 329 and XVII, 565 it is the "iron heaven." In *Odyssey* I, 52–54 we read of "Atlas, who holds the big pillars that divide heaven and earth." The brazen heaven also in Theognis (1979: 146, line 869–870), and in Pindar (*Olympian Odes*, ed. Race 1997: 360, line 27).

Hesiod describes the heaven that arches over the flat earth with these words: "Gaia's first child was Ouranos, starry heaven, just her size, in order that he should encircle her completely" (*Theogony* 126–127, my translation).[4] In Greek mythology, the giant Atlas, who bears the heaven, plays a role similar to that of Shu in the Egyptian world picture, preventing the heaven from falling upon the earth (*Theogony* 519–522; cf. Fig. 1.12). Aristotle refers to this mythological notion: "This is why there is no need to believe in the ancient mythological explanation according to which it (sc. the heaven) thanks its preservation (σωτηρία) to one or another Atlas" (*On the Heavens* 284a19 ff.). On the theme of the separation of heaven and earth and the danger of the heaven falling down, see also Chaps. 14 and 18.

When the celestial bodies are thought of as attached to the firmament that arches over the flat earth as a kind of cupola, then the biggest distance from the celestial bodies to the earth cannot be more than half the diameter of the earth, as clearly follows from Fig. 1.1. Let us take that the diameter of the earth was estimated at about 5000 km[5] and imagine ourselves to be at the center of the earth (Delphi). Then, the sun, moon, and stars are at a distance of no more than 2,500 km, and for the regions that are at the outskirts of the earth, the celestial bodies are even closer, as we can see, again, in Fig. 1.1. Were it not for the Ocean encircling the earth, we could touch them with our own hands. So the archaic universe is, to our understanding, rather small in extent. This is a simple statement, but it is of the greatest importance to appreciate how people in archaic cultures experienced the cosmos. Nevertheless, it is hardly ever mentioned in the literature. An exception is Balls: "Wie bei anderen Völkern wird auch bei den Griechen der Himmelsraum ursprünglich als verhältnismäßig sehr klein vorgestellt" (1949: 231). Since the celestial bodies in the archaic world picture were at relative little distance from us, their size, too, must be relatively small. People even could have estimated the sizes of sun and moon. Since the apparent diameter of the sun and moon is about 0.5°, it would take 720 suns to make up the full circle of the sun around the earth. Accordingly, the diameter of the sun is $(2 \times \pi \times 2{,}500)$: $720 = \pm 22$ km. As we see in Chap. 16, a calculation like this presupposes the conception of the sun making a full orbit around the earth, which is an insight that does not belong to the archaic world picture. Nevertheless, a reasonable estimation would have resulted in a similar figure as well. The sizes of the stars and planets would have been estimated as not much bigger than a large stone (if people envisaged the celestial bodies as stones at all), not unlike the meteorite that fell from heaven in the year 467 B.C. at Aegospotamoi, which was as big as a cart-load (see Fig. 14.2). Even if people were not yet able to calculate it exactly, this must have been the impression that the celestial vault made upon archaic men. The archaic universe was the conveniently arranged shelter in which people could find themselves safe against the dark powers of

---

[4] As regards the reading of the text, I follow Solmsen (1990): ἵνα μιν περὶ πᾶσαν ἐέργοι. Von Schirnding (1991) reads: ἵνα μιν περὶ πάντα καλύπτοι, and translates: "damit er sie völlig umhülle."

[5] See Fehling: "Nun betrug die größte Entfernung innerhalb der damals bekannten Erde (von den Säulen des Herakles bis Babylon) ca. 5000 km" (1985b: 210).

Fig. 1.12 Atlas and Prometheus on a Greek bowl (photograph by the courtesy of Art Resource)

the surrounding Chaos. The goddess of the heaven, Nut, who arches as a protectress over the earth is the ultimate symbol of this world picture. When one realizes how small the universe was for archaic man, one can also understand why astrology played such an important role in human life. If people could almost touch the celestial bodies, it is not strange to think that they could also influence a person's life on earth.

Archaic astronomy consisted of the observation and description of what is enacted on the firmament that arches over the earth. Egyptian astronomy as a whole was not very impressive.[6] The Babylonians and Assyrians, on the contrary, made considerable achievements in this field. Their interest, just like that of the Egyptians was, so to speak, two dimensional: they studied the movements and mutual positions of the celestial bodies on the screen of the heaven. What they developed was a kind of descriptive astronomy. This was what they needed because for them astronomy was subordinate to the demands of religion, astrology, and the making of calendars. They studied meticulously the constellations and the movements, as well as the times of the rising and setting of the celestial bodies at the horizon. They paid special attention to particular phenomena such as lunar and solar eclipses. Apparently, however, the cultures of Egypt and Mesopotamia were not able to imagine a dimension of depth in the universe. Or perhaps better, they were simply not interested in it, as this kind of knowledge had no religious or astrological importance and was not needed in the making of calendars.

In the archaic world picture, what happens with the celestial bodies when they disappear under the horizon is a problem. Especially regarding the sun, one may find different representations. In ancient Egypt, Nut was not only the goddess who symbolized the celestial vault but also the one who gave birth to the sun in the morning, to swallow it again in the evening (see Fig. 1.7). This means that overnight the sun travels back through her body from west to east. In an analogous way, it was thought in India that the sun, during the night, followed the opposite direction along the celestial vault (from west to east) but that it had turned over so that it showed its unlighted side toward the earth (see Keith 1917: 5–6). In Egypt, we also meet very often another representation of the sun, as being transported by a boat through the waters under the earth. In Homer and Hesiod, sun, moon, and stars arise from the ocean in the east in the morning and plunge into it again in the west in the evening. But, as Dreyer says: "What happens with the heavenly bodies between their setting and rising is not stated, but since Tartarus is never illuminated by the sun, they cannot have been supposed to pass underneath the earth" (1953: 7). As regards the one passage where Homer seems to say that the sun passes under the earth (ὑπὸ γαῖαν, *Odyssey* X, 191), I agree with Dicks' judgment: "This expression means no more than that it disappears from view below the horizon (...), and it is quite illegitimate to assume from this a knowledge of the sun's course round and under the earth" (1970: 31–32).

[6] Thurston says that the Egyptian achievements in the field of astronomy, compared with those of the Babylonians, were negligible (1994: 83). A similar judgment in Dicks (1970: 43).

This was the world picture that Anaximander encountered. Before exploring how he made a fundamental break with these archaic conceptions, I insert a number of chapters that deal with the astronomy and geography of the archaic world picture. However, there is one element of the archaic world picture that survives not only in Anaximander's cosmology but even in that of Anaxagoras, Democritus, and other Presocratics, viz., the conception of a flat earth. That is why Presocratics are dealt with in these chapters as well. We get insight into what the archaic world picture was like when we study the problem of the celestial axis on a flat earth, which bothered thinkers as late as Democritus. Although the same holds to a certain extent for Anaxagoras' proof of the flatness of the earth and the calculation of the size of the sun when one thinks that the earth is flat, these subjects are dealt with in part III, where the debate on the shape of the earth is treated.

As a matter of fact, even today elements of the archaic world picture play a role in colloquial language. We say that the sun, moon, and stars rise and set, instead of saying that the earth turns. We say that the moon shines and not that the moon reflects. We speak of the firmament and constellations, as though we do not know that there is no such a thing as the dome of the heavens or that the stars are light-years behind each other. Astrology is still based on the archaic world picture, and even astronomers do not mind using expressions like "the nebula in the Orion."

Chapter 2
Archaic Astronomical Instruments

The oldest astronomical instrument is the *naked eye*, with which the courses of the celestial objects were observed. Since time immemorial, people have noticed that the celestial bodies rise at the eastern horizon and set at the western horizon. They have also noticed that some stars never set and that all stars circle around a fixed point in the northern sky (at least on the northern hemisphere, where the oldest civilizations were. See Fig. 2.1). Already in ancient times, this point was conceived of as the end of the celestial axis. More on the celestial axis in Chap. 5.

The rhythm of day and night is determined by the appearance and disappearance of the sun. That in summer the sun is higher and a longer time in the sky than in winter is governed by another rhythm, that of the year. One can observe that every day the sun rises and sets at another point at the horizon, in summer in the eastern, respectively, and western sky further to the north, and in winter more southward (always on the northern hemisphere, where the ancient civilizations were). The sun reaches its northernmost position on the first day of summer, when the day is longest. This is called the summer solstice. The southernmost position of the sun is reached on the first day of winter, when the day is shortest. This is called the winter solstice. Twice a year, the points of sunrise and sunset lie exactly opposite to each other, due east and due west, respectively. On these dates, night and day are of equal length. These days are called the equinoxes. The circles of the daily orbit of the sun stand perpendicular to the celestial axis. The inclination of the plane of the sun's daily orbit in relation to the horizon varies from place to place, according to the location of the observer. The farther to the south, the higher the sun, and the farther northward, the lower (always on the northern hemisphere. See Fig. 2.2). More on this phenomenon, as seen from a flat earth, in Chap. 5.

The moon has a rhythm of its own, characterized by its monthly passing through the subsequent stadia of new moon, waxing crescent, first quarter, waxing gibbous, full moon, and back again to waning gibbous, last quarter, waning crescent, and new moon. The moon too, stands high or low in the sky in a monthly rhythm that is called the tropical month. Sometimes the strange phenomena of partial or total lunar and solar eclipses take place. The stars are always in the same and fixed mutual positions. From ancient times, people have divided the celestial vault into constellations, which made the topography of the sky easier. The velocity with which the stars move along the sky differs from that of the sun and the moon, making us see

D.L. Couprie, *Heaven and Earth in Ancient Greek Cosmology*,
Astrophysics and Space Science Library 1, DOI 10.1007/978-1-4419-8116-5_2,

Fig. 2.1 Stars circling around the celestial axis on a recent time exposure (photograph by Antonio Fraga, composition, and Gabriel Vazquez, circumpolar stars)

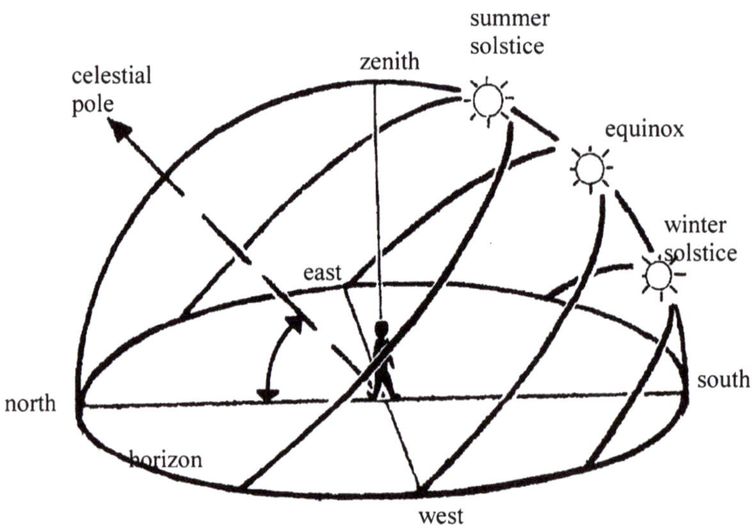

Fig. 2.2 The daily orbit of the sun in the different seasons

different constellations as the seasons pass. Only the constellations that never set remain visible during the whole year. Part of the unchanging starry sky is occupied by a capriciously shaped and softly shining belt, the Milky Way. Seven or eight celestial bodies (depending on whether the morning star is taken to be identical with the evening star) do not have a fixed position on the celestial vault. These are the so-called planetary stars, or planets. They are not at arbitrary places in the sky, but move within the limits of a belt of constellations, the Zodiac.

As a result of careful and regular observations of the sky over many generations, all kinds of regularities have been recorded, for instance, the way lunar and solar eclipses appear. An example can be seen in Chap. 3, Table 3.2, which shows the lunar and solar eclipses that Thales could have observed during his lifetime in Miletus. These eclipses move like garlands through the calendar. Another example is the cycle of Meton. This is the cycle of 235 synodic months (the time between two subsequent new moons), which is approximately 19 years. At the end of this cycle, the sun and the moon, in relation to each other and to the stars, are in virtually the same position as at the start. This cycle is named after Meton of Athens, who introduced it in 432 B.C. to improve the calendar.

From ancient Mesopotamia, we possess an enormous amount of observations, descriptions, and predictions of the rising and setting of the celestial bodies and their courses along the firmament, preserved on clay tablets. The Babylonians were well-versed observers and had achieved excellent results. In the second century A.D., Ptolemy, in his *Almagest*, used the systematic Babylonian registrations of the movements of the sun, moon, and planets, dating as far back as the time of Nabonassar (747 B.C.).

From the Presocratics, on the contrary, we hardly know of any observations of this kind. One of the few exceptions is a report by Pliny of an observation, made by three ancient observers, Hesiod, Thales, and Anaximander, of the time the Pleiades set (*Naturalis historia* XVIII: 213, see also DK 12A20). According to Hesiod, it was the day of the autumnal equinox, which was 30 September in Hesiod's time.[1] According to Thales, it was 25 days, and according to Anaximander, it was the 31st day after the autumnal equinox (which was 29 September in their days).[2] As regards Hesiod, Pliny's source is a lost book on astronomy that is ascribed to him. In his *Works and Days*, Hesiod says no more than the following: "When the Pleiades set at the end of the night, then it is the right time to plough," and: "When the Pleiades, Hyades, and Orion set, remember that it is the season to plough" (*Works and Days* 681–682 and 432–433). Pliny adds the observations of two later astronomers: according to Euctemon, the Pleiades set the 44th day after the autumnal equinox (which was on 28 September), and according to Eudoxus on the 48th day after the autumnal equinox (which was on 28 September). If we take the year 700 B.C. for

[1] Before 1582 A.D., the dates of the equinoxes and solstices shift about one day per 128 years on the Julian calendar. This was corrected by Pope Gregory's calendar reform, which resulted in an error of only one day in about 3,000 years. Moreover, to eliminate the 10-day error that had developed since the church council of Nicea, in the same year, 1582 ten days were passed over so that 4 October was followed by 15 October. This is why Table 2.1 differs from that in Couprie (2003: 181), where 23 September was taken as the date of the autumnal equinox throughout.

[2] White reads for Anaximander: "on the 29th [day from the equinox]" (2002: 10). This makes, however, only a few minutes difference: on 28 October 560 B.C., the sun rose at 4:33 h, and the Pleiades set at 4:17 h.

Hesiod and his birthplace Ascra in Boeotia (23°07′E, 38°23′N) as his observation post, 580 B.C. and Miletus (27°15′E, 37°30′N) for Thales, 560 B.C. and Miletus for Anaximander, 430 B.C. and Athens (23°44′E, 38°00′N) for Euctemon, and Athens and 350 B.C. for Eudoxus, then the times of sunrise and the true setting of the Pleiades are as indicated in Table 2.1.[3]

Table 2.1 Ancient dates of the autumnal setting of the Pleiades

Observer	Place	Date of autumnal equinox	Date of observation	Sunrise (universal time)	Pleiades set (universal time)
Hesoid	Ascra	30 September 700 B.C.	30 September 700 B.C.	4:18 a.m.	6:18 a.m.
Thales	Miletus	29 September 580 B.C.	24 October 580 B.C.	4:27 a.m.	4:30 a.m.
Anaximander	Miletus	29 September 560 B.C.	30 October 560 B.C.	4:33 a.m.	4:07 a.m.
Euctemon	Athens	28 September 430 B.C.	11 November 430 B.C.	5:02 a.m.	3:42 a.m.
Eudoxus	Athens	28 September 350 B.C.	15 November 350 B.C.	5:07 a.m.	3:30 a.m.

The last column in Table 2.1 shows the different time of the true setting of the Pleiades. Because of the light of the rising or setting sun, stars are invisible some time before and after sunrise and sunset. Therefore, the ancient astronomers noted the first and the last moment of visibility of a certain rising or setting star. In relation to the rising sun, these are called the heliacal rising and the cosmical setting. The data of the cosmical setting are (with an insecurity margin of 2 days to both sides) as follows: Ascra 6 November 750 B.C., Miletus, 7 November 580 B.C., Miletus 7 November 560 B.C., Athens 8 November 430 B.C., and Athens 9 November 350 B.C. Euctemon's and Eudoxus' figures seem to refer to the cosmical setting of the Pleiades.[4] They correspond rather well to the duration of the astronomical dawn, which is about 1 h and a half for latitudes between 36° and 44° (see Neugebauer 1922: 21, Table 11). Wenskus has tried to explain the data for Thales and Anaximander, which are, respectively, "um etwa zehn Tage zu früh" and "etwa eine Woche zu früh," by suggesting that in the case of Thales we have to read 35 days after the autumnal equinox instead of 25 and that in the case of Anaximander

[3] This table is made with the help of the computer program *Redshift 5.1* (2005), and compared with Neugebauer for the days of the equinox (1922: 49, Tafel 19).

[4] Information from USHA-member Rob van Gent, according to the computer program *Planetary, Lunar, and Stellar Visibility 3.0*. Pannekoek, discussing Hesiod, gives on one and the same page the dates for the cosmical setting of the Pleiades as 12 and 3 November, the last the same as Wright (Pannekoek 1961: 95; Wright 1995: 18). Dicks (1970: 36) has 5–11 November; Bickerman (1980: 112) has 3–5 November for latitude 38° and the years 500–300 B.C.; Wenskus (1990: 250) has 4–6 November for 700–300 B.C. (see also p. 49), and elsewhere: "Ende Oktober – Anfang November" (1990: 176). White has November, and remarks: "the extended size of the cluster makes its rising and setting impossible to determine precisely" (2002: 10).

we must suppose that he had very sharp eyes and was able to see the Pleiades set less than half an hour before sunrise (1990: 53, see also 52 and 60). Although the notion of "cosmical setting" is somewhat vague and depends on the sharpness of sight of the observer, it is certainly impossible to see stars of that magnitude set half an hour before sunrise, even if one takes into account that the sky at the western horizon is still rather dark when the rose-fingered dawn announces the rising sun in the east. I tend to think that Thales and Anaximander, on the contrary, were not concerned with the cosmical setting but tried to fix the precise moment at which the rising of the sun and the true setting of the Pleiades at the western horizon coincide. If Pliny's report is right, they must have been able to calculate in one way or another the elapsed time from the last moment the Pleiades were visible until their true setting, when they are no longer visible. We may conclude that, according to Pliny, Thales' account was better than that of Anaximander. The resulting date for Hesiod, however, is rather strange because at the autumnal equinox the Pleiades set almost 2 h after sunrise. Perhaps he was not yet able to fix the date of the equinox or to calculate the time the Pleiades actually set. As regards the strange figure given for Hesiod, Wenskus has to admit that "die Hesiod zugeschriebene Angabe ungenau" is (1990: 51–52).

Already in ancient times, observers have tried to improve the accuracy of the results achieved with the unaided eye. The first tool is the *human body* itself. The altitude of a celestial body above the horizon, or the angular distance between two stars, can be measured by means of the finger (the *digit*), the thumb (the *inch*), the fist (the *palm*), the stretched fingers (the *span*), and the forearm from the elbow to the middle fingertip (the *cubit*), and so on. Ptolemy cites observations like these: "In the year 82 of the Chaldeans, Xanthicus 5, Saturn was two digits below the Virgin's southern shoulder"; "In the year 75 according to the Chaldeans, Dius 14 in the morning, Mercury was half a cubit on the upper side of the southern Balance" (*Almagest* XI 7 and IX 7). We may assume that the ancient Greeks also measured distances on the firmament in this way. The person on the left of Fig. 2.3 shows that this method was still used much later. Figure 2.4 shows other examples, redrawn after a more recent handbook.

The Egyptians used the human body in yet another way to locate a point in the sky. In the tombs of Ramesses VI, VII, and IX, a kind of star clock is painted, consisting of 24 sitting men (one for the first, and one for the 16th day of each month), above whom stars are drawn, as in Fig. 2.5. The sitting person, obviously an aide of the observer, had to sit all night with his back to the south (or better: the line between the observer and his assistant had to be the north–south line – we would say the meridian of that place). The rows on the right indicate the hours of the night, and the columns behind the sitting man the various positions of the stars, for instance "the star of Sothis (Sirius) on top of the left shoulder." This method of determining a position has become obsolete, not only because of its inherent inaccuracy but perhaps also because it must surely have been a nuisance for the aide of the astronomer to have to sit still all night (see Clagett 1995: 64–65). Bruins has made the interesting suggestion that "the 'target figure' of the star clocks is *not an assistant* of the observing astronomer, but the astronomer *himself*! The painter depicted the seated astronomer and *what he sees* is, independently, drawn 'behind

Fig. 2.3 Several manners of measuring distances at the firmament on an engraving from 1533 A.D. (picture by the courtesy of Adler Planetarium & Astronomy Museum and Cambridge UP)

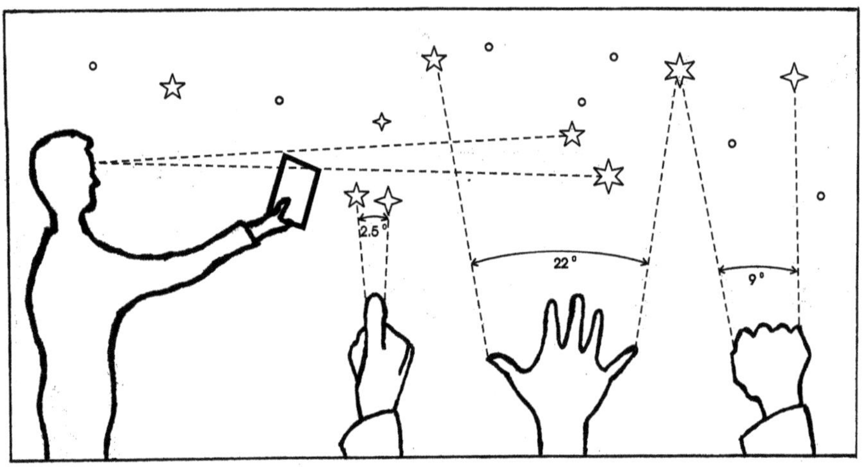

Fig. 2.4 Some more examples of measuring angles between stars (freely after Klepešta and Rükl 1969: 70)

him' in the charts" (1965b: 173). However, his interpretation of the indications "opposite the heart," "on the left shoulder," etc., as "mnemotechnic expressions" sounds not very convincing (cf. 1965b: 174).

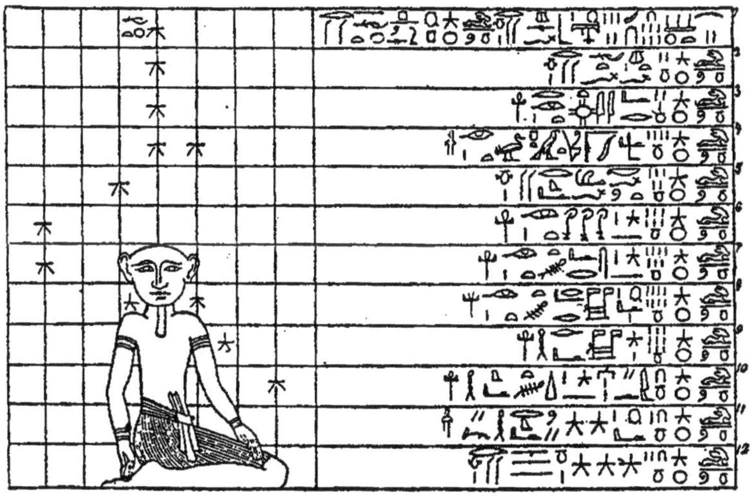

Fig. 2.5 Part of the star clock in the tomb of Ramesses IX (Sloley 1931: 169, Fig. 3)

In the following description of genuine instruments, by which I mean *manufactured tools*, I confine myself not only to the instruments that were used by the ancient Greeks, or with which they could have been acquainted, but also to the period before the discovery of the sphericity of the earth. The instruments that were developed after that have been described sufficiently elsewhere and are of less interest for the scope of this book. It is strange that this subject is scarcely treated in studies on archaic astronomy. Perhaps one reason is that although the Babylonians were experienced observers of the heavens, we do not know whether they made use of any instrument other than the gnomon. Observing instruments are neither mentioned in the texts nor found in the excavations, not even a water clock (see Steele 2008: 45). Although Egyptian astronomy is generally said to be poor as compared with that of the Babylonians, at least three kinds of astronomical instruments have been found in Egypt, as we will see. However, this may be, most authors start with the instruments of Ptolemy. In this respect, even the standard work of Kelley and Milone (2005) has a serious lacuna.[5] Moreover, Dicks' article and Gibbs' chapter on ancient astronomical instruments, in spite of their titles, give less than one would expect (Dicks 1954: 77–85; Gibbs 1979: 39–59). The same holds for Thurston's chapter on the astronomer's tools where mainly later and more sophisticated (and especially Chinese) instruments are treated (1994: 26–44).

The person on the left in Fig. 2.4 uses a little *piece of board* to compare the angular distances of stars. A similar instrument is described by Simplicius (*In Aristotelis De caelo commentaria* 504.16 ff.), explaining how one can easily see that the moon does not always have the same angular diameter. A disk held at a certain distance from our

[5] Kelley and Milone use another definition of "Archaeoastronomy" than in this book, namely, "the practices of pretelescopic astronomy" (2005: vii).

eye sometimes needs a diameter of 11 in. to cover the moon but 12 in. at another time of the year. This is not to be confused with the well-known phenomenon that the moon looks bigger at the horizon than in the zenith, and which can be shown by the same instrument to be an optical illusion. In one of the following sections, we see how a similar tool can also be used to measure the angular diameter of the sun.

One is well advised not to study the sun by direct observation because of the danger of eye damage. To observe the sun, and solar eclipses in particular, people used the reflection on *the surface of a liquid*, for instance olive oil, poured into a bowl, as described by Seneca (*Naturales Quaestiones*: 1, 11.3–12.1). The reader can find out for himself that it is possible, after some time of eye accommodation, to observe a distinct reflection of the sun disk, even at noon. I used this method myself, with perfect results, to observe the partial sun eclipse on 1 August 2008 at Maastricht. In his allegory of the cave, Plato hints at this method when he says that the prisoner, who is freed from the cave and arrives at the surface of the earth, sees the sun "without using its reflections in water or another medium" (*Republic* 516b). Elsewhere he speaks of the risk people run to injure their eyes when looking at a solar eclipse, "unless they study its image in water or something like it" (*Phaedo* 99d). Another way to observe the sun, and in particular a solar eclipse, is with a *camera obscura*, where the light of the sun is captured through a little hole, throwing an reversed image on the opposite wall. Aristotle seems to hint at it somewhere, but it is doubtful whether the Presocratics were already acquainted with this method (*Problems*, book xv, Chap. 6). Thales could have used one of these methods for his observations that led to the prediction of a solar eclipse.

To avoid the unevenness of the real horizon, people may have used an *artificial horizon*, like the little circular wall in Fig. 2.6. With this device they could, for instance, determine the north. From the center of the circle, the observer notes where a certain star rises above the wall, and he puts a mark there. In the same way, he puts a mark where the star sets. The bisector that divides the angle from the observer to the marks into two equal parts will point to the north.

Somewhere – though not in a book on astronomy – Aristotle mentions the *sighting tube* (*Generation of Animals*: 780b 19–2 and 781a 9–12). A sighting tube (Greek: διόπτρα, but Aristotle speaks of an αὐλός) is a hollow tube, put on a stand, a kind of telescope without lenses (see Evans 1998: 33 and 34). A sighting tube facilitates the observation of stars at daybreak or in the evening twilight, by keeping out the atmospheric light from view. Also at night the observation of stars is improved by the use of a sighting tube (thus Eisler 1949: 314). Observing from the bottom of a deep pit or shaft has the same effect. Even during daytime, the stars are visible with this method, says Aristotle.[6] In the Arabic and European Middle Ages, deep pits were said to be used as observation wells (see Sayili 1953: 149–155). Perhaps this makes sense of the story of Thales falling into a well while looking at the stars. He may have descended into a well on purpose with the intention to make use of its sighting tube function (see Eisler 1949: 324, n. 13).

[6] See also Strabo, *Geographica*, ed. H.L. Jones (Strabo 1923, vol. II: 10).

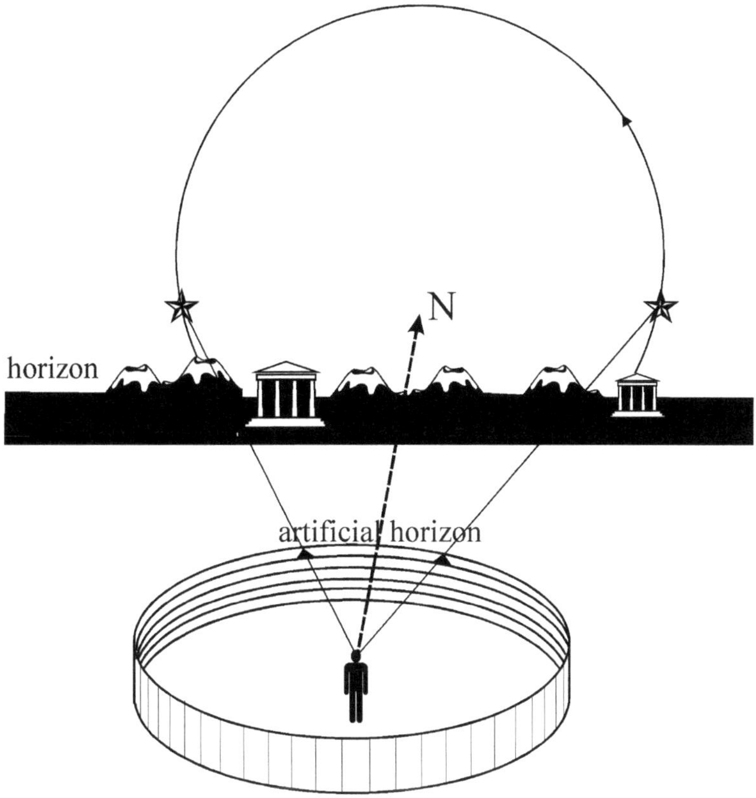

Fig. 2.6 Finding north with the help of an artificial horizon

The sighting tube made it possible to find the north in another way as well. If one points the tube toward an arbitrary star, this star will disappear from sight after a certain time, because of the turning of the celestial vault. If one points it toward the Polar star, this will remain visible all night. Since the Polar star, because of the precession, was in archaic times further removed from the actual celestial pole than today, another star was nearer to the pole. In Anaximander's time, a star just at the limits of human visibility stood almost at the north celestial pole (FK3037, magnitude +6.00, at about 89°27′), but the ancients probably preferred Kochab in the Big Dipper (magnitude +8.00, at about 83°09′). To find the north, one would have to point the sighting tube – of suitable size and fitted on a stand – in such a way that that star described a small circle in the visual field of the instrument. The center of this circle is the north pole of the heaven (Eisler 1949: 313). Figure 2.7, the original of which dates from more than three and a half centuries before the invention of the telescope in The Netherlands, shows that in the Middle Ages the sighting tube was still in use. In Chap. 16, we discuss a rather spectacular measurement with the help of such a sighting tube.

Fig. 2.7 The personification of astronomy with a sighting tube. Drawing (c.1880 A.D.) after medieval manuscript (1241 A.D.). For the original, see http://www.manuscripta-mediaevalia.de/hs/katalogseiten/HSK0523b_a040_jpg.htm. (© Hermann, www.editions-herman.fr)

Diogenes Laertius relates that Thales maintained that the diameter of the sun equals 1/720th part of its orbit (and that the same holds for the moon).[7] This is confirmed by Apuleius (DK 11A1(24) and 11A19). In other words, 720 suns placed in a row add up to the full circle of its daily orbit. This is correct because the angular

[7] Gobry, who reads this text as "Selon Thalès (...) la course de la lune est le cent vingtième de celle du soleil", is twice mistaken (2000: 172).

(or apparent) diameter of the sun (which is the angle between our eye and the both ends of the sun's diameter) equals approximately half a degree (see Fig. 2.8). As with all doxographical accounts on Thales, we have to be careful about the truthfulness of this statement. To measure the angular diameter of the sun implies that Thales would have developed the idea that the celestial bodies pass underneath the earth during their daily course along the firmament, making a full circle. This idea, however, is not consistent with his world picture, at least as far as we are acquainted with it. As we discuss in Chap. 4, Aristotle says that according to Thales, the earth floats on water like a piece of wood. This representation is difficult to combine with the idea of celestial bodies making full circles around the earth, which implies that the earth hangs freely in space instead of floating on water. The unsupported earth is, as we see in Chap. 8, Anaximander's conception. If the account on the measurement of the angular diameter of the sun is based on truth, then Anaximander probably has to be credited with this achievement rather than Thales.

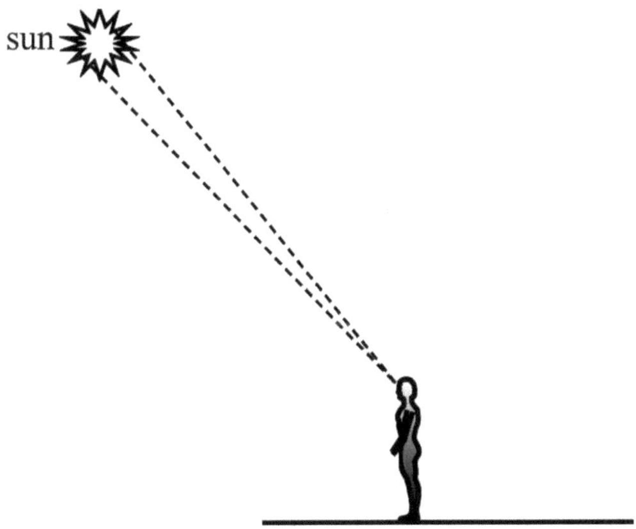

Fig. 2.8 The angular diameter of the sun

He could have made this measurement with the help of a *water clock*, the so-called *clepsydra* (κλεψύδρα, "water thief". See Fig. 2.9). This instrument was already used by the Egyptians about 1350 B.C. and even earlier (see, e.g., Lull 2006: 137–139; Clagett 1995: 65–82 and plates III. 21a–35). The principle of a *clepsydra* is the same as that of an hourglass. Its use is described by Cleomedes in the second century A.D. The picture shows a primitive Greek *clepsydra*, consisting of two containers, one placed above the other. The upper vessel is continuously filled with water that slowly drains away into the lower vessel in a steady stream. When the lower vessel is full, it is replaced by an identical empty one, so that a measurement of time in equal units is achieved. Cleomedes describes the experiment as follows: "During the time from the first appearance of the sun above the eastern horizon until the time the whole sun is visible above the horizon, one vessel of the water-clock will be filled. When one lets the water stream out day and night, until the next sun rises above the horizon, about 750 vessels will have been filled. Therefore, the diameter of the sun equals 1/750th part of its entire orbit" (*De motu circulari corporum celestium* 2.75, at p. 136).[8] Cleomedes' result differs from the 1/720th part mentioned by Diogenes Laertius. This, however, will be due to the intrinsic inaccuracy of the measuring method used (see Dicks 1954: 84: "it was liable to constant error").[9]

Fig. 2.9 A simple Greek *clepsydra* (water clock) (photograph by the courtesy of Agora Excavations, The American School of Classical Studies at Athens)

It is interesting to compare Cleomedes' *clepsydra* with an Egyptian specimen of 14 in. height, found in Karnak and dating from about 1400 B.C. This *clepsydra* was supposed to empty in one night. On the inside of the vessel, inscriptions are made that indicate the water level for the hours of the night at different times of the year (see Fig. 2.10). The length of the night at Karnak varies during the seasons between 610 and 820 min. Although we would say that it does not matter whether one

[8] Wasserstein tries to make acceptable that Thales would have used another method than that with the *clepsydra*, since his result differs from that of Cleomedes (1955: 114–116). Thales' result of 1/720, Wasserstein says, is obviously inspired by the hexagesimal system, in which the circle is divided in 360°. His argumentation, however, is not convincing. Given the inaccuracy of the measuring method, Thales – or whoever performed the calculation – could very well, for instance for aesthetic reasons, have brought his results in line with the hexagesimal system. Moreover, Wasserstein gives no indication of what other method Thales should have used.

[9] The clepsydra on Fig. 2.9 is in the Athenian Agora Museum. It is said to be used to control the length of a testimony in the Dikasterion. When the water stopped flowing, everyone yelled "sit down" to the speaker (information by Robert Hahn). Of course, this does not exclude the possibility of using the clepsydra for astronomical purposes as well.

measures the hours of the day or those of the night, an inscription on the *clepsydra* tells that it was meant to measure the hours of the night. Perhaps its primary use was to let the priests know the exact time to say prayers.[10] When we compare this instrument with Cleomedes' *clepsydra*, we may conclude that it emptied about 375 times faster than the Karnak *clepsydra*, which is an amazing difference. The Egyptian *clepsydra* must have had a very small aperture of less than one millimeter in diameter. The Egyptians probably put a metal (golden) orifice with the desired diameter into the aperture of the *clepsydra* (Cotterell and Kamminga 1990: 62; Sloley 1924: 45, n. 5). Given the volume of the Karnak *clepsydra* of about 22 l, and the volume of a droplet of 50 μl, one may calculate that ideally speaking, the water had to drop out at a steady rate of about ten drops per second (Turner 1984: 46; Sloley 1924: 45). It would have been a question of trial and error to find the diameter that produced the right speed of the outflowing water.

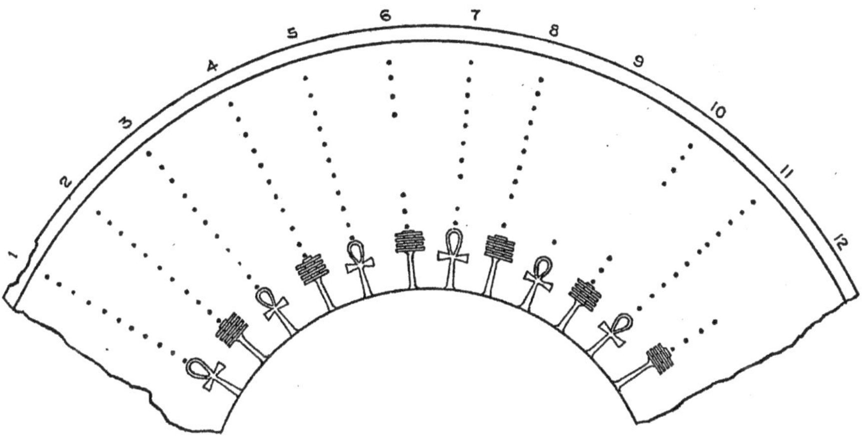

Fig. 2.10 The month scales on the inside of the Karnak *clepsydra* (Sloley 1924: 46)

Another problem was that the velocity of the water flow decreases as the water level drops. The Egyptians tried to solve this problem by making the vessel conical, the lower diameter being smaller than the upper diameter, so that the lower water pressure was compensated by the smaller amount of water flowing. As the water pressure is reduced to zero when the water-level reaches the bottom of the vessel, they made the aperture a bit above the bottom. Recent calculations and experiments have resulted in the conclusion that "the clock would have been accurate as far as the Egyptians were concerned" (Cotterell and Kamminga 1990: 63, and especially Fig. 3.10).

According to White in a recent handbook, Anaximander could have measured the apparent diameter of the sun in another way, using a tool similar to the piece of board already described above. He writes: "if the disk of a cup or plate held at arm's length covers the rising sun, then the disk can be used to measure the horizon by counting how many of the diameters of the cup or plate it spans" (2008: 109).

[10] Information by Robert Hahn.

In order to prevent eye damage and to get a better picture of the sun disk, one has to better wait until the sun is somewhat clouded. White recommends using an artificial horizon like that in Fig. 2.6 to provide for a perfect circle. However, in what follows, he makes the procedure needlessly complicated by suggesting a pointer turning along the horizon. The easiest way would be, instead of holding it in the hand, to put a disk of appropriate size on an artificial horizon while standing at its center and divide the circumference of that horizon by the diameter of the disk. Just like with the use of the *clepsydra*, this procedure avoids calculating with π. On the other hand, it presupposes some knowledge of the laws of perspective to elucidate that the angular diameter of the sun has to its circular path around the earth the same ratio as the diameter of the disk to the circle of the artificial horizon.

Apparently, White did not perform the experiment he describes himself. Otherwise, he would have noticed that not "the disk of a cup or plate," but a vitamin pill (with a diameter of about a quarter of an inch) held at one arm's distance will cover the sun. Probably, this is the reason why he states that "the results are bound to be wildly inaccurate" (2008: 109). On the contrary, they are acceptably accurate, as the reader can easily check by performing the experiment. They remain within an acceptable range, comparable with that in measuring with the *clepsydra*. In a somewhat older handbook, it has already been stated that "(the) value of ½° (...) can be ascertained by the most simple observation" (Pannekoek 1961: 120). The error definitely cannot be a factor 4, as White surmises when he conjectures that Anaximander took the angular diameter of the sun to be 2° instead of a half degree. White makes a curious mistake when he writes that the result is impoverished because "the atmosphere makes the sun appear larger on the horizon than in the sky" (2008: 109). As already explained above, this mistake can be exposed with the help of the same tool. The disk that covers exactly the sun high in the sky will be seen to cover exactly the sun at the horizon as well. Moreover, it is not the atmosphere that causes this illusion. This is a misunderstanding introduced by Aristotle (*Meteorologica*: 373b12–13), as can be read in any book on optical illusions.[11]

The *gnomon* (γνώμων) is usually considered as the most important instrument of archaic astronomy. On the operation of sundials, many books have been written.[12] Usually, however, relatively little attention is paid to the simplest and oldest sundial, the upright gnomon. A gnomon is nothing but a stick or staff put vertically into the ground, the shadow of which can be studied. Any other vertical object, an obelisk for instance, or even the upright human body itself may function as a gnomon as well.[13] Diogenes Laertius says that Anaximander invented the gnomon (DK 12A1(1)). This report must be false, since the gnomon had been in use for centuries all over the world, for instance in Mesopotamia. The oldest records of Babylonian observations with the help of a gnomon, dating from 687 B.C., are preserved in a number of clay

[11] For quick information, see the article "Moon illusion" in Wikipedia.

[12] A good introduction still is, for instance, Mayall and Mayall (1938). A survey of ancient sundials can be found in Gibbs (1976).

[13] See, e.g., Pliny's description of the obelisk that was erected on the Campus Martius in *Naturalis historia* XXXVI, 72.

tablets that are called ^{MUL}.APIN, after their first words. They contain, among other things, tables indicating when the shadow of a standard gnomon has a certain length.[14] Herodotus says somewhere (*Histories* II 109) that the Greeks learned the use of the gnomon from the Babylonians. Probably, then, we have to explain Diogenes Laertius' report in this way that Anaximander introduced the gnomon from Mesopotamia into Greece.[15] Diogenes Laertius and others also report that Anaximander had erected in Sparta an instrument for measuring the hours, and that he used the gnomon not only to measure the time but also to determine the solstices and the equinoxes (DK 12A1(1), DK 12A2, and DK 12A4).

Usually it is said that the gnomon is in the first place an instrument for telling the time of the day. So, for instance, Van der Waerden says: "Der Hauptzweck des Gnomons ist, aus dem Gnomonschatten die Tageszeit zu erkennen" (1965: 254).[16] It may be doubted, however, whether this is as simply true as it sounds. Imagine that you walk around with a stick and want to know the hour of the day. You put your stick perpendicularly in the sand and study the length of its shadow. What does it tell you? All you know is that at different times of the day the shadow has different lengths and that the length of the shadow varies with the seasons. I think you had better throw your stick away, remember where the south is and look directly where the sun stands. Before we discuss a method to handle the problem of telling the time with an upright gnomon by fixing it at one place, we deal with the use of the gnomon as an astronomical instrument.

Pedersen and Phil say: "Even with a simple gnomon it is possible to perform a large number of measurements fundamental to astronomy" (1974: 42).[17] Nevertheless, inspite of its various uses, the gnomon remains a rather limited astronomical instrument because it is, so to speak, the instrument of the day, whereas ancient astronomy is mainly the science of the night sky. Local noon is the only time of the day that can be determined with a gnomon rather precisely, and in different ways, as we see in due time. When employed to find out local noon, the gnomon functions not only as a time indicator but also as an astronomical instrument because it determines the north–south line, since at local noon the sun is at its highest and stands exactly in the south. The first method is to study carefully the shadow of a gnomon during the day and note its smallest length. At that moment, the gnomon's shadow lies exactly on the meridian of the observer, which is the circle that runs through both poles of the earth. Of course this is something the ancients could not know because it presupposes knowledge of the sphericity of the earth. This method, however, is too inaccurate, as differences in length of the shadow are very hard to

[14] ^{MUL}.APIN means as much as "the Plough star." It is a small constellation, consisting of our constellation Triangulum and the star δ Andromedae.

[15] In a recent study, Haase has held the somewhat strange opinion that Herodotus' text must be read in the sense that Anaximander "im Unterschied zum altorientalischen Verständnis dieses Messtechnischen Instruments den Gnomon erstmals *als Medium* begriff" (2008: 18, my italics).

[16] I did it myself in Couprie (2003: 185).

[17] See also Sarton (1959: 174): "A relatively large amount of precise information could thus be obtained with the simplest kind of tool."

perceive during a considerable time around noon, and especially in winter, when around noon during more than an hour the differences in altitude of the sun are no more than 1°. This handicap bothers all instruments that are based on an upright pointer, such as the Egyptian *merkhyt* that is treated hereafter.

The second method to determine noon is more precise and consists of bisecting the angle between an arbitrary morning shadow and an evening shadow of the same length. This is shown in Fig. 2.11, where G indicates the point where the gnomon is put into the ground, GA the morning shadow, GB the evening shadow of equal length, and CG the bisector of the angle AGB. An extra check can be made, as the lines CG (north–south) and BA (east–west) must be perpendicular to one another.

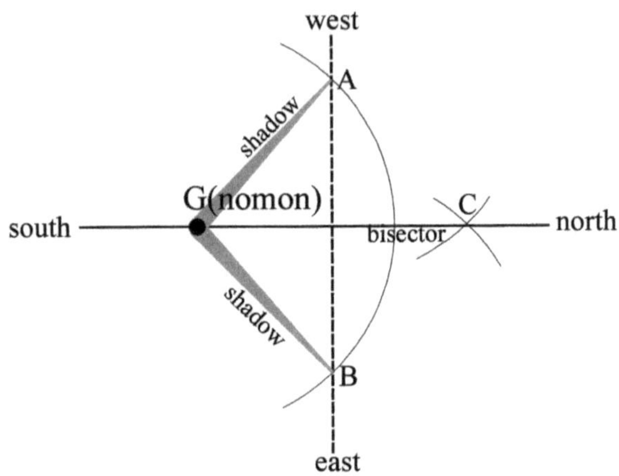

Fig. 2.11 The determination of the noon line with the help of a gnomon

This method is the complement in the day time of determining the north with the help of a rising and setting star, as visualized in Fig. 2.6. However, when you walk around with a stick and want to know when it is noon, these methods will not help you, as noon will be already past when you have finished your observations. With the first method you will notice, when the shadows become longer again, that some time ago it must have been noon, and with the second method you will have to wait for the afternoon shadow to see that some hours ago it was noon. When used to find out noontime, the gnomon functions not so much as a time indicator, but rather as an astronomical instrument determining the north–south line. In addition to Figs. 2.6 and 2.11, at the end of this chapter we discuss another method of finding north.

In a nice little article, Neugebauer has shown how the Egyptians could have used a similar method to orientate their pyramids exactly north–south (1980: 1–3). All they had to do was to take a small but accurately shaped pyramid RSTU with top P (for instance the *pyramidion*, the top of the pyramid itself), and put it roughly on a north–south orientation on a completely flat and horizontal base, where the actual pyramid had to be built (see Fig. 2.12). Then, they had to wait till the morning shadow SMR of the small pyramid was an as-exact-as-possible continuation of the western base UR of the pyramidion to measure the length UM. The same procedure had to be performed in the afternoon, at the time when the shadow RAS was an as-exact-as-possible continuation of the eastern base TS of the small pyramid, and TA could be measured. If after this procedure UM and TA proved not to be of equal length, they had to turn the small pyramid somewhat, and repeat the procedure the next days, until the shadows were of equal length and the big pyramid could be aligned and orientated. The best measurements can be obtained during the winter months, when the sun is lower on the horizon and the shadow of the pyramid is sufficiently long. However, as it is not so easy to construct a perfectly shaped *pyramidion*, nor a perfectly horizontal floor, and as it is rather difficult to determine whether the shadows are exactly equal in length, this method may suffer from inaccuracies.

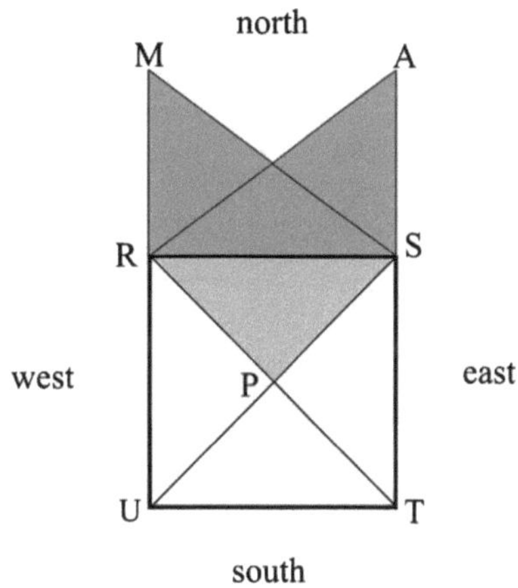

Fig. 2.12 The north–south orientation of a pyramidion (somewhat adapted after Neugebauer)

Doxographical reports tell us that Anaximander observed the (dates of) the solstices and equinoxes. On the equinoxes, 27 March and 29 September, respectively, in the days of Thales and Anaximander, day and night are equally long. At the summer solstice (29 June in Anaximander's days), the noon shadow of the gnomon is at its shortest, and at the winter solstice (26 December in Anaximander's days), it is at its longest. These dates could only approximately be established, according to Dicks "probably to an accuracy of at best some five or six days" (1966: 29). The reason is that during some days around the solstices, there is hardly any difference in the shadow length at noon.

The angle made by the top of the gnomon and the end of its shadow at the time of the solstices can be measured and will show to be about 47° (see Fig. 2.13). This angle equals twice the inclination of the ecliptic (which is the sun's yearly orbit around the starry sky) in relation to the celestial equator (which is the projection of the earth equator on the sphere of the sky). Acquaintance with the obliquity of the ecliptic presupposes knowledge of the sphericity of the earth. This knowledge, however, is not required for measuring the angle between the shadows of the summer and winter solstices with a gnomon. As Sarton says, speaking about Anaximander: "It was possible (. . .) from the observations he made with a gnomon (. . .), to measure the obliquity. Yet, even if Anaximander measured the obliquity, one could hardly say that he understood it" (1959: 292).

On the days of the equinoxes, day and night are of equal length. On these days, the sun rises exactly in the east and sets exactly in the west. With the gnomon, the

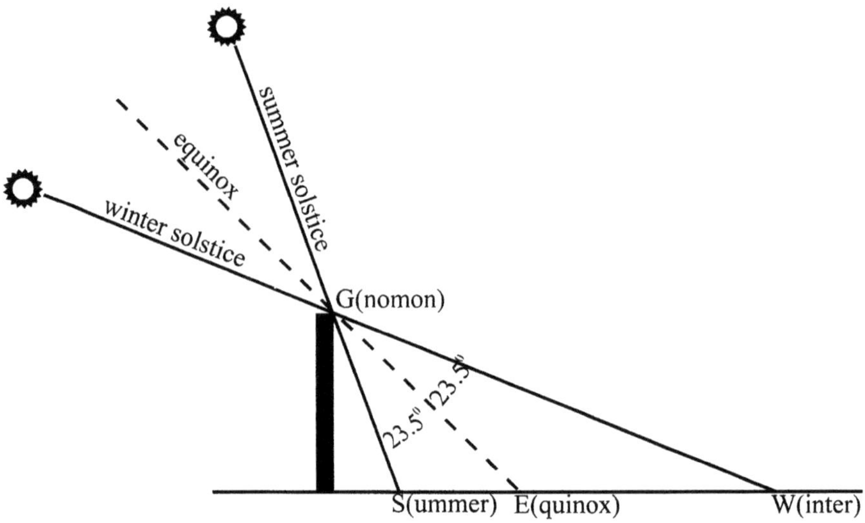

Fig. 2.13 Fixing the shadow points of the equinox and the solstices, and measuring the obliquity of the ecliptic with the help of a gnomon

equinoxes can be found in various ways. The first method is to bisect the angle of the shadows thrown by the gnomon at the summer and winter solstices and to note the day when the shadow reaches the point on the ground found in this way (E in Fig. 2.13). This method is necessarily not very exact, because of the difficulty of measuring the angles at the top of the gnomon and the insecurity of fixing the exact dates of the solstices. The second method is to note on which calendar day the earliest morning shadow and the latest afternoon shadow are just opposite one another. This method too, is not very precise, as it requires a completely smooth horizon on both sides. The third method, which is better, consists of observing on which calendar day the top of the shadow of the gnomon describes a straight line during the day. This line is, for instance, marked on the plate of a Roman sundial (see Fig. 2.16). Contrary to what is sometimes said, none of these three methods presupposes knowledge of the sphericity of the earth, or the idea of a celestial sphere, on which the equator, tropics, and ecliptic are projected.[18] The curves of the

[18] Dicks is wrong when he writes: "the equinoxes cannot be determined by simple observation alone" (1966: 31). And also elsewhere: "The concept of the equinoxes is a more sophisticated one, involving necessarily the complete picture of the spherical earth and the celestial sphere with equator and tropics and the ecliptic as a great circle" (1966: 30). It is also not right to say that "these concepts are entirely anachronistic for the sixth century B.C." (1966: 30; see also 1970: 45). Of course, the ancient ways of fixing the equinoxes and solstices did not possess the grade of accuracy we would expect nowadays. See also, for instance, Fotheringham: "The determination of the exact date of a solstice remained a difficulty throughout the whole course of ancient astronomy. Even Ptolemy deduced from his own observations a date 38 h later than the true date for the summer solstice" (1919: 168).

shadow of the gnomon top during any day, with the exception of the equinoxes, are hyperbolas, the extremes of which are those of the two solstices. That they are hyperbolas was, of course, not yet known, as is clear from the way in which they are rendered in Fig. 2.16. This does not alter the fact that any ancient observer could observe and draw them.

The gnomon can also be used to determine the observer's latitude by measuring the angle of the shadow at the top of the gnomon at an equinox (∠BGE in Fig. 2.14; the latitude depicted is that of Miletus). Of course, this figure makes sense only if one is acquainted with the earth's sphericity.

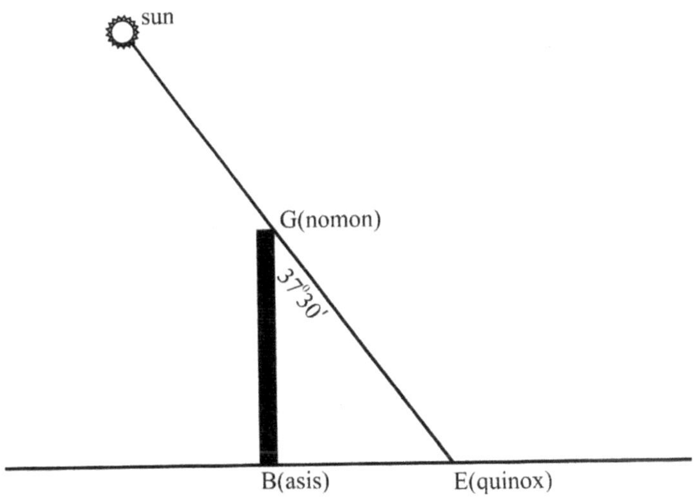

Fig. 2.14 Measuring the observer's latitude with the help of a gnomon

Another possibility is to determine the azimuth of the sun at any time of the day, as in Fig. 2.15. The azimuth is the bearing of an object measured as an angle around the horizon eastward starting from north as the zero point. As is clear from the drawing, one has to determine a north–south line first; the angle between this noon line and the shadow of the gnomon indicates the azimuth. As the stars do not throw shadows, the method at night is somewhat different. To determine the azimuth of a star, you will have to place the gnomon at a certain distance and notice the moment that the star is hiding behind it. Then, the angle of azimuth between the line from the observer to the gnomon and the north–south line can be measured. Combined with measuring the altitude of the star above the horizon with the methods of Figs. 2.3 and 2.4, a rather acceptable determination of the star's position can be obtained. I do not know whether the ancients really used this method. The Egyptians, at least, seem to have preferred the much less precise method of the above-mentioned Rammessian star clocks (see Fig. 2.5), whereas the Babylonians identified the position of the moon and planets by indicating their distances to the so-called Normal Stars (see Steele 2008: 42–44).

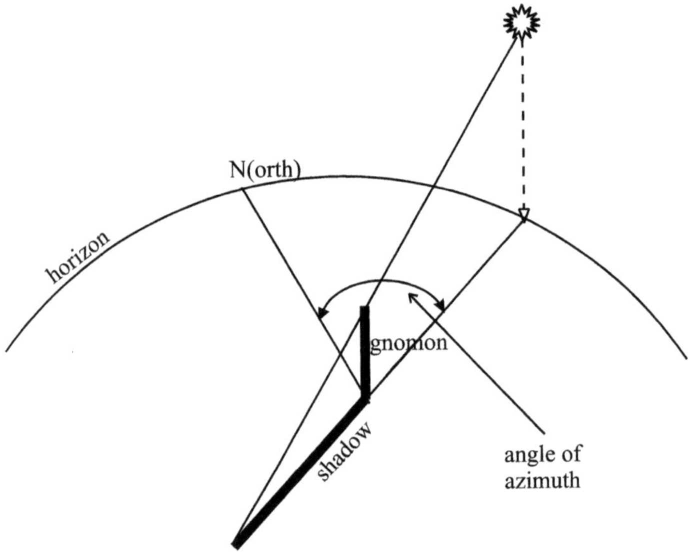

Fig. 2.15 Measuring the azimuth with the help of a gnomon

Now, let us return to the problem of making the gnomon a time indicator. If you cannot use your gnomon as a time teller when you are traveling around, you may decide to put it permanently somewhere, for instance, at the marketplace. Then, you can construct converging hour lines that indicate the time of the day in different seasons, like on the ground plate of the Roman sundial in Fig. 2.16. The black spot on that picture is the place where a vertical gnomon was erected. The idea is that the tip of the shadow of the gnomon touches the same hour line at different points, depending on the time of the year. This is the way it is described in Kirk et al.: "the ground near the gnomon was calibrated so as to give the time of day" (2009: 103). During the day, the tip of the shadow describes a curve. The outermost curves, drawn at the days of the solstices, are indicated (although not as curves but as broken lines) in Fig. 2.16. On the days of the equinoxes, the shadow of the tip of the gnomon does not show a curve but a straight line, as also indicated in Fig. 2.16 between the two solsititial curves. The doxography tells us that Anaximander erected a gnomon in Sparta to observe the solstices and equinoxes and to measure the hours (DK 12A1(1), DK 12A2, and DK 12A4). If these reports can be trusted, the simplest way to understand them is to suppose that Anaximander drew a pattern of lines similar to that of the Roman sundial on Fig. 2.16.

To construct the hour lines, Anaximander could have proceeded as follows. First, at the day of an equinox, he marked the point of a morning shadow of the top of his gnomon that fell neatly within the ground plate of his sundial (cf. the right end of the equinox line in Fig. 2.16). On the same day, he marked the point of the evening shadow of the same length at the other end of the equinox line. With the help of a clepsydra, he divided the equinox line between these points into equal time portions (let us assume ten, as in Fig. 2.16), called "hours" (which do not coincide with our hours of 60 min). He observed that equal time portions did not result in equal distances on the equinox line. Subsequently, at the time of the summer solstice, he marked on the curved line of the summer solstice the point of the shadow at noon and then, after the lapse of five successive afternoon "hours" (measured by the clepsydra), the points of the afternoon shadows. He mirrored these points to get the morning hours. Finally, he connected the same hour points on the curve of the summer solstice and on the line of the equinox and thus found the hour points on the curve of the winter solstice (the first and last of which are, on Fig. 2.16, outside the circle of the ground plate). In this way, the hour lines resulted. Now, at whatever day of the year the point of the shadow of the gnomon fell on, e.g., the second hour line in the morning, it was said to be the second hour in the morning.

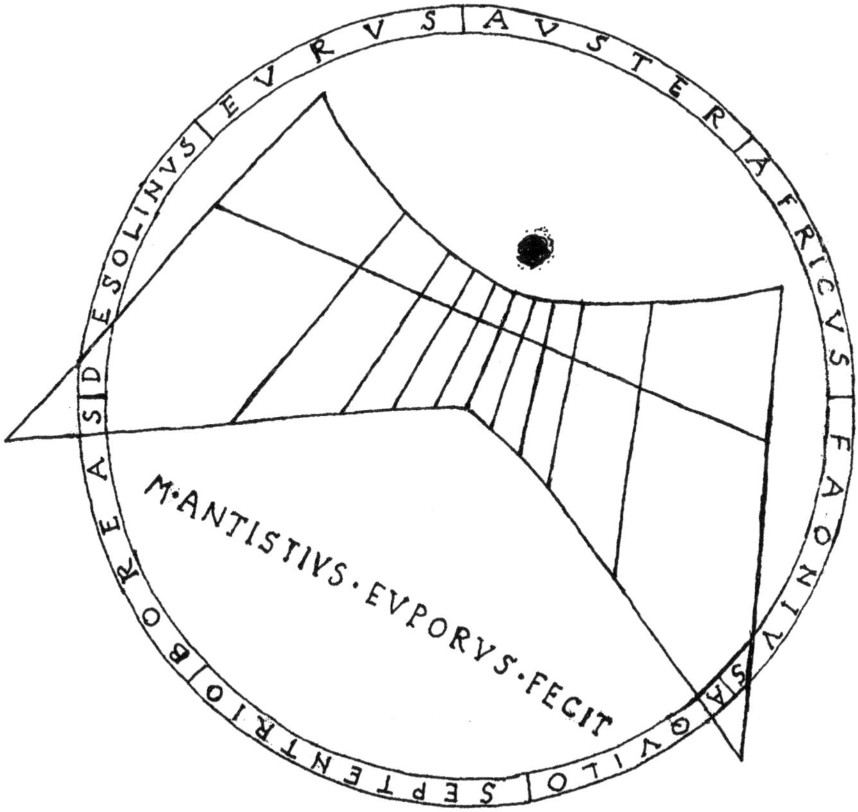

Fig. 2.16 Floor of a Roman sundial (first or second century B.C.), with hour lines, equinoctial line and solstitial curves (drawing by Hans Exterkate)

As a commentary on Kirk's lines quoted above Dicks wrote: "there can be no question of the calibration of 'the ground near the gnomon... to give the time of day'." This is, as he says, "owing to the fact that the altitude and azimuth of the sun are continually altering, no one set of markings applicable all the year round can be formulated to indicate the division of the day into parts" (1966: 29). Against the background of the reconstruction attempted above, this verdict is too harsh. The division of the day into equal parts ("hours") as shown on Fig. 2.16 would have been sufficient for practical purposes in Anaximander's time. An obvious handicap of the sundial as represented in Fig. 2.16 is that it does not show the early morning and late afternoon hours in summer, when the days are longer. This is because its calibration starts from the equinoctial hours. Drawing more intermediate curves and constructing more hour lines for that season could solve this problem. But then, another problem arises, as the resulting morning hour lines lie before what was called "the first hour." Another evident difficulty is that you will always have to run to the gnomon on the marketplace (or wherever it stands) when you want to know the time of the day. When you are at a certain distance of the marketplace you had better spare you the trouble and simply look at the sun to know approximately what time it is.

The problem of telling the time while walking around with a stick still bothered people as late as the eighteenth century A.D. This is shown in an English almanac of the year 1712 A.D., in which for every single month of the year tables of shadow lengths in southern England with their corresponding morning and afternoon hours were published (see Isler 1991b: 170–171). Borchardt mentions an Egyptian table that, however, is so fallacious that he is not even able to conclude from it in which month the summer solstice must be placed. Another table from Taifa in northern Nubia is so inaccurate that it may only function as a very rough rule of thumb (1920: 27–32). After all, we may not suppose that the ancients used to carry around such tables to translate the length of the shadow of their gnomon into the time of the day.

Notwithstanding the above-mentioned proviso, Dicks is basically right when he writes: "observations of the shadow of a gnomon can give only the roughest indication of the time of day, unless the gnomon is so placed that its axis is parallel to the axis of the earth" (1966: 29). The habit of placing the gnomon at an angle, parallel to the earth's or celestial axis (which amounts to the same), however, was developed much later, according to some, in the first century A.D. (Mayall and Mayall 1938: 15). This is the way the gnomon can still be seen on numerous sundials today. When the gnomon is placed parallel to the celestial axis, one reads the shadow of the entire gnomon (not only its top) on a scale.

There is another way of using the gnomon, which is ascribed to Thales, and which at first sight has nothing to do with astronomy. It will, however, appear to have consequences for archaic cosmology, as is shown in Chap. 16. Plutarch tells us that Thales used a gnomon to measure the height of a pyramid. To illustrate his description, I have inserted capitals in his text corresponding with those in Fig. 2.17: "You set up a stick (GH) at the end of the shadow cast by the pyramid, so that by means of the sunbeam that touches both the top of the pyramid and that of the gnomon, you have made two triangles (AGH and APQ). Then you have shown that the ratio of the one shadow (of the pyramid, PA) to the other one (of the stick, GA) is the same as that of the (height of the) pyramid (PQ) to the (length of the) stick (GH)" (DK 11A21, my translation). Thales probably tried to measure the height of the Great Pyramid of Giza (Cheops' pyramid) that is neatly oriented north–south, as we saw.

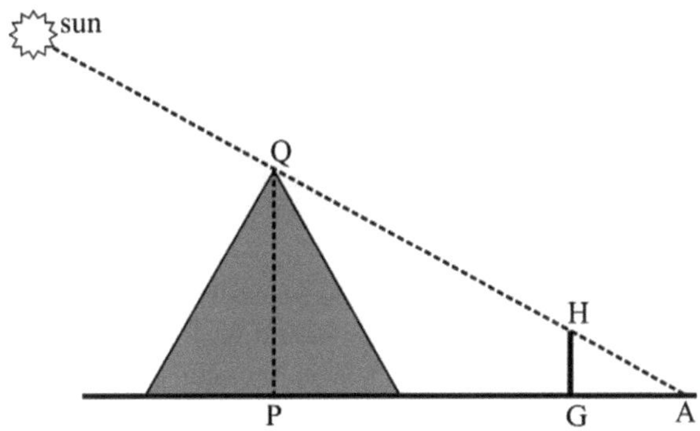

Fig. 2.17 How Thales measured the height of a pyramid

Thales would have had to solve two other problems, before he could measure the height of the pyramid.[19] The first problem was that he had to measure the distance AP, whereas P is hidden in the center of the pyramid. To measure this distance (and taking for granted that the pyramid had an exactly square base), Thales would have had to put his gnomon right in front of a point halfway the side of the pyramid, opposite to the sun at noon (the pyramid is, as we have seen, aligned north–south). In Fig. 2.18, which is in plan view, G is the base of the gnomon and P the hidden center of the pyramid, right below its top. Then, the line GP is perpendicular to SR, which it cuts into two equal halves. SCP is an isosceles right-angled triangle, from which follows that SC = PC. Now, the total length of the shadow of the pyramid is the addition of two lines of known length. SC (=PC) + CG (in Fig. 2.18) + GA (in Fig. 2.17).

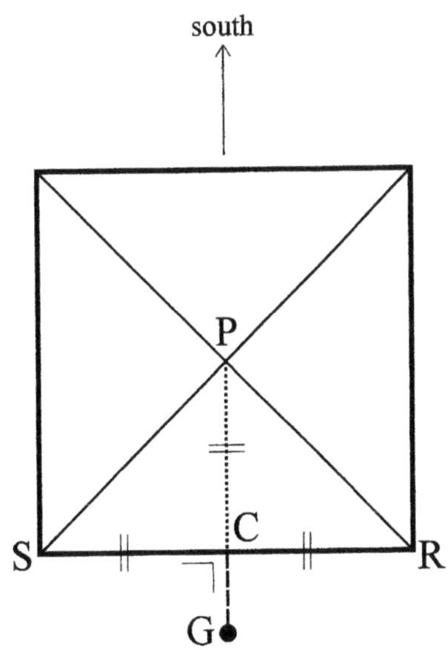

Fig. 2.18 Measuring the distance GP

[19] These were the problems Carlo Rovelli's students were confronted with when he asked them to repeat Thales' measurement.

If SR in Fig. 2.18 is the northern base of the pyramid, the line GC points to the south, the direction of the sun at noon. However, if you try to measure the Great Pyramid's shadow, the second difficulty is that during a considerable part of the year the pyramid does not cast a shadow at noon. This is because the angle of its sloping sides is about 52° to the horizontal. Since the Great Pyramid is at 30°N, the sun at the equinoxes is 60° above the horizon. At the summer solstice, at noon, the sun even gets as high as 83.5° above the horizon. At the winter solstice, the altitude of the sun at Giza is about 36.5°. So Thales had better perform his measurement in winter. Another possibility would be for him to face the west or east side of the pyramid and watch the sun in summer a few hours after its rising or before its setting, when the sun is due east or due west, and not too high in the sky.

An easier way to measure the height of a pyramid is to wait until the shadow is exactly equal to the size of the gnomon. Then, the shadow of the pyramid is also equal to its height. According to Burch this method fails because "a pyramid with a 45° slope (and the Egyptian pyramids are nearly that) casts no shadow at all under the circumstances required by the rule" (1949–1950: 139). Burch is too pessimistic, as the slope of all important pyramids is 50° or more, except one (the north or red pyramid of Snefru) that is 43.5°.[20] The slope of the Great Pyramid is, as we have seen, about 52°. In Thales' time (600 B.C.), the transit altitude of the sun at Giza was 45° on 14 February and 7 November. This means that at those days the shadow at noon fell far enough outside the pyramid to be measured, whereas the length of the shadow of the gnomon was equal to the length of the gnomon, and accordingly the length of the shadow of the pyramid (measured from right beneath its top, hidden inside the pyramid) was equal to its height.

Let us return to Fig. 2.17. This picture invites us, as it were, to draw yet another line from the sun downward to the flat earth, and to measure the distance of the sun. To be able to do so we first need to calculate the distance from A to the point on earth where the sun is right above our head (in the zenith). How this problem can be solved, we will see in Chap. 16. Another possible application of the gnomon is to outline the shape and boundaries of the inhabited part of the earth (the οἰκουμένη) on a map of the flat earth, as will be discussed in Chap. 6 and is shown in Fig. 6.1.

A later development is to place the gnomon vertically in the center of a hemispherical bowl with its top in the plane of the bowl's rim. Such an instrument is called a σκαφή ("bowl"). The bowl creates an inverted celestial vault. The shadow of the gnomon's tip draws curves on the inner side of the bowl that mimic those of the sun in its daily track along the celestial vault (see Fig. 2.19). The oldest σκαφή dates to the fourth century B.C. (Pedersen and Pihl 1974: 47).

As we have seen, when you are walking with a stick it is not a very helpful to use it as a gnomon to tell the time of the day. Yet the gnomon can rather easily be used to make appointments, as a kind of portable agenda. Today, we are used to make appointments to the minute, checking our watches. For instance, we will meet at a

[20] According to Clayton 1994: 44, the norm was 51°52′.

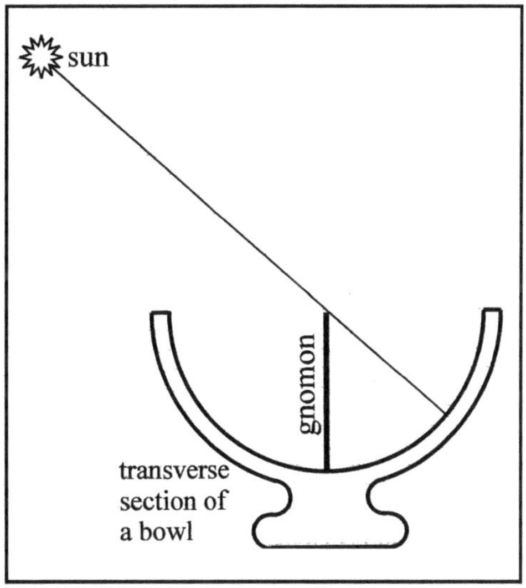

Fig. 2.19 Transverse section of a bowl with an upright gnomon

quarter past five in the afternoon; the train departs at 10.37 a.m. sharp, etc. The ancient civilizations, too, with the steadily increasing complexity of their societies, must have felt a growing need for an instrument that enabled people to make rather precise appointments. The ancients used the *clepsydra* to tell the hours, but this instrument was of course tied to one place and thus of no use for making appointments when the persons involved were at some distance from each other. Two (or more) staffs, used as gnomons, however, were well able to do the job. Mayall and Mayall hint at such a use of the gnomon for making appointments, when they write: "How could the traveler return at a prearranged time? He could carry with him a stick equal in length to the height of the one which had been securely placed in the ground near his cave. No doubt Mrs. Caveman frequently remarked, 'don't forget your shadow pole and return when the shadow's length is one pole'" (1938: 2). However, the authors are too precise, for a serious advantage of the gnomon is that the two persons do not need to carry identical sticks (sticks of the same length). Any vertical stick will do when you arrange to make an appointment like that of Mr. And Mrs. Caveman. This is the only place in the literature that I could find where the possible use of gnomons for making appointments is mentioned.

This method could be easily generalized. Imagine two or more persons carrying sticks with standard marks of, say, one half, one quarter, and one-third of the stick, or even a finer scale. People could then make an appointment when the shadow of the stick was, for instance, $1\frac{2}{3}$ its length. Of course, you will not only have to take into account that the same shadow length will occur twice a day, in the morning and afternoon, but also that the same shadow length will indicate different times of the day according to the season. For instance, in Athens in 500 B.C. around the 9th of March the shadow of a gnomon was equal to its length at noon, whereas 3 months later (9th of June) it had the same length at 8.40 a.m. and at 3.20 p.m. (local time). This would not have caused a big problem, as a daily use of the gnomon would have led to a continuous adjustment of the length of the shadow to make an appointment for approximately the same desired time of the day. Provided all persons involved noticed the shadow length agreed upon, they would all come at about the same time for their appointment. There are some indications that the ancients did it this way. I do not know whether Greek staffs with measuring marks have been found, but some Egyptian staffs seem to bear such marks, as in the statue of Amenhotep II in Fig. 2.20 (Isler 1991b: 174, Fig. 23).

Fig. 2.20 Statue of Amenhotep II holding a staff with a measuring scale on its shaft (Isler 1991b: 174, Fig. 23, by the courtesy of Martin Isler)

Roman indications of calculating with fractions of staff length are in Pliny: "In Egypt at noon on the day of the equinox the shadow of the gnomon measures a little more than half the gnomon itself, whereas in the city of Rome the shadow is one-ninth shorter than the gnomon, in the town of Ancona 1/35th longer, and in the district of Italy called Venezia at the same time and hour the shadow is equal to the gnomon" (*Naturalis historia* I: 182, my translation). Similar remarks were made a century earlier by Vitruvius (*De architectone* IX: 1.1). These observations regard the differences in shadow between different cities, but the point is that the shadow lengths were expressed in terms of parts of the length of the gnomon. As stated previously, the gnomon that is always available is the upright human body with its shadow. Isler remarks somewhere that "the empirical method of telling time by estimating, in paces, the length of a man's own shadow, is ancient and widespread" (1991b: 179). An amusing example is in one of Aristophanes' plays, when a hungry person concludes from the length (in feet) of his own shadow that it is time for dinner (*Ekklesiazusae* 652).[21] Of course, this last method is much less precise than measuring the length of the shadow of a well-scaled staff.

Concluding this section it may be clear that the gnomon, being by far the simplest tool you can think of, and although it was practically confined to use by daylight, was actually a powerful and multifunctional instrument. Moreover, the gnomon inspired the development of computation and measurement, and more specifically stimulated the calculation of angles. If the invention of the wheel stood at the cradle of technology, the use of the staff as a gnomon can be said to have stood at the cradle of the natural sciences. And if it is true that Anaximander introduced the gnomon in Greece, he may also be credited with the introduction of measurement and calculation as scientific tools.

The Egyptians used an instrument, called *merkhyt* (or *merkhet*) that is akin to the gnomon (see Figs. 2.21 and 2.23 right). The *merkhyt* is called after its upright part as a *pars pro toto*. Actually, you may look upon the *merkhyt* as a gnomon with a part of the ground attached to it (the horizontal plank). The *merkhyt* is a rather small instrument,

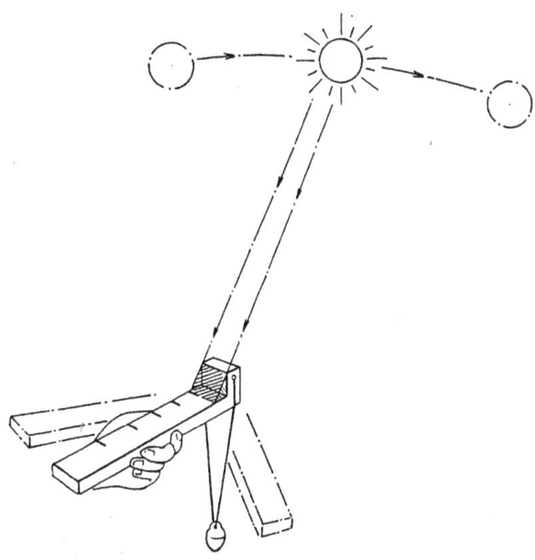

Fig. 2.21 The handling of a *merkhyt* according to Isler (Isler 1991a: 67, Fig. 8, by the courtesy of Martin Isler)

[21] Similar remarks in Menander, fragment 304 (364K) and Eubulus, fragment 119.

which makes it easy to carry along. In modern representations (e.g., recently in North 2008: 31, Fig. 20), it is often depicted with a crossbar on top of its upright part, but this is a fiction resulting, as Isler has convincingly shown, from a wrong reading of an Egyptian text. Moreover, such a crossbar has never been found (See Isler 1991a: 57–59, 1991b: 177–179). A plumb on a line, as in Figs. 2.21 and 2.23, was used to keep the instrument horizontal. Isler lets the observer hold the *merkhyt* in his hand (Fig. 2.21 = Isler 1991a: 67, Fig. 8), but it seems more appropriate to put it on something like a wall or table. When it is turned toward the sun, the shadow of the short upright part, thrown on the horizontal piece, can be read on a scale.

In the description of a *merkhyt* found in the cenotaph of Seti I (±1280 B.C., see Fig. 2.22), the mark that is nearest to the upright part is obviously the noon mark at the equinox, as the angle of the shadow at the upright part is about 30°, corresponding to the latitude of northern Egypt. The plank is divided according to the numerical indications 3, 6, 9, and 12, given in the text (Fig. 2.22, columns 8 and 9). Nowhere is indicated which unit has to be taken 3, 6, 9, or 12 times. I take it that the counting unit is the distance between the upright part and the noon mark (which we may call "a") and that the counting starts from the noon mark, although this is not well represented in the drawing. This results in a distance of 3a between the noon mark and the second mark, a distance of 6a between the noon mark and the third mark, and so on.[22]

A similar counting method is used on the *merkhyt* that is preserved in the Ägyptisches Museum in Berlin (see picture in Von Bomhard 1999: 68–69, Abb. 49), although this one show marks in a rising sequence (1:2:3:4:5). On this specimen the noon mark is so close to the upright part that the instrument must have been calibrated for the summer solstice. Henceforth, I confine myself to a discussion of the *merkhyt* in the cenotaph of Seti I, but, *mutatis mutandis*, the same holds for other *merkhyts* as well.

A main problem is that the way in which the marks are branded on the plank makes no sense as an indication of hours or other time units. The noon mark, for instance, is valid only on the days around the equinoxes. In other times of the year, the shadow at noon is either shorter or (much) longer. This entails that the marks in different seasons indicate different times of the day. Moreover, the equal distances between the marks do not correspond to equal time units. As Clagett puts it: "Even if these marks correctly measured equal hours at the equinoxes (which they did not), they would not have accurately marked the lengths of those hours at other times of the year in view of the changing declination of the sun throughout the year" (1995: 86).

Nevertheless, in the text, the marks are said to indicate the hours of the day. The word "hours," then, is used here in a rather loose way. The instrument neglects the first two hours in the morning and the two last afternoon hours, as is explicitly mentioned in the text in columns 12 and 13: "It sums at [only] eight hours, for two hours have passed in the morning before the sun shines [on the shadow clock] and

[22] The text on top may be translated as "knowing the hours of day and night, starting from fixing noon", as I will defend in a forthcoming article.

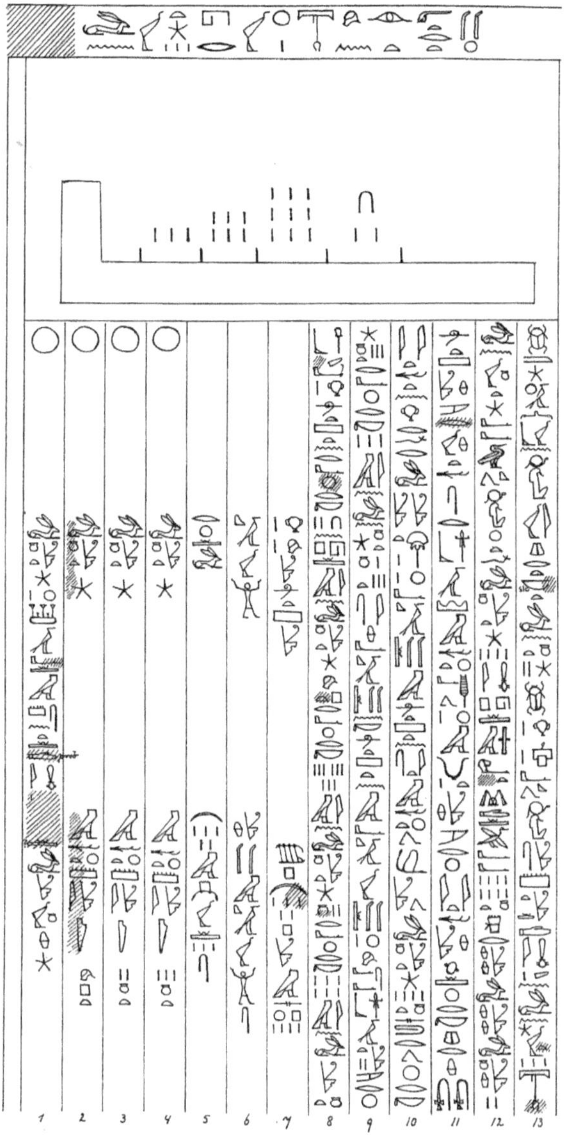

Fig. 2.22 Description of a *merkhyt* from the cenotaph of Seti I (Frankfort 1933: Plate LXXXIII)

another two hours [will] pass after [which] the sun enters [the Duat]" (transl. Clagett 1995: 466).[23] Consequently, the mark that is farthest away from the upright part marks the end of the second hour in the morning (and of the fourth hour in the

[23] Strictly speaking this holds only for the time between the autumnal equinox and the vernal equinox, when the noon shadow falls either on the first mark (at the equinoxes) or somewhere in between the first and the second mark. In the other half of the year, the shadow falls somewhere between the upright part and the first mark, thus creating an extra "hour."

afternoon), although in column 8 it is called (the mark of) the first hour. By this last expression is meant, accordingly: the first "hour" indicated on the instrument.

All this taken together results in the conclusion that the *merkhyt* has, to say it friendly, a very limited use as a time teller. The distances of the marks on the instrument are apparently not meant to indicate precise hours of the day, but chosen in a way that should make them easy to reproduce in order to get exact copies. This feature leads to an interpretation of the use of the *merkhyt* analogous to that given above for the gnomon: the instrument was perfectly apt to make appointments or to fix the moment of, say, a certain ceremony. As far as I can see, scholars have always tried to give an interpretation of the use of one *merkhyt* at a time, whereas nobody has bothered about the use of two or more identical *merkhyts*, used by different persons. When two or more persons had a copy of a *merkhyt*, made according to identical instructions, they could easily agree to meet when the shadow had reached, e.g., the second mark in the morning, or start a celebration when it had reached the fourth mark in the afternoon, and so on. No matter the season of the year, they would all come at the appointed place at the right time. As the marks do not indicate exact times of the day, it does not matter very much where exactly they are drawn, provided they are identical on the *merkhyts* of the persons who make the appointment. Summarizing, three features make the *merkhyt* into a rather practical instrument for making appointments: (1) that it was portable and thus easy to carry with you, (2) that the shadow could be read on the instrument itself instead of on the ground, and (3) that it was easy to reproduce, especially when its marks were at regular distances, so that more people could handle identical instruments. I do not discuss here later developments of this instrument with tilted hour scales, as this would take too much space.

In the literature, the *merkhyt* is often mentioned in combination with another instrument, called the *bay* (which is called *merkhyt* as well by some authors).[24] The *bay* and the *merkhyt* seem to belong together, as at least one set has been found with the name of the same priest on both instruments. The *bay* is a stick with a split upper end. The length of the *bay* in Fig. 2.23 is 52.5 cm. Perhaps it is noteworthy that the *merkhyt* was written as an ideogram in hieroglyphs (see Isler 1991a: 63), but that this is not the case with the *bay*. How this instrument was used is a much discussed question. Borchardt was the first to describe its supposed use, with the following words: "ein Visirstab, der vertical dicht vor das eine Auge zu halten ist, während man das andere schließt" (1899: 14, see also Borchardt 1920: 53–54). Other authors repeat this alleged use of the bay, suggesting that "it would concentrate the vision and so give a sharper image" (West 1982: 121). I am not able to understand, however, what the advantage would be of looking through the split end of a stick held before the eye.

As the *bay* and the *merkhyt* seem to belong together, several authors have tried to imagine what their combined use could have been. Sloley figured out that the observer and his aide were sitting on a north–south line, the first holding a *bay* in one hand and a *merkhyt* with a plumb line in the other, whereas the aide holds the plumb line of his *merkhyt* above his head (see Fig. 2.24). The observer is supposed

[24] E.g., Sloley (1931: 169 and Plate XVI, 4).

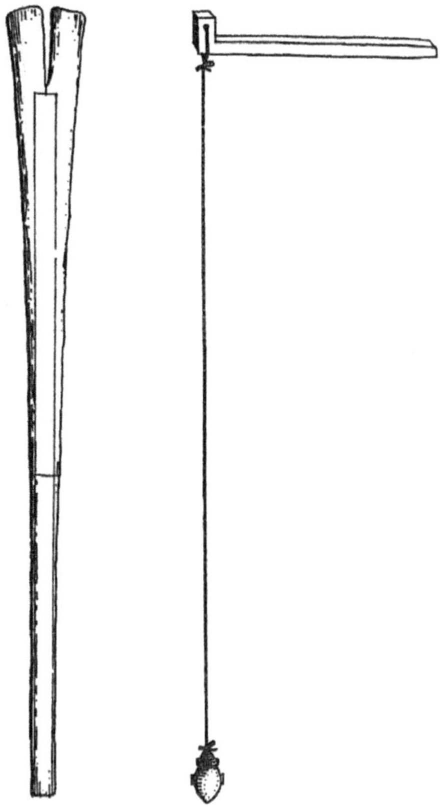

Fig. 2.23 Two Egyptian instruments, a *bay* (*left*), and a *merkhyt* with plumb line (Borchardt 1899: 10)

to look through the split end of the *bay* and along the plumb line of his *merkhyt* and that of his aide to mark the position of a star.

Other attempts to comprehend the combined use of the *bay* and the *merkhyt* are derivatives of Sloley's picture but usually have only one *bay* and one *merkhyt*. Mostly, they have one observer hold the *bay* before him, while an aide holds the *merkhyt* in his hand. (e.g., Ronan 1971: 56). The observer is supposed to look through the split end of the *bay* and along the plumb line of the *merkhyt* to mark the position of a star. Lull inverts the order and lets the observer look along the plumb line of the *merkhyt*, whereas the aide holds a kind of stick (2006: 296, Fig. 98, and 299, Fig. 100, here reproduced as Fig. 2.25).

The trouble with all these alleged methods is that even if the supposed observers could manage to hold their hands still enough to make any observation possible, this looks like a clumsy way of observing a star. Neugebauer and Parker already remark: "That two persons, sitting opposite each other, cannot resume exactly the same position night after night is clear. To fix accurately the moment of transit, when even very small motions of the eye of the observer will displace the apparent position of a star, is impossible" (1960–1969, Vol. II: x). Probably for this reason,

Fig. 2.24 The use of *merkhyt* and *bay* according to Sloley (1931, plate XVII,1 between 170 and 171)

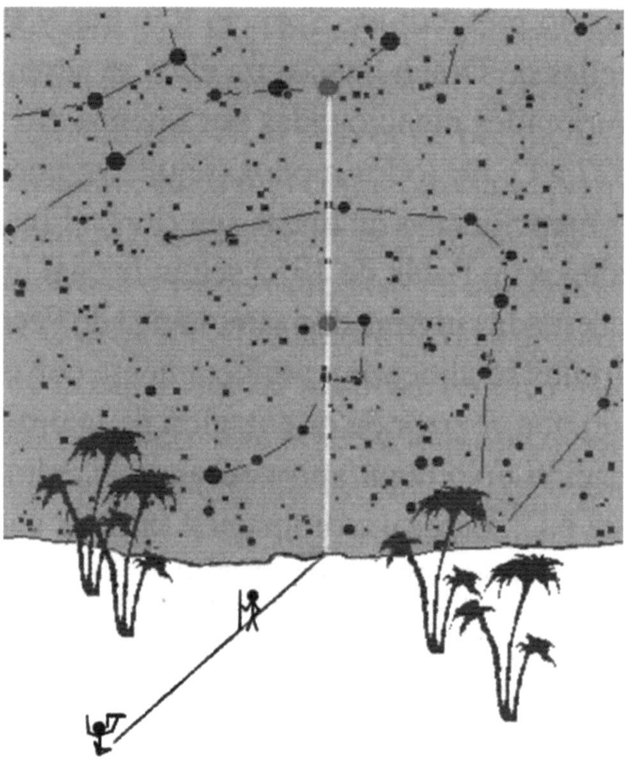

Fig. 2.25 The use of *merkhyt* and stick according to Lull (2006: 299, fig. 100)

Pecker makes the *merkhyt* the cross-beam of a gallows, on which a plumb line hangs (2001: 31, Fig. 1.13). The observer is thought to stand behind a board with a vertical slit that is provided with a scale, and to look through this slit along the plumb line to determine the apparent height of a star above the horizon. Needless to say that such a board with a slit and a scale is no more than a product of Pecker's fantasy.[25] None of the proposed methods of using the *bay* seems to me convincing.

Of course the *merkhyt* itself is not necessary for the use of its plumb line as a kind of sighting instrument. A mere *plumb line*, its upper end tied to something like the gallows mentioned above, will do the job as well. It is well known that the ancient Egyptians were interested in the culmination of other stars, especially the 36 so-called decans.[26] To watch these culminations, the observer needed a permanent and dependable north–south line. I think he could obtain such a line by using the *bay* as a calibration device. The procedure would look like this: The observer sets himself south of a rather long plumb line that hangs down from a stake and, always looking with one eye to prevent parallax problems, he waits until he can move so that he can see stars culminate when passing the plumb line. Then, he lets his aide put a *bay*, with the split end on top, perpendicularly in a holder between himself and the plumb line, so that he sees the plumb line exactly in the split of the *bay*, As soon as he has achieved this, he asks his aide to fix the holder on that spot. Now, he has made sure that every time he will return to the same place and put his *bay* into the holder, he will provide a perfect north–south line by setting himself south of the *bay* so that the plumb line is caught in the split of the *bay*. In other words, he has made a simple but convenient observatory, by means of which he can observe the culmination of a star, say Sirius, or another of the 36 so-called *decans*. Mark that the *bay* is not held close to the eye, as Borchardt supposed, but at a certain distance, because the observer uses the *bay* only to make sure that he will sit in the right place. The observatory is shown in Fig. 2.26. This reconstruction of the way the instruments were used is of course also a fruit of fantasy, but at least it makes sense.

The procedure just described can be used for stars on the northern sky. For the observation of culminations in the southern sky, the observer, having drawn a north–south line on the ground, simply has to change his position to north of the plumb line and to look southward, making sure that the plumb line is seen in the split of the *bay*. In Fig. 2.26 I made use of the fact, exposed by Spence, that in 2467 B.C. the imaginary line between two stars, Mizar (ζ of the Big Dipper) and Kochab (β of the Little Dipper), ran through the pole (2000: 320–324). Lull did the

[25] Isler proposes still another use, quite different, of the *bay*. He lets the observer put it upside down (with the split end under) at the top of the shadow of a gnomon "to help clarify a shadow by reducing surface reflection" (1991b: 162, Fig. 9; cf. 1989: 198, Fig. 5; see also Lull 2006: 292, Fig. 96). Moreover, Isler shows all kinds of forked and curved sticks that could function as a gnomon, but none of them looks exactly like the *bay* in Fig. 2.23.

[26] The so-called decans were stars that were used by the ancient Egyptians for marking the hours of the night. More on this subject in Von Bomhard (1999: 50–65), and especially in Leitz (1995).

same in Fig. 2.25 above. According to Spence, this datum was used to align the pyramids. Spence concluded that, with an uncertainty of 5 years, the pyramid of Cheops must have been built in 2467 B.C. Spence's article has met with severe criticisms that need not bother us here.[27] As already said, any culminating star could have been used to set up the observatory as in Fig. 2.26.

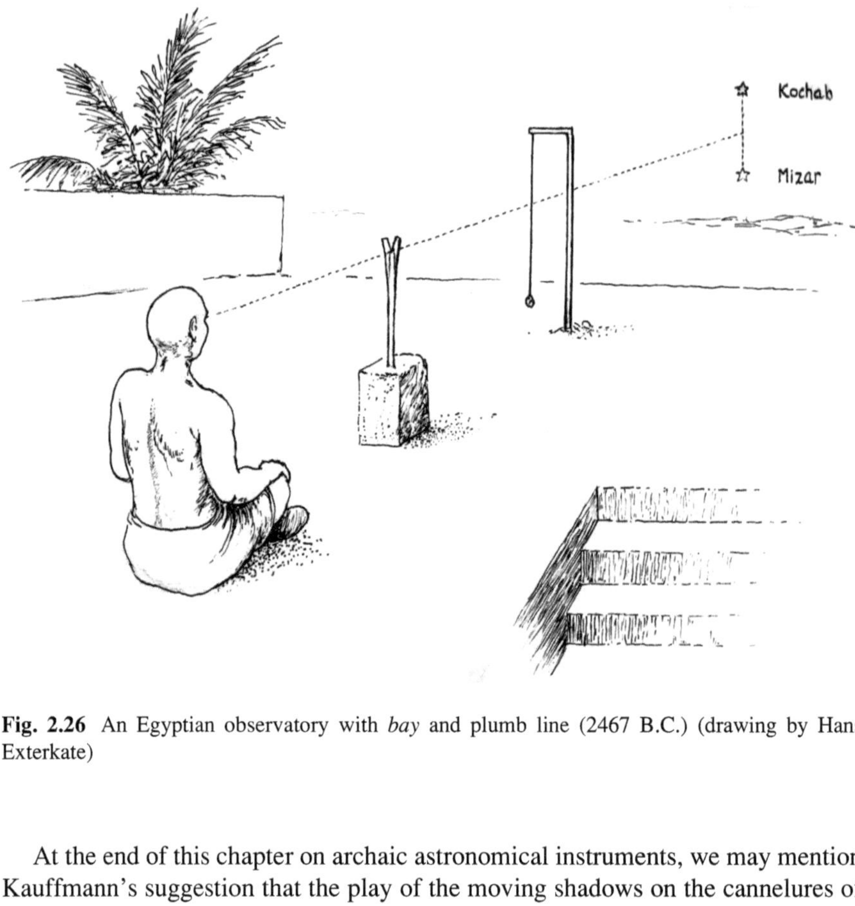

Fig. 2.26 An Egyptian observatory with *bay* and plumb line (2467 B.C.) (drawing by Hans Exterkate)

At the end of this chapter on archaic astronomical instruments, we may mention Kauffmann's suggestion that the play of the moving shadows on the cannelures of temple columns functioned as a sundial (1976: 28). There are, however, no ancient sources to confirm this hypothesis (Fig. 2.27).[28]

[27] See e.g., the discussion in Gingerich (2000: 297–298), Rawlins and Pickering (2001: 699), Spence (2001: 699–700), Bauval (2001: 320–324), and Lull (2006: 299–300).

[28] See for some critical remarks Couprie and Pott (2001: 47).

Fig. 2.27 The play of light and shadow on the cannelures of temple columns (photograph by Victor Abrash)

Chapter 3
How Thales Was Able to Predict the Solar Eclipse of 28 May 585 B.C.

In the year 467 B.C. at Aegospotamoi, a stone fell from the heaven. It is said that Anaxagoras, thanks to his knowledge of astronomy, was able to predict the fall of this famous meteorite (DK 59A1(10), DK 59A6, DK 59A10, and DK 59A11; also in Ammianus Marcellinus' *History of Rome* 22.8.5, not in DK, but see remark on p. 9, line 12). Nobody, however, has ever been concerned to find out the method he could have used to predict this event. Taking for granted that it is impossible to predict the fall of a meteorite, the attribution of that prediction to Anaxagoras has been dismissed as an instance of the habit of crediting the Presocratics with all kinds of discoveries. In his *Life of Dion*, Plutarch reports that during Plato's visit to Sicily, Helicon of Cyzicus predicted a solar eclipse. When his prediction came true, he was admired and rewarded by the tyrant (*Life of Dion* 19, see also *Life of Lysander* 12). Nobody has ever tried to determine the method Helicon might have used to accomplish such a remarkable achievement. Many scholars have, by contrast, tried to reconstruct the method that Thales might have used in predicting a solar eclipse, with which he is credited by ancient authors.[1] The story is told by Herodotus: in the 6th year of the war between the Lydian king Alyattes and the Medes under Cyaxares, during a battle, all of a sudden the day became night (*Histories* I 74 = DK 11A5). The reports of other authors depend on his account (see DK 11A1, DK 11A2, and DK 11A5). The eclipse of 28 May 585 B.C., which was almost total at Miletus, is generally accepted as the one predicted by Thales. This date also matches best Pliny's testimony that the eclipse took place in the 170th year after the foundation of Rome, 753 B.C. (*Naturalis historia* II, 53 = DK 11A5; see also Stephenson and Fatoohi 1997: 280).

O'Grady writes: "modern astronomy confirms that the eclipse took place and that it was total" (2002: 128). However, all sources I consulted confirm that this is not right. The relevant map in Kudlek and Mickler indicates that both Nineveh and Miletus lay outside the central zone of the eclipse (1971: 192, no. 1, 507, 900). So does the map in Stephenson and Fatoohi, although they remark that the eclipse

[1] Much has been written on Thales and the solar eclipse. Three of the most important articles are the following: Hartner (1969: 60–71), Panchenko (1994a: 275–288), and Stephenson and Fatoohi (1997: 279–282).

D.L. Couprie, *Heaven and Earth in Ancient Greek Cosmology*,
Astrophysics and Space Science Library 1, DOI 10.1007/978-1-4419-8116-5_3,

"would probably be total at Miletus, where Thales lived" (1997: 281). Hartner speaks of "the nearly total eclipse" (1969: 68, Fig. 1). Lull's picture of the path of the eclipse also shows that the central zone past north of Miletus, and even north of Ephesus, Mitilene, and Sardes (Lull 2005: Fig. 6). The computer program *Redshift 5.1* (2005) shows that the moon's shadow passes north of Miletus and that at that city a small sickle of the sun remained visible. For the making of Fig. 3.1, I consulted NASA's *Five Millennium Catalog of Solar Eclipses* on the internet.[2]

Fig. 3.1 The path of the central zone of the solar eclipse of 28 May 585 B.C.

Although the subject of this chapter has to do with the solar eclipse predicted by Thales, I treat lunar eclipses as well because the cycles of lunar eclipses play a certain role in the prediction of solar eclipses. For these cycles, the notion of "synodic month" or "lunation" is important. This is the period of about 29.5 days between two consecutive new moons. Moreover, one has to keep in mind that Thales indeed could observe solar and lunar eclipses but that he had no knowledge whatsoever of the true nature and cause of these phenomena. As we will see in Chap. 14, it was probably Anaxagoras who discovered this. When ancient sources honor Thales because of his knowledge of the eclipses and maintain that he understood the causes of eclipses (DK 11A3, DK 11A5, DK 11A17, DK 11A17a, DK 11A19 and in the *Oxyrhynchus Papyri*, 3710, ed. Haslam 1986: 106, quoted by O'Grady 2002: 142), they must be regarded as apocryphal because it makes no sense within the context of his conception of the earth floating on water, nor within the historical context of the opinions of his immediate followers.

A solar eclipse occurs when the moon in its orbit around the earth comes between the sun and the earth. Solar eclipses only occur at new moon, when the moon, seen from the earth, is in line with the sun. Yet, not every new moon yields a solar eclipse. This is because the moon's orbit is somewhat oblique with regard to the plane of the ecliptic. If the moon moved precisely along the ecliptic, there would be a solar eclipse

[2] More specifically http://www.eclipse.gsfc.nasa.gov/SEsearch/SEsearchmap.php?Ecl=05840528.

every month. A solar eclipse is not visible everywhere on earth, but only on a relatively small strip where the shadow of the moon falls upon the earth (Fig. 3.2). This is one of the reasons why it is difficult to predict a solar eclipse for a specific place.

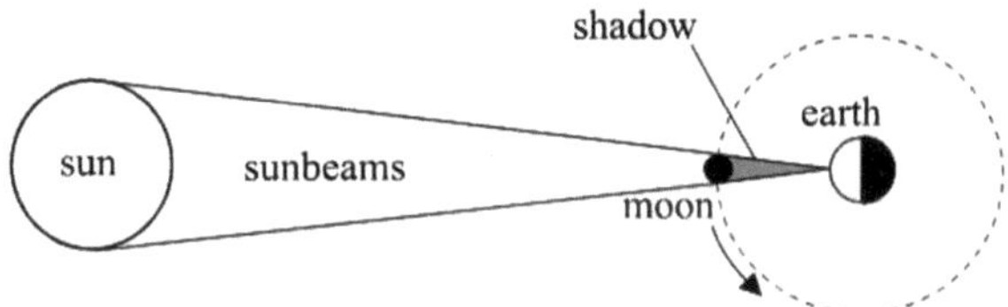

Fig. 3.2 How a solar eclipse originates

A lunar eclipse occurs when the moon passes through the cone of the shadow of the earth. A lunar eclipse can be observed everywhere on the hemisphere of the earth that is turned away from the sun. At the time of an eclipse of the moon, the earth is between the sun and moon, which means that it is a full moon. Yet not at every full moon does a lunar eclipse occur. This is because the orbit of the moon is inclined with 5° as regards the plane of the ecliptic. If the moon moved precisely along the ecliptic, there would be a lunar eclipse every month. A lunar eclipse can be total or partial depending on the part of the moon that is covered by the shadow of the earth. When the moon does not go through the kernel shadow of the earth (*umbra*), but through the half-shadow (*penumbra*), a penumbral eclipse takes place (Fig. 3.3). I leave this kind of eclipses out of consideration as they would probably not have been observed by Thales. The observation of penumbral eclipses is rather difficult and presupposes, moreover, knowledge of the true nature of eclipses that Thales did not possess. Even the late Babylonian astronomical texts do not contain any observations of penumbral eclipses (see Steele and Stephenson 1997: 123). The lists in Kudlek and Mickler (1971) only mention umbral eclipses and partial eclipses, but no penumbral eclipses.

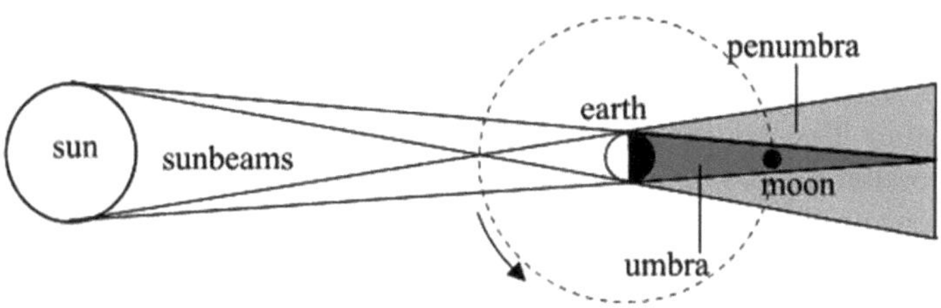

Fig. 3.3 How a lunar eclipse originates

According to White, Thales' discovery was that "if the moon is new, a solar eclipse is possible. Such a claim made early in the sixth century and vindicated in 585 would readily inspire the story that Thales 'foretold' an eclipse" (2008: 101). However, as 12 or 13 new moons occur within a year, this would have been a very tiny basis for the prediction of a solar eclipse. Moreover, that eclipses of the sun always occur at new moon is such a basic observational datum that it hardly can be called a "claim." Accordingly, several scholars who have tried to explain Thales' prediction have focused on one or another kind of cycle that they believe he learned from foreign (preferably Mesopotamian) astronomers. The so-called Saros cycle of 223 lunations is famous as a method for predicting eclipses. However, the main reason why the Saros cycle was of little help for Thales in predicting solar eclipses for Miletus is that an eclipse in the next cycle of one Saros takes place nearly 120° of longitude further westward.[3] This means that a theory for the prediction of solar eclipses has to take into account the sphericity of the earth, with which Thales was not acquainted, as we will see in Chap. 4. It also means that a triple Saros, the so-called Exeligmos, which is an eclipse period of 669 lunar months or 54 Julian years and 1 month, theoretically promises the best predicting results for a given location, says Miletus. Accordingly, the Exeligmos turns out to be far and away the best of the 15 cycles compared by Hartner (1969: 62–64). This is why Panchenko tried to apply the Exeligmos cycle, but unfortunately this method fails for the famous eclipse of 28 May 585 B.C. (Panchenko 1994a: 280).[4] The eclipse of 21 September 582 B.C., on the other hand, which is Panchenko's favorite, "may well have passed completely unnoticed," its magnitude at Miletus not exceeding 0.85, as Stephenson and Fatoohi explained in their rejoinder (1997: 281). This date therefore conflicts with Herodotus' report that during Thales' eclipse "the day turned into night." After all, Panchenko's attempt to discover Thales' method for predicting an eclipse appears to have failed as well. For a critical discussion of O'Grady's recent attempt, I refer to my article on Thales and the solar eclipse where it is shown that this one fails as well (Couprie 2004b: 323–328).

According to Aaboe, "Babylonian lunar theory of the Seleucid period, when it has reached its ultimate stage, was certainly capable of predicting *lunar* eclipses" (1972: 106, my italics). The Seleucid era began in 312 B.C., about 300 years after Thales. We may wonder, then, how much Babylonian theory for the prediction of *solar* eclipses existed in Thales' times. Most Babylonian observational material on solar eclipses dates from after Thales. According to Steele and Stephenson, Babylonian texts contain details of solar eclipses from 350 to 50 B.C., a considerable time after Thales (1997: 119). Britton deals with a text in which 38 solar eclipse possibilities are described from 473 to 456, which is also more than a century after Thales (1989: 29ff. and Table 8 on p. 34). Steele discusses solar eclipse possibilities of the late Babylonian period from 474 B.C. to 49 A.D., starting roughly one century after Thales (2000: 444–448, Table 4). Stephenson mentions two Assyrian records of solar eclipses before Thales' time (those of 27 May 669 B.C. and 15 June 763 B.C.), and concludes that "at this early period, Babylonian

[3] More information on the *Saros*-cycle and the problems connected with it in (among others) O'Grady (2002: 129–133), and in Pannekoek (1961: 57–62). Hartner examines 15 different cycles, among which the Saros and the Exeligmos, according to their usefulness in the prediction of solar eclipses (1969: 60–71).

[4] A misprint in Panchenko's table on the same page 280 should be rectified because it plays a role in his argumentation: the date of the solar eclipse no. 7 at Nineveh in Table 1 has to be 650 Nov 21 instead of 650 Feb 21.

astronomers (. . .) attained very poor success in anticipating eclipses for a given location" (1997: 125 and 343).

As far as I know, the cuneiform evidence has delivered a treasure of astronomical observations but not a single explicit system or theory for predicting future solar eclipses. Against this background, one may wonder why scholars, and most recently O'Grady, have tried in the case of Thales to identify a theory that even the most distinguished observers of his time did not possess. If Thales really did have a successful method for predicting solar eclipses, his followers would presumably have applied it as well, and we would possess scores of predictions of eclipses in ancient Greece. This, however, is not the case. How little the mechanism of eclipses was understood even in Herodotus' time is demonstrated by his report that has puzzled many scholars. Herodotus writes that Thales foretold the eclipse "fixing it within the year in which the change did indeed happen." Obviously, Herodotus did not realize, as Panchenko says, that "if one can predict an eclipse at all, one can predict it to the day" (1994a: 275, see also already Martin 1864: 189). Commenting on Thales' alleged prediction, Burch writes: "Now to predict the year of a total eclipse is no very great achievement, in view of the fact such eclipses occur almost every year; the trick is to predict the exact time and place" (1949–1950: 138, n. 5). Burch is right, although he is too modest, as solar eclipses occur at least twice a year somewhere on earth. Herodotus, however, obviously meant a prediction for some particular place: Miletus.

Nevertheless, although it is not explicitly documented, the observers in Mesopotamia must have used one method or another, however simple, for their reports show that they *expected* eclipses and *announced* the events (see Pannekoek 1961: 45). The conclusion of Britton's thorough study of Babylonian lists of eclipses prior to 456 B.C. is that "in general (. . .) the empirical basis for the eclipse theory (. . .) is very simple and limited" (1989: 46). Statements by Assyrian observers, writing to the king (who had asked them whether a solar eclipse was to take place), typically run like this: "Now since it is the month to watch the sun." And elsewhere: "[Eclipses] cannot occur [dur]ing certain periods." And again: "It is (indeed) the month for the watch of the sun" (Parpola 1993: 170, 71, and 45).

These reports show that the Assyrian astronomers were acquainted with the simple rule of thumb that not at every new moon, but only once in 6 (or sometimes 5) months, the *possibility* of a lunar or solar eclipse arises. This is what O'Grady calls "the six month–five month rule of thumb" (2002: 139). This rule, which is "the most basic scheme for calculating eclipses," states that "eclipses can be predicted by moving 6 or occasionally 5 lunar months from the preceding eclipse possibility" (Steele 2000: 423. See also Aveni 1993: 47: "one of the simplest rules of eclipse prediction"). Consequently, "the interval in months between consecutive *visible* eclipses (. . .) is always a multiple of 6 months or a month less than such a multiple" (Britton 1989: 5). The Assyrians already knew that the interval sometimes counts 5 rather than 6 months. Thus, for instance, on 12 December 671 B.C. the court astronomer Balasî wrote to the king: "We will observe twice, at the 28th of *Marchevan* (the 8th month) and on the 28th of *Kislev* (the 9th month)" (Parpola 1993: 45). Note that they did not need to know the cause of this phenomenon but could have extracted this rule from a long series of observations. Whether an eclipse really occurs on such a date is far from certain, especially in the case of solar eclipses, but they knew that those were the days for observing the sky.

The interval of 5 days instead of 6 between the dates of possible eclipses does not occur at random but with certain regularity. As only *possible* eclipses are at stake, prolonged observations over many years or even over generations of observers are needed to discover this kind of regularity. Whether this was the case before or after Thales is not quite clear. Steele has shown that it is acceptable to assume that the Assyrians made this discovery (2000: 421–454). The scheme is as follows: we start with the date of a possible eclipse after an interval of 5 months, then the next seven occur, each of them with an interval of 6 months, then after 5 months no. 9, then the next six with intervals of 6 months, then after 5 months no. 16, then the next seven with intervals of 6 months, then after 5 months no. 24, then the next six with intervals of 6 months, then after 5 months no. 31, and finally the next seven with intervals of 6 months, whereupon the whole series starts all over again. In Table 3.1, this scheme is represented, in which the numbers 1–38 stand for possible eclipses, and (5m) and (6m) stand for an interval of 5, respectively, 6 months between two possible eclipses.

Table 3.1 The system of alternating intervals of possible lunar eclipses

(5m) 1	(6m) 2	(6m) 3	(6m) 4	(6m) 5	(6m) 6	(6m) 7	(6m) 8
(5m) 9	(6m) 10	(6m) 11	(6m) 12	(6m) 13	(6m) 14	(6m) 15	
(5m) 16	(6m) 17	(6m) 18	(6m) 19	(6m) 20	(6m) 21	(6m) 22	(6m) 23
(5m) 24	(6m) 25	(6m) 26	(6m) 27	(6m) 28	(6m) 29	(6m) 30	
(5m) 31	(6m) 32	(6m) 33	(6m) 34	(6m) 35	(6m) 36	(6m) 37	(6m) 38

(5m) 1	(6m) 2	(6m) 3	(6m) 4	(6m) 5	(6m) 6	(6m) 7	(6m) 8
(5m) 9	(6m) 10	(6m) 11	(6m) 12	(6m) 13	(6m) 14	(6m) 15	
Etc.	Etc.						

Whether Thales had discovered this scheme is not relevant for my argument in what follows. As a matter of fact, we do not know whether Thales had any knowledge of foreign astronomical observations and theories. Perhaps for this reason, Hartner has investigated whether Thales himself, on the basis of observations of solar eclipses at Miletus, could have discerned a cycle enabling him to generate his prediction. Hartner presupposes the existence of Milesian records of solar eclipses since as early as 710 B.C. This highly speculative supposition is based on nothing more than the fact that since 775 B.C., the Greeks had kept records of the names of Olympic victors (1969: 5). The result of his investigation is that "the unexpected 'eclipse of Thales' came as a surprise to the Sage" (1969: 69). We will see how this conclusion can be challenged.

Let us suppose, just like Hartner, that Thales was not acquainted with any observations or theory regarding the occurrence of lunar or solar eclipses whatsoever, neither from abroad nor from former Milesians. Instead, let us confine ourselves to observations that he could have made himself. In other words, let us assume that he had at his disposal the dates of lunar and solar eclipses that were observable and observed at

Miletus during his lifetime. We may readily imagine how he became interested in astronomy when, at the age of about 30, he observed three solar eclipses, namely those of 30 September 610 B.C. (the first solar eclipse visible at Miletus for 25 years), 13 February 608 B.C., and 30 July 607 B.C.[5, 6] One may suppose that Thales was also acquainted with the simple observational datum (without necessarily understanding its cause) that lunar eclipses always occur at full moon. He could either have heard or have concluded himself from his observations that solar eclipses always occur at new moon. He would surely have known too that the length of a lunar (synodic) month is 29½ days – a primary observational datum on which ancient calendars (with the exception of the Egyptian calendar) are based.[7]

After having observed solar eclipses of different magnitudes, Thales probably surmised that very small eclipses could occur too, but this does not necessarily mean that he had observed them. Nowadays, we know beforehand exactly when and where to look for even the slightest eclipse. But all Thales knew was that at an unknown time of the day at a certain new moon, a solar eclipse of unknown magnitude might perhaps occur. Stephenson's table of changes in light intensity during solar eclipses shows how light intensity drops spectacularly at eclipses <0.9 (1997: 53, Fig. 3.5). This means that most partial solar eclipses will not have been noticed at all, unless special measures were taken to improve the observation. As stated earlier, even an eclipse of 0.85 may well have passed unnoticed by Thales' fellow citizens. However, although there are no records of the methods used, we may assume that observers like Thales regularly watched the reflection of the sun on a liquid surface, as described in Chap. 2. This method notably increases the visibility of solar eclipses, especially of those that are fairly high in the sky, whereas solar eclipses near the horizon are easier to observe with the naked eye (see Panchenko 1994a: 288 n. 36. See also Stephenson 1997: 54, who quotes Seneca, *Naturales Quaestiones* 1.11). On the basis of all this, it seems fair to suppose that solar eclipses in which less than half of the solar surface was eclipsed (i.e., <0.5) could have escaped Thales' observation, unless he had a special reason to expect one. We will see in due time what this special reason could have been.

Let us also suppose that Thales, at the end of 586 B.C., had ordered his observational data of the past 24 years in the shape of a calendar table like that shown in Table 3.2. This table is based on possible observations only, without

[5] Apollodorus gives the 35th Olympiad as Thales' birth date, which is 640 B.C. (DK 11A1). Many authors take it, on the authority of Diels (1876: 16), that one should read here 39th Olympiad, which is 624 B.C. Fehling (1985a: 100–102) has convincingly shown that Diels' emendation of the text is unjustified. However, if Diels is right, this means that in 610 B.C. Thales was 14 years old, which hardly injures my argument.

[6] F. Richard Stephenson drew my attention to the solar eclipse of 17 April 611, which reached its maximum (0.49) after sunset (at 18.6 h., altitude $-3°$). Given these data, and taking into account the fact that the undermost part of the sun was eclipsed, I seriously doubt whether this eclipse was observable at all at Miletus. It is listed neither by Hartner nor by Panchenko.

[7] The real length of a synodic month, which is the interval of time between two new moons, being $29^{d}\ 12^{h}\ 44^{m}\ 3^{s}$,

Table 3.2 Lunar and solar eclipses visible at Miletus from 610 to 586 B.C.

Year B.C.	Jan	Feb	Mar	Apr	May	Jun	Jul	Aug	Sep	Oct	Nov	Dec
610									1 ○ ●	30	28	28
609	27	24	26	25	24	23	22	21	○ 19	18	17	16
608	15	● ○	15	14	13	12	12	10 ○	9	8	7	6
607	4	3	5	3	3	1	1 ●	29	28	27	26	25
606	○ 24	22	23	22	21	20	○ 19	18	17	16	15	15 ○
605	13	12	12	10	10	8	7	6	5	4	3	3 ○
604	31		2 31	29	29	27	28	25	24	23	22	22
603		19	21	19	*	17	16	14	13	12 ○	11	11
602	10	9	10	9	8	6	6	4	2	2 31	30	30
601	28	27	28	○ 26	26	24	24	22	20	20	18	18
600	17	16	17	16	15	14	13	12	10	9	8	7
599	6	5 ○	6	5	4	3	2	1 ○31	29	29	27	26
598	25	23	25	24	23	22	21	20	18	18	17	16
597	15	13	13	12	11	10	●	8	7	6	5	5
596	3	2	2	1	1 30	*	28	27	25	25	24	○ ●
595	22	21	22	20	20	○ 18	17	16	14	14	13 ○	13
594	12	10	12	10	* ○	8	7	5	4	3	2	2 31
593	30	29	30	28	27	26	25	23	22	21	20	19
592	18	17	19	○ 17	17	15	15	13	11	11	9	9
591	7	6	8 ○	7	6	5	4	3	1 30	30	28	28
590	26	25	26	25	25	23	23	21	20	19	18	17
589	16	14	15	13	13	12	11	10	8	8	7	6
588	4 ○	3	4	3	2	1 30	●	28	27	27	25	25
587	○ 24	22	23	22	21	○ 19	19	17	16	16	15	●
586	13	12	13	11	11	9	8	7	5	5	4	3

presupposing any theory of eclipses and without presupposing any knowledge of the sphericity of the earth, nor of the true nature of the celestial bodies or even of eclipses, nor of what sun and moon do when they are beyond the horizon. By "possible observation" is meant the whole of observations that could have been made by a diligent observer under favorable weather conditions. Perhaps one might object that to write down data in a table like this looks rather anachronistic. In defense might be argued: "that either Thales or Anaximander compiled a comprehensive astronomical calendar (. . .) rests on strong circumstantial evidence" (White 2008: 96). However this may be, for the conclusions reached this table is not necessary; it simply visualizes some of them and so in a sense makes them easier to observe. Table 3.2 is a kind of luni-solar calendar, the basis of which is our calendar B.C. Of course, Thales would have used a different kind of calendar than ours (based on lunar months), but that would have shown an analogous picture. Table 3.2 shows the calendar from 610 until 586 B.C., which is from the year of the first solar eclipse Thales could have observed until the year before the eclipse he is said to have predicted. In this table, the consecutive data of new moon are indicated. This makes it possible to count easily intervals of months. Lunar eclipses that were visible at Miletus are marked "○". For reasons already mentioned, no penumbral lunar eclipses are listed. Solar eclipses >0.5 visible at Miletus are marked "●," and solar eclipses <0.5 are marked "✱." These last two marks stand in place of the corresponding dates of the new moon.[8]

Looking carefully at Table 3.2, three relevant features catch the eye. The first is that lunar eclipses show a certain pattern as they occur in a kind of garlands throughout the calendar. The effect is even more apparent if one wraps this table around a suitable cylinder. Thales would probably not have noticed all the eclipses in the list, because of weather conditions or other inconveniences, but the pattern is so manifest that an incidental loss of an observation does not affect the overall result of observations registered over many years.[9] Thales could have learned from his observations that the interval between consecutive lunar and solar eclipses is always a multiple of 6 lunar months (sometimes minus one), as explained above. The garlands in Table 3.2 are, so to speak, a visualization of this six month–five month rule of thumb, shown in Table 3.1. Table 3.2 shows that Thales did not have to appeal to foreign knowledge. All he had to do was to register the eclipses, note the pattern of garlands, and count the months elapsing between consecutive eclipses.

The second relevant feature is that solar eclipses fit into the same pattern of garlands, always occurring at new moon. Of the 12 or 13 new moons in a year, the only ones that remain as real candidates for possible solar eclipses are those that fit into the same pattern, always being half a month before or after a possible lunar

---

[8] This table is made with the help of *Redshift 5.1* (2005), and checked with the data in Kudlek and Mickler (1971). Nowadays, a registration of lunar and solar eclipses as shown in the table is not unusual. It is, for instance, used by Ekschmitt (1969: 15) and Herrmann (1969: 45).

[9] Note the frequent remarks such as "(very) overcast," "a lunar eclipse that was omitted," "I did not watch," "clouds were in the sky," "rain shower," and even "coughing and a little *rišûtu*-disease," in the observational reports collected in Sachs (1988: *passim*).

eclipse. These two features would have been sufficient for Thales to extrapolate his data and notice that 28 May 585 B.C. could be a possible date for a solar eclipse. There were, however, other candidates, for instance, 21 November of the same year. Moreover, these were only dates of *possible* eclipses, and, as the table shows, not every possible date will give rise to a real solar eclipse. To deliver a real prediction, more was needed.

The third relevant feature of Table 3.2 is that the observed solar eclipses seemed to occur in clusters within a few years of one another. The first cluster was formed by those of 30 September 610, 13 February 608, and 30 July 607. The intervals between these eclipses were 17 and 18 lunar months so that the third eclipse came 35 months after the first. If we ignore solar eclipses <0.5, namely, those of 18 May 603 (magnitude 0.49) and 28 June 596 (magnitude 0.35), which could, as explained, have escaped Thales' attention, then the second cluster started with the eclipses of 9 July 597 and 23 December 596, the interval between them being 18 months. Perhaps Thales surmised that another eclipse would again take place 35 lunations after the first one in this cluster. In that case, he might also have noticed the eclipse of 9 May 594, even though it was relatively small (0.46). If he had seen this last eclipse, he could also have concluded that the clusters apparently consisted of three consecutive eclipses, so that between the first and the second eclipse 17 or 18 lunations elapsed, and between the first and the last eclipse 35 lunar months. When, 77 lunations later, the eclipse of 29 July 588 occurred and was followed, 17 lunar months later, by that of 14 December 587, the suggestion that these eclipses were the first two of a new cluster of three eclipses within 35 lunations must have had an obvious appeal. How, then, after he had observed the eclipse of 14 December 587, could Thales have resisted the temptation to make a bold guess, saying to his fellow citizens: "Look carefully at the sun on 28 May 585 B.C. (or rather, of course, whatever date corresponded to this date on his calendar), then you will possibly see something amazing, a solar eclipse?" A pronouncement such as this must have sounded to his fellow citizens as a stroke of real genius. When on that date a solar eclipse actually did take place, and was, by chance, almost completely total, everyone would have claimed that Thales had predicted it.

Table 3.3 shows the groups of solar eclipses that Thales could have observed, as explained above. In the third column, the number of lunar months elapsing between the eclipses is given. The number of months elapsing between the eclipses of one and the same cluster is printed in boldface.[10] The last line shows the eclipse that Thales is said to have predicted.

My hypothesis about the way in which Thales could have predicted a solar eclipse was triggered by a remark in Panchenko's article: "The eclipse of 28 May 585 was separated from the two previous eclipses by intervals of 18 and 17 lunations respectively. But this sequence had taken place once before! The eclipse of 30 July 607 was also separated by intervals of 18 and 17 lunations from the two

[10] For the figures of the magnitudes of the solar eclipses at Miletus, F. Richard Stephenson kindly placed his calculations at my disposal.

Table 3.3 Clusters of solar eclipses at Miletus from 610 to 585 B.C.

Date of solar eclipse	Maximum phase	Lunations since last eclipse
30 Sept 610	0.59	371
13 Feb 608	0.76	**17**
30 July 607	0.63	**18**
9 July 597	0.73	123
23 Dec 596	0.61	**18**
9 May 594	0.46	**17**
29 July 588	0.88	77
14 Dec 587	0.75	**17**
28 May 585	0.97	**18**

preceding eclipses (. . .). Thales therefore could have discovered that not only one interval previously encountered had been repeated, but a series of two intervals between consecutive eclipses. Such a 'regularity' might have allowed him to venture a prediction" (1994a: 283). The reason why Panchenko does not follow his own suggestion and fails to see the third cluster of intervals of 17 and 18 months results from another curious mistake, this time in his table of the solar eclipses at Miletus (1994a: 281, Table 2; for the other mistake, see Note 4). There we read that between the eclipses of 23 December 596 B.C. and 9 May 594 B.C. an interval of 23 months takes place, whereas the number has to be 17. The origin of this slip is that Panchenko takes the number of 23 months from Hartner's table. Hartner, however, obtains an interval of 23 months because he forgets to mention the eclipse of 23 December 596 B.C. (1969: 66, Table 3, column 5). This should be a reminder that one always has to be careful in taking for granted another man's data.

If Thales made his prediction of the solar eclipse of 28 May 585 B.C. in the manner explained above, his achievement should not be called a real prediction. Of course, the alleged pattern of three solar eclipses within a time span of 35 lunar months, the second occurring halfway between the first and the third, was quite coincidental, and depended on eclipses that were actually observed, taking into account the above mentioned assumption that those solar eclipses in which less than half of the solar disk was eclipsed, and for which he had no special reason to expect one, escaped Thales' attention. A real method of predicting eclipses presupposes more knowledge – about the sphericity of the earth, for instance, and of the exact movements of sun and moon – than Thales possessed.

These considerations also explain why no other instances of solar eclipses according to this method are recorded. As a matter of fact, the alleged method of predicting eclipses collapsed right after the "predicted" solar eclipse of 28 May 585 B.C. had appeared. On 21 September 582 B.C., 41 lunations after the previous one, another eclipse occurred (magnitude 0.85), which was followed after six lunations by another one on 26 March 561 B.C. (magnitude 0.55).[11] Although the

[11] The solar eclipses of 18 May 584 B.C. and 1 October 583 B.C. were too small to be seen at Miletus – the last one actually at its maximum being beneath the horizon.

latter was not impressive, it could easily have been observed (weather conditions permitting), since it occurred not too high above the morning horizon. Neither of these two eclipses fitted into the alleged system of groups of eclipses within 35 lunations. If Thales really had used the method explained above for predicting a solar eclipse, he must have realized afterward that his method was wrong, although it had, by accident, enabled him to predict a solar eclipse. Thales' prediction, then, was an early instance of the not uncommon situation in science that a right conclusion may be based upon a false presupposition.

We will never know whether Thales argued in the way described above. However, this chapter may give an impression of the kind of observations and conclusions drawn from them that are possible within the framework of the archaic world picture.

At the start of this chapter, I mentioned the eclipse predicted by Helicon of Cyzicus at Sicily. If we take that to be the annular eclipse of 12 May 361 B.C., then his prediction is most easily explained by the Saros, as 223 synodic months before that date, on 2 May 379 B.C. a partial eclipse could have been recorded in Sicily. However, Helicon was as lucky as Thales because a Saros is not an absolute guarantee for an eclipse being visible at the same location, and he certainly could not have known that he would see an annular eclipse.

Chapter 4
The Shape of the Earth According to Thales

In the archaic world picture, the earth is flat and usually conceived of as a round disk. First of all, it is important to bear in mind that almost all the Presocratics, of which we have reports about what they are supposed to have said about the shape of the earth, believed that the earth is flat. It is reported by several sources that according to Anaxagoras, the earth is flat (πλατεῖα) (DK 59A1(8), DK 59A42(3), DK 59A47, and DK 59A87). Archelaos is said to have conceived of the earth as somewhat concave in the middle (μέσον δὲ κοίλην), but generally speaking he too may be considered to have conceived of the earth as flat (DK 60A4(4)). The ambiguous word στρογγύλος in the doxographical report on Diogenes' earth can here only mean circular, not spherical (DK 64A1). The earth is said to be drum-shaped (τυμπανοειδῆ) according to Leucippus, and the surface of the earth is said to be disk-shaped (δισκοειδῆ μὲν τῷ πλάτει) and concave in the middle (κοίλην δὲ τῷ μέσῳ) according to Democritus and Anaxagoras (DK 67A1(30), DK 67A26 and DK 68A94). The concavity of Democritus' earth does not prevent Aristotle from calling it flat (πλατεῖα) (*On the Heavens* 294b15 = DK 59A88). Elsewhere, however, Democritus' earth is said to be oblong (ἡμιόλος) (DK 68B15). Diels' translation "nicht rund, sondern länglich" is tendentious because the words "nicht rund, sondern" are not in the Greek text. As will be explained in Chap. 6, and drawn in Figs. 6.1 and 6.2, I hold this rectangle to be the shape of the οἰκουμένη on an otherwise disk-shaped and flat earth.[1] That Empedocles too conceived of the earth as flat may be concluded from a doxographical report, in which the earth is called "circular" (κυκλοτερής) (DK 31A56). Guthrie rightly remarks: "The word is κυκλοτερής, not the ambiguous στρογγύλος" (1965: 192 n. 4). Against this simple evidence, Burkert's argument that the idea (in DK 31A56) that the sun is the earth's reflection implies the sphericity of the earth is not very convincing (1972: 305 n. 30).

[1] See Heidel: "It is quite certain that the continental mass, not to speak of the οἰκουμένη, was not circular, though the map probably was in the earlier times" (1937: 12, note). Heidel argues that the frame of the ancient maps, such as Anaximander's, was rectangular as concerns the inhabited earth (οἰκουμένη), but circular as concerns the surface of the earth as a whole (γῆ). More specifically on Democritus: Heidel 1937: 100.

D.L. Couprie, *Heaven and Earth in Ancient Greek Cosmology*,
Astrophysics and Space Science Library 1, DOI 10.1007/978-1-4419-8116-5_4,

However, Diogenes Laertius mentions four Presocratics who are credited with the discovery of the sphericity of the earth: Hesiod, Anaximander, Pythagoras, and Parmenides (DK 12A1, 28A1, and 28A44). The first three cannot be taken seriously: Hesiod is certainly too early for such a claim, Anaximander is known to have conceived of the earth as a column drum, as will be explained in Chap. 8, and the ancients ascribed all kinds of "discoveries" to Pythagoras. The truth is perhaps that in the circle of the Pythagoreans, the idea of a spherical was advanced, for instance, by Archytas, as some scholars suggest (see Chap. 17). The case of Parmenides is more complicated and much-discussed. Diogenes Laertius credits him twice with the discovery of the sphericity of earth. In DK 28A1, he says, without mentioning a source, that Parmenides was the first to call the earth spherical, using the word σφαιροειδῆ. In DK 28A44, he says that according to Theophrastus, it was Parmenides, whereas in the same text he gives the credits to Pythagoras (on the authority of Favorinus) and Hesiod (on the authority of Zeno) as well. Here, he uses the ambiguous word στρογγύλην ("round").

Frank and more recently Fehling have argued rather convincingly that at least two mistakes are here at stake (Frank 1923: 198–200; Fehling 1985b: 202–206. See also Dicks 1970: 51). In the first place, Diogenes Laertius seems to have "confused with the earth the 'one being' which was spherical" as Morrison expresses it with consent (1955: 64). In the second place, he seems to have misunderstood Theophrastus' "round" as "spherical," and accordingly changed στρογγύλος into σφαιροειδής. Nevertheless, other authors tend to defend Parmenides' claim. Some argue that it is hard to disbelieve the authority of Theophrastus and that Diogenes Laertius' στρογγύλος can be considered as just a loose rendition of the more precise term σφαιροειδής, used by Theophrastus (Kahn 1994: 115; Burkert 1972: 304). It is also argued that the division of the earth into zones is ascribed to Parmenides and that this can only be reconciled with a spherical earth (DK28A44a. See Burkert 1972: 305–306, and recently Panchenko 2008: 189–193). One has to consider, however, that this division of the earth into zones is also ascribed to Thales and Pythagoras (DK 11A13c). I tend to think that Frank c.s. have the stronger arguments, and my personal guess is that Oenopides has to be credited for the proposal of a spherical earth, as will be explained in Chap. 13. However this may be, from Anaxagoras onward, we may infer that there existed a discussion about the shape of the earth, for he put forward an argument that intended to prove its flatness (see Chap. 15).

Thales, who is known as one of the Seven Sages and who is said to be the first Greek philosopher, still taught, according to Aristotle, that the earth floats on water (*On the Heavens* 294a28–b6 = DK 11A14; *Metaphysics* 983b20–21 = DK 11A12). Usually, it is taken for granted that Aristotle implies that Thales meant that the earth is flat and floats like a leaf or a piece of wood on the surface of the water that surrounds and supports it. The conclusion from these texts is that "he (Thales) was ignorant of the sphericity of the earth," as Guthrie puts it (1962: 47). Recently, however, a monograph on Thales has been published, written by Patricia O'Grady (2002: 87–107), who, contrary to the generally held opinion, defends the view that already Thales has said that the earth is spherical and floats free in space.

O'Grady proposes an original but also peculiar interpretation of the text in which Aristotle says that the earth, according to Thales, floats on water. Aristotle, she says, has completely misunderstood Thales' meaning, and, put on the wrong track by his authority, all commentators after him until now. What Thales meant to say was not, according to O'Grady, that *the Earth* (Greek: γῆ) *as a whole*, but that *earth* (in Greek also: γῆ) rests on water. What Thales wanted to say is that the great land masses of the earth are, as it were, big floating islands. As an example, O'Grady mentions the floating island in Egypt that is described by Herodotus (*Histories* II 156). The earth as a whole, on the contrary, has no need of support "or any kind of attachment, securing the Earth to anything below it," she writes (2002: 93).

Against O'Grady's hypothesis that Aristotle has misunderstood Thales, several objections can be raised. To begin with, nowhere in the ancient sources can be found that the Presocratics have been worried about the question what it is that the land masses of the earth rest on. If Thales really had purported such a theory, one would expect it to have initiated a discussion among his immediate successors. More important is that O'Grady's interpretation that not *the Earth*, but *earth* floats on water implies, as she says, that to Thales has to be ascribed the notion of the earth hanging free in space as well. If the earth does not float on water, it must be thought of as completely unsupported, she argues. The notion of an unsupported earth, however, is the corollary of the notion of the celestial bodies making full circles around the earth, as we will see in Chap. 8. O'Grady credits Thales with this conception: "He (Thales) *surely* considered that when the sun disappeared below the horizon it proceeded on its journey along a dome beneath the Earth to appear again at sunrise" (2002: 99, my italics). How can she be so sure? Nowhere in the doxography is mention made of Thales defending such a revolutionary conception of the celestial bodies. *To us*, it might be obvious that the celestial bodies make full orbits around the earth, but this was not the case in those ancient times. First of all, it does not concern an observation, but a conclusion that the sun in its daily course goes under the earth cannot be observed by someone who does not travel around the earth. Moreover, in the Presocratic tradition, the notion of a free-floating earth, which is the consequence of the conception of the celestial bodies making full circles, is linked with the question why the earth does not fall. Nowhere in the tradition, however, is reported that Thales asked that question, let alone that he answered it.

After having disposed of Aristotle's texts that are thought to refer to Thales' flat earth, O'Grady concludes that there is not a single piece of textual evidence for the commonly agreed opinion that the earth according to Thales is flat. However, the textual evidence for her own hypothesis that Thales should have been acquainted with the sphericity of the earth is also rather thin. It consists, all in all, of only one statement by Aëtius that sounds: "Thales and the Stoics and those who followed them state that the earth is shaped like a ball" (Aëtius 3.10, not in DK, see Diels 1879: 376; O'Grady 2002: 95). This is a dubious text, as it mentions Thales in one breath with the Stoics and does not mention others like Plato and Aristotle who also taught the sphericity of the earth. O'Grady also quotes two texts in which Seneca writes: "For he (Thales) says that this *round of lands* is sustained by water, and is carried along like a boat," and "Thales of Miletus judges that the whole earth is buoyed up and floats upon liquid that lies underneath. (...) The *disc* is supported by this water (...) just as some big heavy ship is supported by the water which it presses down upon" (Seneca, *Naturales Quaestiones* 3,14.1–2 = DK 11A15, and *Naturales Quaestiones* 6,6.1, not in DK, but see *Nachtrag zum ersten Band*: 486, line 35; my italics; O'Grady 2002: 101–102).

O'Grady maintains that these lines suggest that according to Thales, the earth was a sphere (2002: 107, n. 53). The Latin expression *orbis terrarum* ("ring of lands"), however, is nothing but the usual description of the then known lands around the Mediterranean Sea, without any connotation as to the shape of the earth. Moreover, Seneca explicitly says that Thales' earth floats upon liquid that lies underneath, whereas according to O'Grady Thales taught that the earth floats unsupported.

The objection that it is very improbable that Thales knew of the sphericity of the earth, whereas his direct successors Anaximander and Anaximenes still thought that the earth is drum-shaped or flat, O'Grady regards as of little relevance. We have to do here, she says, with one of those "outstanding examples of backward steps 'that sometimes occur in the history of the natural sciences'" (2002: 100, quoting Burkert 1972: 303 and 305). I think this is an example of bad reasoning. That sometimes such a step back occurs is not to say that we may postulate it whenever it suits us. Moreover, O'Grady does not take into consideration that even Anaxagoras and Democritus, contemporaries of Socrates, still believed that the earth is flat. If I do understand her intentions well, we have to accept that Thales conceived of the earth as a big and unsupported sphere of water, on the surface of which the continents float. One really wonders, which Presocratic thinker would have ventured such a conception.

Over and above her curious reading of Aristotle, O'Grady presents another set of arguments. The first is that Aristotle would have had Thales in mind when he said, talking about the shape of the earth: "some think that it is spherical" (*On the Heavens* 293b33–294a1; O'Grady 2002: 95). This is an *argumentum e silentio*, and as such it is not very strong. All we can say about it is that unfortunately Aristotle did not mention the names of these "some." O'Grady's second argument is that the doxographical reports, which say that Thales should have settled the solstices and determined the size of the sun and the moon, must mean that he made a great number of observations and that "these observations could have played a role in his reasoning" about the shape of the earth (2002: 95; DK 11A1 and DK A17). However, if these testimonies were trustworthy at all (which they are not), then they still do not as such say anything about the spherical shape of the earth. At least O'Grady does not explain *how* this kind of observations should have led to, or perhaps presuppose, the discovery of the earth's sphericity. As we will see in Chap. 16, it is possible to calculate the size of the sun, even when one thinks that the earth is flat (and moreover, using a type of mathematical reasoning that is ascribed to Thales, as we saw in Chap. 2). Such a calculation, however, has nothing to do with the knowledge or ignorance of the solstices. The third argument, to which O'Grady pays most attention, is that Thales could have done all kinds of observations. She mentions that at lunar eclipses the border of the eclipsed part is always part of a circle, that stars, which can be seen at a given location, become invisible when we go north- or southward, whereas new stars arise above the horizon, and that, standing on the shore, one sees a ship gradually descend below the horizon (2002: 98–99). In short, O'Grady supposes that Thales could have adduced similar arguments as mentioned 200 years later by Aristotle for the first time in written history. There is, on the contrary, not the slightest indication that Thales actually made such observations, let alone that he drew from them the right conclusion.

O'Grady's way of arguing on the basis of observations Thales *could have made* is an anachronistic fallacy. *To us*, who know that the earth is spherical, it seems obvious that these observations can be adduced as conclusive arguments. For people, however, who are raised within the paradigm of a flat earth, these observations and the possible inferences from them are not in the least obvious. Astronomers (and especially such versed observers as the Assyrians and Babylonians, and later the Chinese) and sailors have for centuries watched this kind of phenomena without relating them to the shape of the earth. Dreyer thinks anachronistically as well, when he supposes that this kind of stories shows that "some people must have been able to perceive the consequences of the earth being a sphere" (1953: 39). Speaking of these observations, Guthrie rightly concludes: "But all they show is that these phenomena were observed. It does not follow that they were correctly explained" (1962: 294 n. 1).

Sometimes O'Grady's argumentation makes further commentary superfluous. For instance, when she in support of her conviction that Thales did understand that the earth is spherical adduces that the horizon looks curved (2002: 98). This is no indication of the sphericity of the earth at all. On the contrary, it seems more likely that when people, especially at full sea, looked at the full circle of the horizon around them, this could have contributed to the image of a circular flat earth, as in Homer. It is rather the other way round: the fact that the rising or setting sun seems to be cut off by the horizon with a straight line is used by Anaxagoras to prove that the earth is flat, as we will see in Chap. 15. We may conclude that there are no good arguments to throw overboard the traditional interpretation that Aristotle understood very well Thales' intention when he said that according to Thales the earth floats on water and that he meant to imply that according to Thales the earth is flat.[2]

[2] It is strange that O'Grady in her article "Thales" in *The Internet Encyclopedia of Philosophy*, section "The Earth Floats on Water", nowhere mentions her own hypothesis that not "the earth", but "earth" is meant.

Chapter 5
The Riddle of the Celestial Axis

The tilting of the axis of the heavens must have been one of the big riddles for the ancients who studied the skies. Why does it look as if the stars turn around a point in the northern region of the heavens, and not around the zenith? In other words, why is the axis of the heavens tilted?[1] One would expect myths to be told about this phenomenon. However, I know of only one, from ancient China, where, just as in Presocratic Greece, people believed that the earth is flat. The legendary hero Kung-kung struggled with the also legendary emperor Chuan Hsü about the sovereignty of the empire. In great anger, he threw a mountain that shattered the pillars of the heavens. Since that time, the heavens are inclined toward the north-west (see Needham 1959: 21). It is strange that the heavens are said to tilt in the direction of the north-west instead of the north, as we might expect. Perhaps one source gives a reason for this anomaly, as there it is added: "the earth does not fill the South-East, so the rivers and the rain floods find their home there" (Graham 1990: 96). Broadly speaking, one might say that the great Chinese rivers flow to the south-east. Here, the same identification of the tilting of the heavens with a dip of the earth as in Presocratic Greece, which I discuss below, seems to be the case.

The Presocratic philosophers too pondered about the problem of the tilting of the heavens. As Furley writes: "Greek cosmologists (...) had to face the problem that the axis of the whirling stars is visibly not perpendicular to the horizon (in Greece). They commonly explained this by the *ad hoc* assumption that the heavens tilted somehow, after the formation of the earth" (1989: 12 n. 32). Unfortunately, the doxographical reports on this tilting are rather confused, and so are most modern commentaries. This chapter is my attempt to disentangle the ways in which the phenomenon of the tilted axis of the heavens has been misrepresented and misinterpreted.

First of all, it is important to bear in mind, as seen in the preceding chapter, that all the Presocratics, of which we have reports on what they should have said about the tilting of the heavens (Anaxagoras, Empedocles, Archelaos, Diogenes,

[1] At the time of the Presocratics, due to the precession, the Polar star was much farther away from the north pole of the heavens than in our times. Around 500 B.C., Kochab, one of the stars of Ursa Minor, was the nearest star to the pole, at about 7°, whereas in our days Polaris is at no more than 41′ distance from the pole.

D.L. Couprie, *Heaven and Earth in Ancient Greek Cosmology*,
Astrophysics and Space Science Library 1, DOI 10.1007/978-1-4419-8116-5_5,

Leucippus, and Democritus), still believed that the earth is flat. In other words, just like Anaximander, as we will see in Chap. 8, they all preserved this important feature of the archaic world picture. For those who think that the earth is flat, the axis of the heavens, around which the stars revolve, must be thought of as going through the center of the earth. We will have to realize that all thinkers mentioned above lived on what we call the northern hemisphere. The ancient Greeks considered Delphi as the navel of the earth, as Agathemerus remarks just after having mentioned Anaximander's map of the earth (DK 68B15). Accordingly, the celestial axis must be thought to run from the celestial pole to Delphi. As Delphi lies at 38.5° N, the tilting of the celestial axis must be 38.5° with regard to the northern horizon, and 51.5° with regard to the zenith, as is shown in Fig. 5.1.

Fig. 5.1 Celestial axis and earth's axis on a flat earth

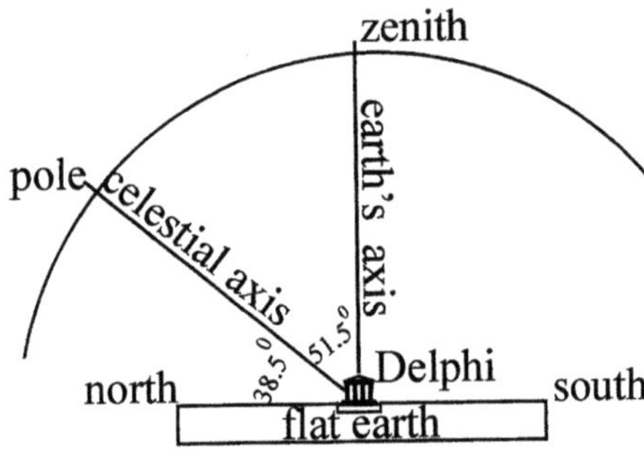

Of course, the ancient astronomers must have been aware of the phenomenon that the celestial pole stands higher on the firmament as one moves to the north, and lower as one moves to the south. The only way in which they would have been able to explain this is by inferring that the stars are not far away, as is shown in Fig. 5.2.[2] However, they obviously did not recognize that this must imply that on a flat earth the celestial pole is no longer exactly in the north as one goes farther to the east or to the west. The observation that the pole remains exactly north, however far one goes to the east or to the west, should have led to the conclusion that the earth cannot be flat but must be convex, or even spherical (and, accordingly, the stars far away). We, being acquainted with the conception of a spherical earth, would say that this observation and the conclusion drawn from it are obvious. That this conclusion was not drawn shows how hard it must have been to overcome the apparent evidence that the earth is flat and to embrace the counterintuitive concept of a spherical earth. As soon as people realized that the stars are far away indeed, they were forced to admit that the earth is not flat, but spherical. For if the earth were flat and the stars far away, the celestial pole would have to be practically at the same height all over the earth.

[2] See also Rovelli: "la hauteur du soleil varie avec la latitude (...) l'interprétation chinoise: la Terre est plate et la variation est due à la faible distance du soleil" (2009c: 131, subscription of Fig. 18, left).

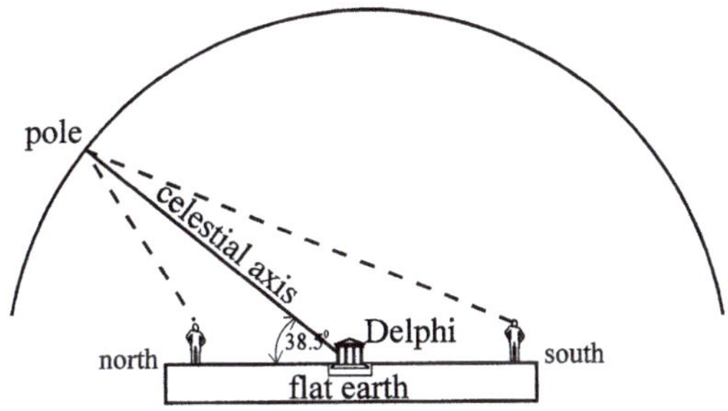

Fig. 5.2 The height of the pole on a flat earth

Most misrepresentations and misinterpretations of the issue of the tilted celestial axis originate from confusing the astronomy of a flat earth with that of a spherical earth. Or, more precisely, they result from confusing the tilting of the celestial axis, as seen from a flat earth, with the obliquity of the ecliptic. On a spherical earth, the celestial axis coincides with the earth's axis, which is perpendicular to the equatorial plane. As the plane of the ecliptic is inclined 23.5° in relation to the plane of the equator, the celestial axis is inclined as well 23.5° in relation to the ecliptic pole (see Fig. 5.3).

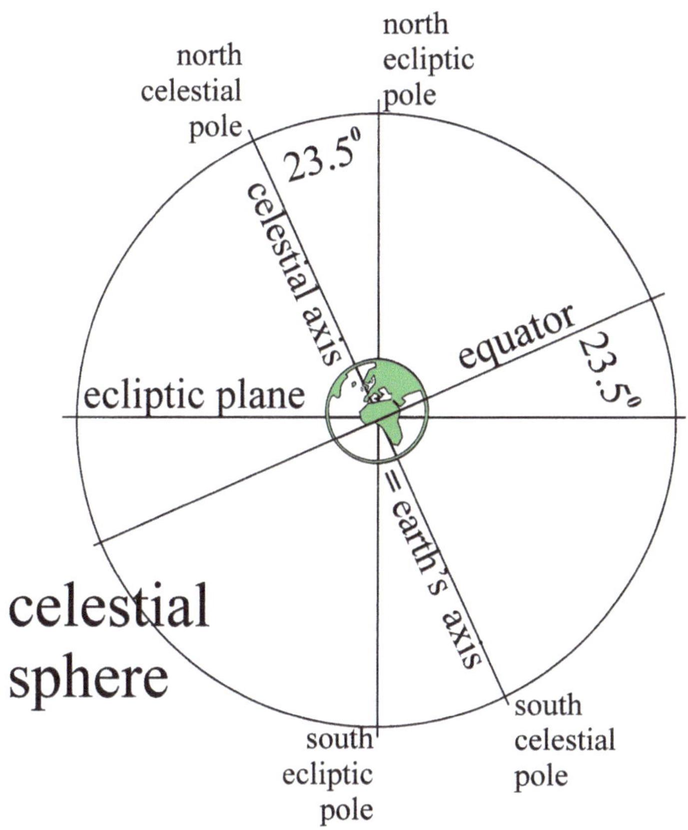

Fig. 5.3 Celestial axis, equator, and ecliptic on a spherical earth

On a flat earth, on the contrary, the celestial axis does not coincide with the earth's axis. Actually, on a flat earth, which is thought to stand still, the conception of an axis of the earth does not make much sense, although one might call the perpendicular from Delphi to the zenith the earth's axis, as done in Fig. 5.1. The amount of the angle of the celestial axis on a flat earth depends on the place where the center of the earth is thought to be, and is, as we have seen, 38.5° with respect to the surface of the earth when you think that the earth's center is at Delphi.

The most straightforward account is that the pole of the heavens originally stood vertically upon the earth, and that later, for some reason, it has tilted. This is explicitly reported by Diogenes Laertius as Anaxagoras' view: "Originally the stars moved as in a dome (θολοειδῶς) so that the always visible pole was right above (κατὰ κορυφήν) the earth" (DK 59A1(4) and DK 59A1(9)). This original state of affairs can be compared with what a person sees who is situated on the north pole of a spherical earth: the celestial pole is right above his head, and the sun turns around the horizon. This is also how Heidel describes it: "the early scientists contended that originally the sun had moved round the edge of the earth-plane" (1933: 122).[3] The reason for the tilting, ascribed to Empedocles, that the air gave in to the violence of the sun is hardly understandable (Cf. DK 31A58). Furley notes to this text: "I do not know what this means" (1987: 93 n. 13). Kingsley's interpretation that according to Empedocles the tilting of the heavens was due to the abundance of fiery stars which made the universe top-heavy, and that as the fire slid down the circular boundary of the universe it accumulated into one concentrated fire, the sun, is not supported by the text (1995: 49–50). Another text says that, according to Anaxagoras: "Later, however, the cosmos assumed a tilt (ἐγκλιθῆναι) all by itself (ἐκ τοῦ αὐτομάτου)" (DK 59A67).

Our sources on the Presocratics use different wordings that, however, all boil down to the same thing: we are informed that according to Diogenes and Empedocles it was the κόσμος that was tilted, according to Anaxagoras the κόσμος, or the ἄστρα, or the πόλος, and according to Archelaos the οὐρανός (DK 31A58, DK 59A67, DK 59A1(9), and DK 60A4(4)). The result of this tilting of the heavens is shown in Fig. 5.4.

So far, so good. Diogenes Laertius and Hippolytus do not indicate the direction of the tilt (DK 59A1(9) and DK 60A4(4)). However, Ps-Plutarch says that according to Anaxagoras, the axis of the heavens was tilted *to the south*: εἰς τὸ μεσημβρινὸν αὔτου (sc. κόσμου) μέρος (DK 59A67). Something seems to have gone wrong here. For the pole of the axis of the heavens has tilted *to the north*, and not to the south, as compared with its original upright position, as is illustrated in Fig. 5.4.

How has this strange mistake originated? In the ancient civilizations in Europe, Asia, and northern Africa, for those who believed that the earth is flat, "south" is defined as the direction in which the sun stands at noon, whereas "north" is the

[3] See also Zeller-Nestle: "Die Gestirne drehten sich anfangs seitlich um die Erde" (1920: 1274; see also Dümmler 1889: 106).

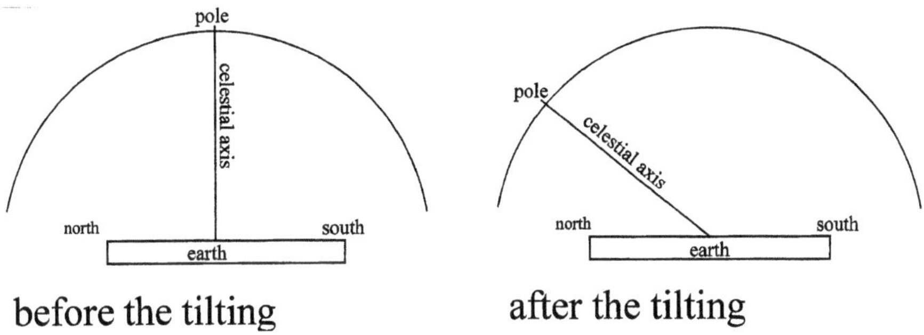

Fig. 5.4 The position of the celestial axis before and after its tilting, according to Anaxagoras c.s.

opposite direction. "East" is the direction where the sun rises (more precisely, at the equinoxes), and "west" is the direction of sunset. We have to realize, however, that Ps-Plutarch lived in a time (second century A.D.) when the sphericity of the earth had already been an established fact. On a spherical earth, "south" and "north" have to be defined otherwise, for only on the northern hemisphere "south" is the direction of the sun at noon, whereas on the southern hemisphere the direction of the sun at noon is "north." On a spherical earth, then, "north" and "south" are defined as the opposite directions of the celestial axis, which coincides with the axis of the earth. Moreover, in Ps-Plutarch's time, the inclination of the earth in relation to the plane of the ecliptic (or, which amounts to the same thing, the inclination of the ecliptic in relation to the celestial equator) was an established fact as well.

Let us assume that in the conception of a *spherical* earth originally the ecliptic was not inclined so that the celestial poles coincided with the ecliptic poles, and the ecliptic plane coincided with the equatorial plane. Then, standing on the northern hemisphere, it makes some sense to say that the axis of the heavens has tilted *toward the south*, as Fig. 5.3 shows. With this image of a spherical earth in mind, Ps-Plutarch may have abusively attributed to Diogenes, Anaxagoras, and Empedocles a tilting of the heavens to the south. In other words, Ps-Plutarch confused the tilting of the celestial axis, as seen from a flat earth, with the inclination of the ecliptic in relation to the celestial equator.

The same Ps-Plutarch, in his account on Empedocles, tells that "both poles have tilted, and hence the north celestial pole has moved *upward* and the south pole *downward*, which occasioned the tilting of the heavens as a whole: ἐπικλιθῆναι τὰς ἄρκτους, καὶ τὰ μὲν βόρεια ὑψωθῆναι, τὰ δὲ νότια ταπεινωθῆναι, καθ' ὅ καὶ τὸν ὅλον κόσμον" (DK 31A58). Here, he seems to be completely confused, as on a flat earth the tilting of the pole is naturally described as *downward*, and not upward.[4] I suppose that Ps-Plutarch has in mind that Aristotle calls the north pole of the celestial

[4] A theoretical possibility would be that originally the celestial axis is thought of as lying in the plane of the surface of the flat earth, and that later on it was lifted up onto its present angle. This, however, would not be in accordance with the sources, which speak of a tilt, and, in the case of Anaxagoras, explicitly of an originally upright position of the celestial axis.

axis "the lowest part" (τὸ κάτω μέϱος), and its south pole "the upper one" (τὸ ἄνω) (*On the Heavens* 284b6–286a2). Heath has explained his meaning by imagining that when you are lying face up along the world's axis, with your head toward the south pole and your feet toward the north pole, east is at your right, west at your left, south is above (sc. your head), and north is under (sc. your feet) (1913: 232). This view is not "unnatural," as Dicks calls it (1970: 72), provided you take as a starting-point west, from where the movement of the heavens start, at your right, as Aristotle says.

Fig. 5.5 According to Aristotle the North Pole is "below" and the South Pole "on top"

So, in Fig. 5.5 (and again, being on the northern hemisphere of a spherical earth) we could say that the celestial axis has moved upward (sc. toward the south). In other words, Ps-Plutarch confuses what would be the case on a spherical earth with what would be the case on a flat earth. Ps-Plutarch's mistake is due, again, to a confusion between the tilting of the celestial axis and the inclination of the ecliptic, which, on a flat earth, are not the same. However, we do not have to do here with Aristotle, who knew that the earth is spherical but rather with Empedocles. To him who believed that the earth is flat, lying on the earth in the above described position did not mean being parallel to the celestial axis, because the celestial axis is at an angle in regard to his flat earth. On a flat earth, it makes no sense to describe the tilting of the celestial axis as an upward movement.

Other texts describe the phenomenon of the tilting of the celestial axis in yet another, and quite curious, way. The atomists Leucippus and Democritus, who also believed that the earth is flat, are said to have held that it is not the axis of the heavens but the earth itself that had tilted. The sources say that according to them the earth has tilted *to the south*: πρὸς μεσημβρίαν, as one source says, εἰς τὰ μεσημβρινὰ μέρη, or κατὰ τοῦτο (sc. μεσημβρινόν) according to another source (Diogenes Laertius in DK 67A1(33), Ps-Plutarch in DK 67A27 and DK 68A96). Obviously, when you describe the phenomenon that the celestial pole is not in the zenith, you may do this by saying that either the heavens have tilted toward the north or that the earth has undergone a dip to the south. So this is a confirmation of what was said above, that those who maintained that the celestial axis has tilted must have expressed that by saying that it has tilted toward the north.

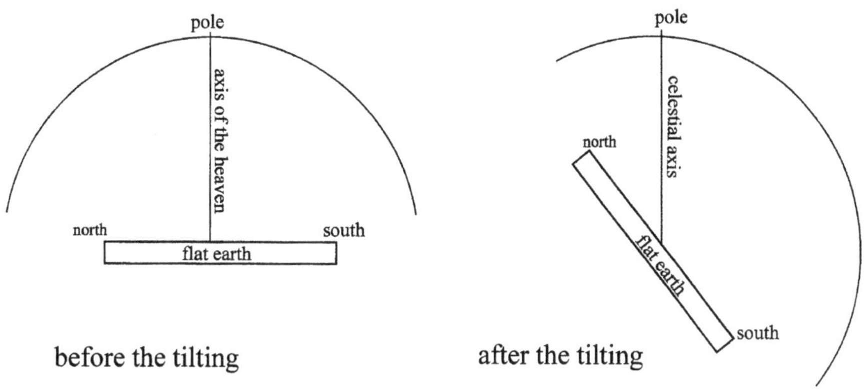

Fig. 5.6 The position of the earth before and after its tilting, according to the doxography on Leucippus and Democritus

The picture above gives an impression of the supposed dip of the earth to the south. Here, too, one has the uncomfortable feeling that something has gone wrong. When the celestial axis has tilted 51.5° in relation to its former upright position (as at Delphi the pole is 38.5° above the northern horizon), then the dip of the flat earth to the south, in Leucippus' and Democritus' version, also amounts to 51.5°, as is shown in Fig. 5.6. But when the earth has tilted that much, how do we manage not to slide off the earth? Or, as already Zeller puts it, why doesn't all the water of the earth accumulate in the southern regions? (1920: 1108, n. 6). In the same sense Wöhrle says: "Die Neigung der Erdscheibe wäre so beachtlich, daß man sich kaum (...) vorstellen kann, wie sie noch auf dem Luftpolster schwimmen könnte und wieso die Bewohner der Erde von dieser Neigung nicht das geringste verspüren" (1993: 75). Aristotle, speaking about Anaximenes', Anaxagoras' and Democritus' flat earth, not only states that it, "covers like a lid the air beneath it," but even that it "does not cleave the air beneath it" (*On the Heavens* 294b16 = DK 13A20). Simplicius repeats that according to Anaxagoras and Democritus "the earth rides on the air below like a lid" (*In Aristotelis De caelo commentaria* 520.28 = DK 59A88). How are these remarks to bring into line with such an enormous dip of the earth?

I think the only way to cope with this difficulty is to consider Aristotle's account as the most reliable, as he was nearest in time to Democritus and wrote a book on him, which Simplicius quotes, but which is now lost (see Simplicius' account in *In Aristotelis De caelo commentaria* 294.33 = DK 68A37, see also Dicks 1970: 82). This would mean that the reports on the so-called dip of the earth must be mistaken, and that the atomists taught a tilting *of the heavens*, just like the other Presocratics. The origin of the mistake in the doxography of the atomists was, I suppose, the same as that indicated above, by which a tilting of the heavens to the south instead of to the north was erroneously reported. When the earth is thought of as spherical, the axes of the heavens and of the earth coincide, as can be seen in Fig. 5.3. And just like, in that case, the celestial axis can be said to have tilted to the south, the same can be stated of the axis of the earth, and thus of the earth itself, as is also shown in Fig. 5.3. The doxographers, however, did not realize that on a flat earth this is not the case. On a spherical earth, the amount of the tilting of the earth is 23.5°, which is the obliquity of the ecliptic in relation to the equator (see again Fig. 5.3). On a flat earth, however, we have to distinguish between, on the one hand, the tilting of the heavens, which is in relation to the zenith (and which is 51.5° at Delphi), and on the other hand the obliquity of the ecliptic. In other words, on a flat earth, the inclination of the celestial axis is not equivalent to the obliquity of the ecliptic (23.5°), and has, *au fond*, nothing tot do with it.[5] Consequently, the ancient Greeks used two different terms: ἔγκλισις for the tilting of the heavens, and λόξωσις for the obliquity of the ecliptic.[6]

In more recent times, most scholars simply echoed the different reports, without even recognizing the twofold inconsistency of explaining one and the same phenomenon by a tilting of the heavens or a dip of the earth, both toward the south. Usually, the misguided reference to the obliquity of the ecliptic is repeated as well. Let us look at a few examples: Dicks mentions the tilting of the cosmos as well as that of the earth to the south, both meant to explain the same kind of phenomena, without noticing any problem (1970: 80, 78, and 59). He also adds that "this seems to be a recognizable attempt to account for the facts that the plane of the ecliptic is inclined to the plane of the equator (...) and that in Mediterranean regions the point in the sky about which all the stars are seen to revolve (i.e., the celestial north pole) is not at the observer's zenith" (1970: 59). As we have seen only the last mentioned is relevant. Panchenko already rightly remarks that Dicks does not take into account that Anaxagoras' earth was flat, which means that "it is likely, therefore, that he had in mind the inclination of the solar path relative to the

[5] On a flat earth, the tilting of the celestial axis will only by accident be the same as the obliquity of the ecliptic, viz., when the center of the flat earth is thought to be at what we would call 23.5° latitude, for instance at Syene. Only in that case the celestial axis on a flat earth will lie in the plane of the ecliptic.

[6] See Dicks: "ὁ λοξὸς κύκλος was in later Greek astronomy a normal expression for the ecliptic" (1970: 71). See also Kahn: "the term for this general tilting is always ἔγκλισις, whereas λοξὸς (κύκλος) is the technical expression for the obliquity of the ecliptic" (1970: 102).

surface of the earth" (1999: 39). Gershenson and Greenberg mix up three different things when they mention under the head "On the inclination of the ecliptic" both "the inclination of the cosmos to its southern (!) part", and "the inclination of the earth" (1964: 340–341, my exclamation mark). They translate the title of Ps-Plutarch's account on Diogenes and Anaxagoras as: "what is the reason the world is inclined to the plane of the ecliptic?," whereas the original has only τίς ἡ αἰτία τοῦ τὸν κόσμον ἐγκλιθῆναι, without any mentioning of the plane of the ecliptic (1964: 118, quoted as Aëtius, *Placita philosophorum* 2.8; not in DK, but see Diels 1879: 337). On the other hand, they maintain that Plato in *The Lovers* 132a3 (= DK 41A2) speaks of "the inclination *of the earth*" (1964: 59), whereas the text has only "some tiltings" (ἐγκλίσεις τινάς). Curiously enough, Dumont translates the same passage as "certains positions *d'écliptique*," which is not in the text either (1988: 477, my italics). Guthrie quotes, without any commentary, the report on Empedocles that describes the tilting of the heavens by stating that "the northern parts were heightened," apparently not noticing that something is wrong (1965: 192). The same holds for his (and Furley's) rendition of the report on Anaxagoras that says that the cosmos has tilted to the south (Guthrie 1965: 305; Furley 1987: 73). Furley, speaking about Democritus, writes that "*the tilt of the earth* relative to the north pole was introduced, as we have seen, to account for the effect that the star-circles are not parallel to the surface of the earth." However, for the clause "as we have seen" he refers to a passage (on Anaxagoras and Empedocles) in which it is stated that *the axis of the heavens is tilted* with respect to the earth, without noticing any inconsistency (1987: 145 and 92–93, my italics). Dreyer declares that according to Anaxagoras "the inclination of the axis of the heavens to the vertical was caused by a spontaneous tilting *of the earth* toward the south," as if he were describing Democritus' alleged ideas instead of those of Anaxagoras, who taught a tilt *of the heavens* (1953: 31, my italics). And speaking of Empedocles, he erroneously explains "the obliqueness of the axis of *the heavens* to the horizon (...) whereby the north side has been elevated and the south side depressed," by saying that "*the earth* (...) was tilted so as to bring the northern end up and the southern end down" (1953: 26), whereas the text of DK 31A58 clearly states that according to Empedocles *the cosmos* was tilted. Last but not the least, Diels tries to emend an obviously corrupted text of Diogenes Laertius (DK 67A1 (33)), by inserting the words τὴν δὲ λόξωσιν τοῦ ζῳδιακοῦ before τῷ κεκλίσ-θαι τὴν γῆν πρὸς μεσημβρίαν. There is, however, no reason to burden Diogenes Laertius, when he reports on Leucippus, with a mistake of Ps-Plutarch. Heath already remarks that Diels' emendation is incorrect, as "the reference must be to (...) the angle between the zenith and the pole or between the earth's (flat) surface and the plane of the apparent circular revolution of a star" (1913: 122 n. 3).

Both the ancient and the modern confusions are due to a lack of understanding of how celestial phenomena appear to those who believed that the earth is flat. My conclusion is that the Presocratics, including Leucippus and Democritus, taught a tilting of (the axis of) the heavens and that the direction of this tilt was to the north, and not to the south. This tilting of the celestial axis was not meant as an explanation for the obliquity of the ecliptic. Moreover, I conclude that the accounts on a dip of the earth are mistaken. This also means that the strange and somewhat childish explanations of this alleged dip (Leucippus: the northern region is always frozen; Democritus: the enveloping air in the south is thinner, which makes that the earth is in the south heavier by bearing fruits), must be regarded as apocryphal as well (DK 68A96, cf. DK 67A27). Dicks (1970: 83), already surmised that Aëtius' account in DK 68A96 is faulty. How Guthrie can read in these texts a confirmation of Aristotle's account that according to the atomists the earth settled on the air like a lid, I do not understand (1965: 424). Probably, the original texts said no more than that on a flat earth it becomes colder as one goes to the north, until one reaches a region that is always frozen, and warmer as one goes to the south, where the fruits grow abundantly. This was erroneously understood by the doxographers as having bearing on the alleged dip of the earth.

Chapter 6
The First Map of the Earth

Several sources attest that Anaximander drew a map of the earth and even that he was the first one to do so (see DK 12A1(2) and DK 12A6). His map has been lost, just like his book. Several scholars, however, have tried to reconstruct it.[1] Studying the possibilities of such a reconstruction, I pay special attention to the astronomical implications of drawing a map when one thinks, like Anaximander, that the earth is flat. In Chap. 8, the shape of the earth according to Anaximander will be discussed more extensively.

The most plausible assumption is that Anaximander's map was circular. The Greeks were familiar with this shape, as it is already found in Homer, where the Ocean ('Ωκεανός) is envisaged as circling around the outskirts of the inhabited earth. Homer mentions "the Ocean, recurring into itself" in *Iliad* XVIII, 399. Herodotus confirms that the early maps were circular in shape "with the Ocean encircling the earth, which is round as drawn with compasses" (*Histories* IV 36). Aristotle also pointed out that the oldest maps were circular (*Meteorologica* 362b12 ff.). Agathemerus, following Eratosthenes, immediately after mentioning Anaximander's map and other ancient mapmakers, says that the ancients drew the inhabited earth as circular with Greece in the middle and Delphi as the navel (ὀμφαλός) in the center (DK 68B15(2)).

In his book on the frame of the ancient Greek maps, Heidel argues, however, that the frame of these maps such as Anaximander's was rectangular, and accordingly his reconstructed map is drawn as a rectangle. What Heidel means to say is that the frame was rectangular as concerns the inhabited earth (οἰκουμένη), but circular as concerns the surface of the earth as a whole (γῆ). The title of Part I of his book is *The Frame in Relation to the Flat Disk Earth*, and he says explicitly: "It is quite certain that the continental mass, not to speak of the οἰκουμένη, was not circular, though the map probably was in the earlier times" (1937: 12 n. 22). Heidel was also the first to suggest the use of the gnomon for drawing a map of the earth when he mentioned "the undoubted relation of the Ionian map to the horizon and the

[1] Reconstructions of Anaximander's map of the earth (or that of his successors, like Hecataeus) in Herrmann (1931: 47), Heidel (1937: 6), Bengtson et al. (1963: 8). Brumbaugh (1981: 22), Robinson (1968: 320), Bunbury (1979: between 140 and 141), Cohen and Drabkin (1948: 153), and Naddaf (2003: 54).

D.L. Couprie, *Heaven and Earth in Ancient Greek Cosmology*, Astrophysics and Space Science Library 1, DOI 10.1007/978-1-4419-8116-5_6,

points that were marked on the dial of the gnomon" (1937: 58). Pédech means something similar when he remarks: "Les Ioniens concevaient le monde par analogie avec l'horizon, où les points extrèmes, lever et coucher du soleil en hiver et en été, c'est-à-dire sud-est et nord-est, sud-ouest et nord-ouest, marquaient les limites coïncidant avec l'espace habitable" (1976: 35).

This idea has been worked out more thoroughly by Hahn in his book on Anaximander and the architects (2001: 205–208). Obviously, a gnomon or sundial is of little use in defining a circular flat earth. On the other hand, a gnomon can be used very well for establishing the shape and limits of its inhabited part (οἰκουμένη). The procedure would be like this: put a circular model of the earth, for instance a column drum, flat on a horizontal surface. Erect a gnomon in the center of the model and mark where the shadow line of the gnomon at sunrise at one of the equinoxes cuts the circumference of the model. If needed, use an artificial horizon as described in Chap. 2. Do the same at the equinoctial sunset (azimuth, respectively, 90° and 270°, see Fig. 6.1). Connect the two points thus found. The line drawn runs from east to west through the center of the model and divides it into two equal halves. Determine the northernmost point of sunrise and sunset at the summer solstice (azimuth at Delphi in 600 B.C., respectively, 58.5° and 301.5°), and draw the line between these two points, parallel to the first line. Repeat this procedure for the southernmost points of sunrise and sunset at the winter solstice (azimuth at Delphi respectively 120.5° and 239.5°).[2] Heidel called the three parallel lines thus drawn the "Ionian equator and tropics," and he concludes: "We have,

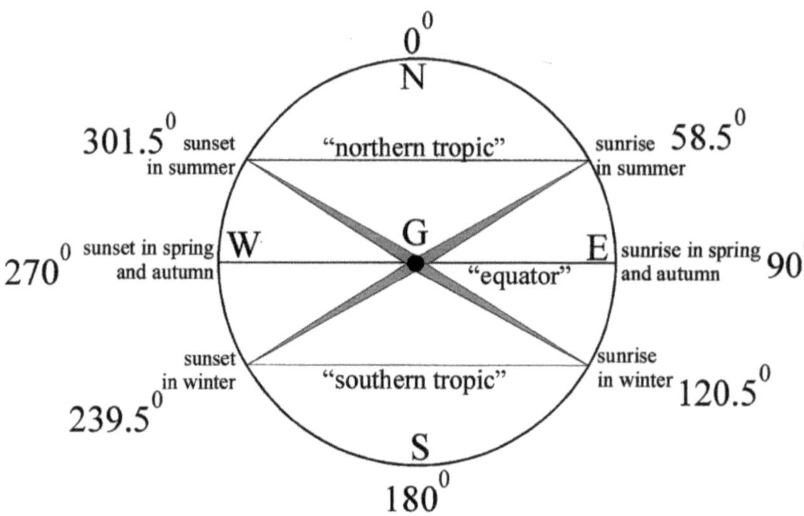

Fig. 6.1 Model of a flat and circular earth with "equator" and frame of "tropics"

[2] Azimuth figures found with *Redshift 5.1* (2005).

then, three clearly indicated parallels on the Ionian maps, corresponding to the tropics and the equator on our maps, but drawn at places where we should not think of locating them" (1937: 20). And elsewhere: "There can be no doubt, therefore, that the Ionian map had an 'equator' based, like its 'tropics', on the appearance of the fixed horizon with the points marked by the rising[s] and settings of the sun" (1937: 54). The oblong area between the two "tropics" is the inhabited earth (οἰκουμένη).

Now, we can start to fill in our model of an ancient Ionian map of the earth. First of all, we mark the center of our circular model and call it Delphi, the navel of the world. Consequently, the "equator" we drew just now runs through Delphi. Hahn, in his above quoted book, does not mention what has to be considered as the center of Anaximander's map. In a recent book, however, he argues that not Delphi (or Miletus, for that matter) but Syene in Egypt must have been the center of Anaximander's map (2010: 157–158). His rather strange argument is that at Syene a gnomon at the summer solstice does not throw a shadow. Moreover, if one tries to sketch a map of the earth with Syene in the center, the hardly believable and for the ancient Greeks probably unacceptable consequence results of Greece being situated on the fringes of the habitable world.[3] Some years earlier, Naddaf also suggested that the center of Anaximander's earth map was in Egypt, although he pleaded for the Nile Delta. One of his arguments is from Herodotus, who "suggests when discussing early Ionian maps that there was a north-south meridian running from the Nile in the south to the Danube/Ister in the north" (2003: 53). According to me, Herodotus is only saying that the Ister (where it flows into the Euxine) and the Nile flow on the same "meridian," without implying that this was the central meridian of the early maps. And even if so, any other place on this "meridian" could have been the center of the map, and not necessarily northern Egypt. Naddaf's other argument is that the Nile Delta is somewhat further to the East, which makes more room on the map for the outermost point of the Persian empire (2003: 55). I think for Anaximander's map no more was needed than the Caspian Sea, a bay of the eastern Ocean, as one of the easternmost points of the map. In other words, I for one see no reason to take another center for Anaximander's map than Delphi.

We will have to realize that on a flat earth the climate becomes colder the farther we go to the north, and warmer as we go southward. This means that the "equator" through Delphi divides the earth in two equal halves, one northern and colder, and one southern and warmer half. Then, we draw the inner periphery of the encircling Ocean. The next site we mark on the "equator" is the strait between the Pillars of Hercules (which is now called the Strait of Gibraltar) that connects the Ocean with the Mediterranean Sea. Just like Polybius, who made use of ancient maps, we situate the Pillars of Hercules at the western end of the "equator" where sun sets at the equinoxes (see Heidel 1937: 54). We may suppose that the daring Milesian sailors had come thus far. As in most reconstructions, we let the "equator" run

[3] For a more extensive discussion of Hahn's ideas in this book, see my review in Couprie (2010).

through Miletus, at an appropriate distance east of Delphi (see Pédech 1976: 35: "L'Asie Mineure (...) se trouvait donc sur l'équateur de la carte"). In reality, the Pillars of Hercules, Delphi, and Miletus are not exactly on the same east–west line (we would say on the same latitude) but the differences are so small that the early mapmakers easily could have overlooked them. Now, we can outline the contours of the Mediterranean Sea and the Euxine. We add some big rivers, the Nile, the Ister (= the Danube), the Euphrates, and the Tigris, and finally we draw the Caspian Sea and the Red Sea as bays of the eastern Ocean (see Heidel 1937: 32). It has to be noted that Eratosthenes (who knew of the sphericity of the earth) still divided his map of the inhabited part of the earth into two parts by a line drawn from west to east, from the Pillars of Hercules through the Strait of Sicily, the southern capes of the Peloponnesus and Attica, and Cilicia (the south-eastern part of Asia Minor), as was done by Strabo (*Geographica* II 1.1; see also the reconstruction of Eratosthenes' map in Cohen and Drabkin 1948: 154).

We may observe that the inhabited earth (οἰκουμένη) with its oblong shape, lying between the two "tropics" in a central moderate zone, consists mainly of the lands around Mediterranean Sea: the Iberian, Italian, and Balkan peninsulas, a strip of land north of them, the Euxine, Asia Minor, Mesopotamia, Syria, Palestine, Arabia, Egypt, and Libya (which is the name of the other North-African countries). The continent north of the "equator" is called "Europa," and that south of the "equator" is called "Asia." In this conception, Libya (which is to say Africa) is a part of the continent Asia. This results in the reconstruction of Anaximander's map as given in Fig. 6.2. This picture is almost the same as Cohen and Drabkin's reconstruction of the map of the world according to Anaximander's successor Hecataeus (1948: 153). The main difference is that they, for some unexplained reason, do not make Delphi the center of the map, but the Hellespont (which is now called the Dardanelles).

The word οἰκουμένη can be translated in two ways. When one translates it as "the habitable earth," then it has to do with climatological issues: it is the moderate zone the north of which is too cold and south of which too warm to inhabit acceptably. When one translates it as "the inhabited earth," then it concerns the region where the civilized peoples live. North of it, mythical peoples like the Hyperboreans live, and south of it live the peoples burnt black by the sun. To quote Pédech again: "Au-delà (de l'espace habitable) le froid et la chaleur extrèmes interdisaient la vie" (1976: 35). Yet, it might snow in these southern countries, for Anaxagoras explained the yearly inundation of the river Nile by the melting of the snow in Ethiopia during the summer (So Diodorus of Sicily, *Bibliotheca Historica* 1.38.4. See also DK 59A91).

There seems to be a misunderstanding in the doxographical report of the above-mentioned Agathemerus, as he maintains that the ancients described the inhabited earth (οἰκουμένη) as round, whereas Democritus was the first to describe the earth (γῆ) as oblong (DK 68B15). This remark would make more sense when the words οἰκουμένη and γῆ would be reversed. It is also strange that Diels without any comment counts this report on Democritus as authentic (in the B-category), whereas Heidel laconically remarks that, of course, the *oikoumene* was meant

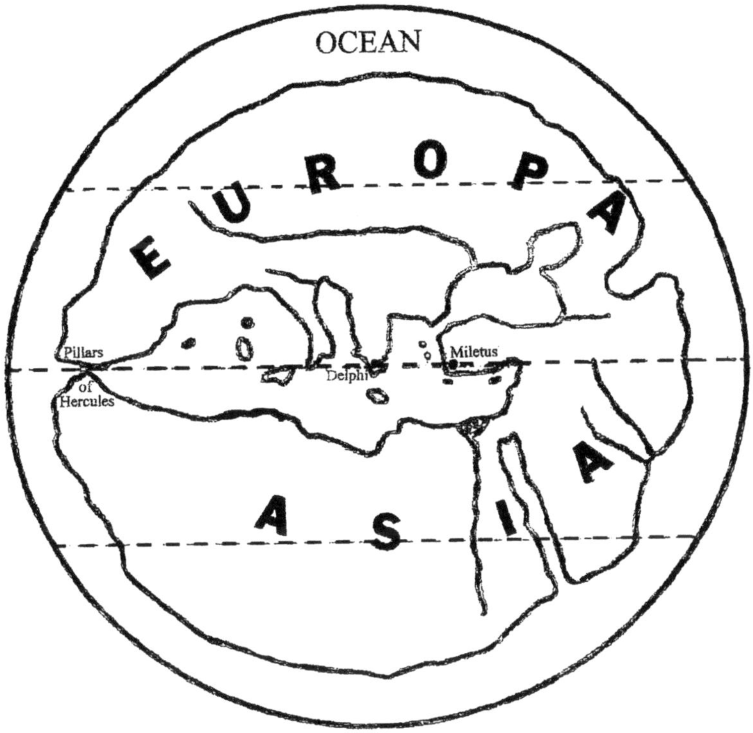

Fig. 6.2 Reconstruction of Anaximander's earth map

(1937: 100 and 110). There is no good reason, I think, to qualify Agathemerus' report as a genuine fragment. It belongs rather to the doxographical testimonies. Another doxographical report, from Eustathius' commentary on the *Iliad*, has it right: it was the οἰκουμένη γῆ (the inhabited part of the earth) that was characterized as oblong by Democritus (DK 68A94).

The conception of the oldest Ionian maps presented above is in concordance with Herodotus, who writes: "I laugh to see how many have ere now drawn maps of the earth, not one of them showing the matter reasonably: for they draw the earth round as if fashioned by compasses, encircled by the river Ocean, and Asia and Europe of like size" (*Histories* IV 36). I follow Heidel, who takes this passage to mean that the oldest maps showed only two continents, "Europe occupying the northern and Asia the southern segment" (1937: 12, see also pp. 22 and 31). And also Berger remarks that the majority of the sources ascribes to the oldest geographers a division of the earth in two continents, and he adds: "Wenn er (Herodotus) aber die Jonier darum tadelt, daβ sie Europa und Asien gleich machen, dann kann er sich nur die kreisförmige Ökumene in einen nördlichen und südlichen Halbkreis zerlegt vorgestellt haben" (1903: 81). Elsewhere Herodotus says, however, that the Ionians discerned three parts of the earth: Europe, Asia, and Libya, the river Nile being the border between Asia and Libya (*Histories* II 16). In his reconstruction of the oldest Ionian map Robinson, followed by Hahn, divides the surface of the earth in three parts (Robinson 1968: 32; Hahn 2001: 210; see also recently Rovelli 2009c: 38). This is, I think, a misunderstanding, as Herodotus is not talking here about the oldest maps, as he did in the above-quoted lines, but about later developments of maps, on which more details of lands were depicted.

The issue of the Ionian "equator" and "tropics" shows how early Greek astronomy and the geography of a flat earth are interrelated. Maybe Anaximander did not go so far as to actually draw the "equator" and the "tropics." But if he did, they must have run as indicated in Fig. 6.2, and their meaning must have been as described above, entirely different from our conception of the equator and tropics. Anaximander must have thought that he lived on or near the "equator" of his flat earth, for he could have observed that at the time of the equinoxes the sun rose due east and set due west. This was the base of the reconstruction of his map with the help of a gnomon. Obviously he did not realize that the same phenomenon is the case at other latitudes as well, for then he would also have realized that the earth cannot be flat.

However this may be, we may state for sure that whatever it was that Anaximander may have drawn, it was something we would have recognized as a geographical map on which we would have been able to distinguish continents, countries, and seas. The reconstruction above is in accord with the doxography in which it is said not only that Anaximander made a geographical map of the earth, but also that he drew the inhabited earth on a map and even that he outlined the contours of the sea and the land (DK 12A6 and DK 12A1). As we saw, Herodotus, who undoubtedly had the oldest Ionian maps before his very eyes, leaves no doubt that they were real geographical maps, although he considered them as ridiculously primitive and speculative. He also tells how Aristagoras, the tyrant of Miletus, about 500 B.C., with the help of a similar map, engraved in a copper plate, tried to persuade the king of Sparta to march against the Great King of Persia. On that map, one could not only discern the contours of the earth, the seas, and the rivers, but it was so precise that one could perceive all countries that one had to go through when going from Miletus to Susa (*Histories* V 49). Anaximander's map will have been much more primitive, but essentially it did not differ from that of Aristagoras. Bunbury expresses the same thought rather apodictically: "There can be little doubt that the bronze tablet, which was brought by the Milesian Aristagoras to Sparta in B.C. 500 (...) was in substance a reproduction of this original map of Anaximander" (1979: 122). What matters is that Anaximander's map was purely geographical, which is to say that it was of a completely different order than the remaining ancient maps of the earth.

A good example of such an old map is the well-known Babylonian map of the earth that is approximately contemporary with Anaximander's (see Fig. 6.3). According to Horowitz, it "can be no older than the ninth century" but "it is likely, however, that the map dates to the late eight or seventh century" (1998: 25 and 26). The hole in the center is probably the place where one leg of the compass was fixed. Above this central hole, an oblong shape is marked "Babylon" with two lines running through it from north to south, probably depicting the two banks of the river Euphrates.

This river takes its rise from the North-west Mountains and runs into an also oblong marsh or swamp in the south. Some cities and lands are indicated by circles. To the right of Babylon, Assyria is situated, and north of it Armenia. Outside the Ocean lie the mythical *Seven* (or *Eight) Islands*, depicted as equal triangles, only two of which are completely preserved.

Fig. 6.3 Babylonian earth map (photograph British Museum, BM 92687)

Another surviving ancient map of the earth is drawn on an Egyptian coffin and dates from the 30th dynasty, that is, from the fourth century B.C. On this picture, the goddess of the heaven, Nut, arches over the world that is carried by two arms making the *ka* sign. We perceive again an encircling Ocean. Inside the Ocean is a ring with the names of the neighboring countries, then a ring with the hieroglyphs of the Egyptian districts, and in the innermost circle the underworld is represented (Fig. 6.4).

These Babylonian and Egyptian maps may be characterized as a mixture of mythological and symbolical or schematic features. The triangular "mountains" outside the Bitter River on the Babylonian map, and the goddess Nut arching over the earth on the Egyptian map are essential features of these maps. However, they clearly do not have a geographical meaning, but instead only a religious or mythological one. Moreover, these maps do not show contours of lands and seas (apart from the encircling Ocean), but only a few schematic lines (on the Babylonian map), or a list of geographical names (on the Egyptian map). The method of

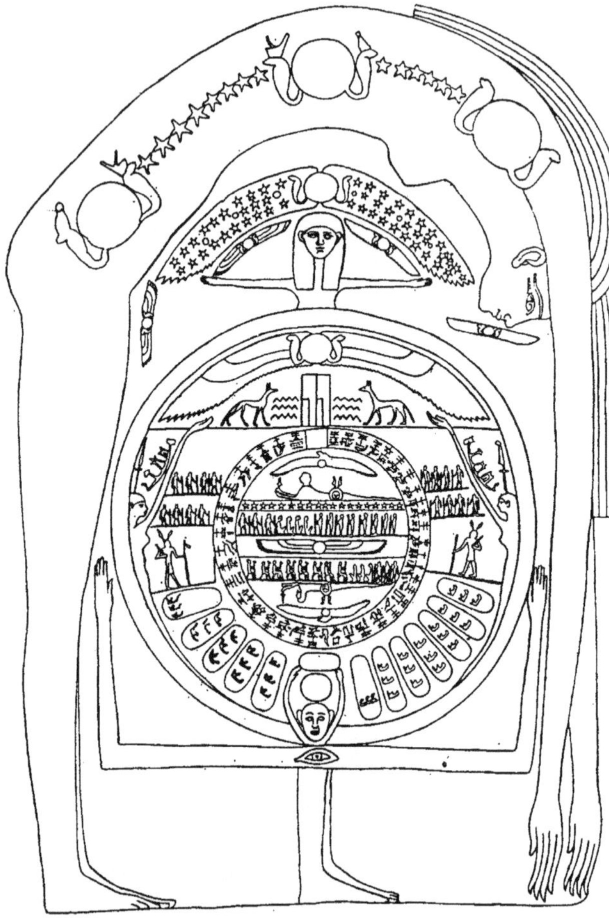

Fig. 6.4 Egyptian earth map (Clère 1958: 30)

mapmaking introduced by Anaximander replaced the Babylonian and Egyptian traditions of making maps of the earth. Although unfortunately we do not possess a specimen of his map, and although his representation of a flat earth still remained within the limits of the archaic world picture, we may conclude that Anaximander created a new paradigm in geographical mapmaking. Or, as Jacob says: "Le rire hérodotéen témoigne donc déjà d'une évolution de la cartographie. De modèle abstrait et *a priori*, la carte est devenue l'instrument d'un savoir 'historique' au sens grec, support de l'enquête ethnographique" (1988: 280).

Part II
Anaximander and the Discovery of Space

Chapter 7
Anaximander: A Survey of His Ideas

The history of Western philosophy begins with Anaximander of Miletus (610–547 B.C.), in Asia Minor, now Turkey. This is how he is treated, for instance, by Karl Jaspers in the first volume of his *Die grossen Philosophen*. Anaximander was the first Greek who wrote a treatise in prose that is referred to in the tradition under the title *On Nature*. Anaximander seems to have reflected on the discovery of writing, or more precise on the letters of the alphabet. It is said that Anaximander maintained that the letters of the alphabet stem from the Phoenicians and were introduced in Greece by Danaüs and not by Cadmus (DK 12C). In this context, the word στοιχεῖα means "letters" and not "elements," as Dumont translates (1988: 40 and note on 1194). Diels qualifies this text as "Zweifelhaftes" and thinks that here another Anaximander is meant, namely, Anaximander the Younger, who lived about 400 B.C. (note at DK 12C). There are, however, reasons to believe that this text refers to the great Anaximander, as Panchenko has argued (2000: 418–420. He refers to arguments of Heidel 1921: 257–260. See also Naddaf 2003: 46).

Only a few lines of Anaximander's book have been preserved, in Simplicius' commentary on Aristotle's *Physics*. Simplicius lived in the sixth century A.D., but he made use of a book by Aristotle's disciple and successor Theophrastus that is now lost. Unfortunately, Anaximander's book itself has been lost as well, but at the time of Aristotle and Theophrastus it was probably still available in the library of the Lyceum. The story tells that in the second century B.C. Apollodorus found a copy of it, probably in the famous library of Alexandria. As already said in the *Introduction*, this story has become more plausible since a recent discovery at Taormina in Sicily (see Blanck 1997a, and somewhat more extensively Blanck 1997b, in which also photographs of his find). This discovery showed that in the second century B.C. Anaximander's book formed part of the library of the gymnasium of that city since a fragment of a catalog was found on which the name Anaximander can be read. In the transcription of the almost completely erased text, we read on the first two lines: ΑΝΑΞΙΜΑΝΔΡΟΣ ΠΡΑΞΙΑΔΟΥ ΜΙΛΗΣΙΟΣ, which is: "Anaximander, son of van Praxiades, from Miletus." The missing part of the second line and the third line can be restored as [ΜΑΘΗΤΗΣ] ΕΓΕΝΕΤΟ ΜΕΝ Θ[ΑΛ]ΕΩ, which is: "who was a pupil of Thales" (see 1997a: 509) (Fig. 7.1).

D.L. Couprie, *Heaven and Earth in Ancient Greek Cosmology*,
Astrophysics and Space Science Library 1, DOI 10.1007/978-1-4419-8116-5_7,

Fig. 7.1 Transcription of a text, found in Taormina (Blanck 1997a: 509, Abb. 2)

We know very little of Anaximander's life. He is said to have led a mission that founded a colony called Apollonia on the coast of the Black Sea. He probably introduced the gnomon into Greece and erected a sundial in Sparta. So he seems to have been a much-traveled man, and that is no surprise as the Milesians were known to be audacious sailors. It is also reported that he displayed solemn manners and wore pompous garments (DK 12A1, DK 12A2 and DK 12A3). Most of the information on Anaximander's works comes from Aristotle and his pupil Theophrastus, whose book on the history of philosophy was used, excerpted, and quoted by many other authors, the so-called doxographers, before it was lost. Sometimes in these texts, words or expressions appear that can with some certainty be ascribed to Anaximander himself. Approximately one third of the reports have to do with astronomical and cosmological questions. These will be treated in the next five chapters.

According to Aristotle and Theophrastus, the first Greek philosophers were looking for the "origin" or the "principle" (the Greek word ἀρχή has both meanings) of all things. Anaximander is said to have called this principle τὸ ἄπειρον, which has been translated as "the boundless," or "the unlimited," or "the nonfinite" (the Greek word ἄπειρος means "that which has no limit"). Already in ancient times Aëtius complained that Anaximander fails to explain what he meant by the *apeiron*: ἁμαρτάνει δὲ οὗτος μὴ λέγων τί ἐστι τὸ ἄπειρον (DK 12A14). More recently, authors have disputed whether the *apeiron* should be interpreted as spatially or temporarily without limits, or perhaps as that which lacks further qualifications, or rather as something inexhaustible. Some scholars have even associated it with the Greek word for "to experience," "to perceive" (περάω), instead of with the word "limit," "boundary" (πέρας), and have defended that ἄπειρον has to be understood as "that which cannot be perceived.[1]" In the next chapter, we will see that some scholars take the *apeiron* to be the infinite space outside the cosmos.

[1] For a more extensive exposition of the etymology and meaning of (τὸ) ἄπειρον, see Couprie (1989: 134–140).

However this may be, the suggestion is almost irresistible that Greek philosophy, by making the *apeiron* the principle of all things, started on a high level of abstraction. On the other hand, some have pointed out that this use of "apeiron" is atypical for Greek thought, which was occupied with limit, symmetry, and harmony. The Pythagoreans placed the *apeiron* on the list of negative things, and for Aristotle, too, perfection became associated with limit (πέρας), and thus *apeiron* with imperfection. Therefore, some authors suspect eastern (Iranian) influence on Anaximander's ideas (see, e.g., Eisler 1910: 162 and 176; Burkert 1963: 106 and 110–111; West 1971: 89–90).

It seems that Anaximander not only put forward the thesis that the Boundless is the principle but also tried to deliver arguments for it. We might say that he was the first who made use of philosophical arguments, meaning arguments that have to do rather with the meaning of words than with empirical evidence. Anaximander's arguments have come down to us in the disguise of Aristotelian jargon. Therefore, any reconstruction of the arguments used by the Milesian must remain conjectural. Nevertheless, the data, provided they are handled with care, allow us to catch glimpses of what the arguments of Anaximander must have looked like. The important thing is, however, that he did not only utter apodictic statements, but also tried to give arguments. This is what makes him the first philosopher in history.

Aristotle reports a curious argument that probably goes back to Anaximander, which says that the Boundless has no origin because it is itself the origin. When we read this argument we should rather say that it looks more like a string of associations and wordplays than like a formal argument. It runs as follows: "Everything has an origin or is an origin. The *apeiron* has no origin. For then it would have a limit. Moreover, it is both unborn and immortal, being a kind of origin. For that which has become has also, necessarily, an end, and there is a termination to every process of destruction" (*Physics* 203b6–10 = DK 12A15). Similar arguments were used later by Melissus and Plato (DK 30B2[9]; *Phaedrus* 245d1–6). The Greeks were familiar with the idea of the immortal Homeric gods. Anaximander added two distinctive features to the concept of divinity: his *apeiron* is an impersonal something (or "nature," the Greek word is φύσις), and it is not only immortal but also unborn. However, perhaps not Anaximander, but Thales should be credited with this new idea. Diogenes Laertius ascribes to Thales the aphorism: "What is the divine? That which has no origin and no end" (DK 11A1(36)).

Several sources mention another argument that is somehow the other way round and answers the question of why the origin should be boundless. In Aristotle's version, it runs like this: "(The belief that there is something *apeiron* stems from) the idea that only then genesis and decay will never stop, when that from which is taken what has been generated, is boundless" (*Physics* 203b18–20 = DK 12A15). In this argument, other versions of which in DK 12A14 and DK 12A17, the *apeiron* seems to be associated with an inexhaustible source. Obviously, it is taken for granted that genesis and decay will never stop, in addition to which the *apeiron* has to guarantee the ongoing of the process, like an ever-flowing fountain.

A third argument is relatively long and somewhat strange. It turns on one key word (in Greek: ἤδη), which is here translated with "long since." It is reproduced by Aristotle in the context of his rendition of the idea of some Presocratic thinkers that there has to be something independent of and preceding the so-called elements (earth, water, air, and fire): "Some make this (viz. that which is additional to the elements) the *apeiron*, but not air or water, lest the others should be destroyed by one of them, being *apeiron*; for they are opposite to one another (the air, for instance, is cold, the water wet, and the fire hot). If any of them should be *apeiron*, it would *long since* have destroyed the others; but now (this is not the case) there is, they say, something other from which they are all generated" (*Physics* 204b25–29 = DK 12A16). The same argument is also used by Plato (*Phaedo* 72a12–b5), but even more interesting is that it was used almost 2,500 years later by Friedrich Nietzsche in his attempts to prove his thesis of the Eternal Recurrence: "If the world had a goal, it would have been reached. If there were for it some unintended final state, this also must have been reached. If it were at all capable of a pausing and becoming fixed, if it were capable of 'being', if in the whole course of its becoming it possessed even for a moment this capability of 'being', then again all becoming would *long since* have come to an end" (1974: 280, n. 36[15], my translation and italics). Nietzsche wrote these words in his notebook in 1885, but already in an unpublished manuscript from 1873 that was not published during his lifetime, he mentioned the argument and credited Anaximander with it (1973a: 315).[2]

The only existing fragment of Anaximander's book (DK 12B1) is surrounded by all kinds of questions. The ancient Greeks did not use quotation marks, so we do not know where Simplicius, who has handed down the text to us, is paraphrasing Anaximander and where he begins to quote him. The text is cast in indirect speech, even the part of which most authors agree that it is an actual quotation. One important word of the text (ἀλλήλοις), here translated by "upon one another," is missing in some manuscripts. Some authors think that not the whole text but only the second half goes back to Anaximander himself. In the analysis of Havelock, even almost nothing authentic is left (1983: 62–65).[3] Simplicius' impression that it is written in rather poetic words has been repeated in several ways by many authors.[4] Therefore, I offer a translation in which some poetic features of the original, such as chiasmus and alliteration have been imitated. In the fourth and fifth line, a more fluent translation is given for what is usually rendered rather cryptically by something like "giving justice and reparation to one another for their injustice":

---

[2] More on the history of this argument in Couprie (1999).

[3] About the text of the fragment Havelock says: "A more extended scrutiny of their vocabulary will enlarge doubt to cover the whole" (1983: 52). Already in a former publication Havelock concludes: "An image of the Milesian original, one suspects, has been compressed into the prose of an epitome, and in the process given abstract formulation" (1978: 65). In my dissertation, however, the conclusion is in favor of the authenticity of Anaximander's fragment (Couprie 1989: 2–41).

[4] A collection of more than hundred different translations in Couprie (1989: 192–211).

Whence things have their origin,
Thence also their destruction happens,
As is the order of things.
For they execute the sentence upon one another
– The condemnation for the crime –
In conformity with the ordinance of Time.

As regards the interpretation of the fragment, it is heavily disputed whether it intends to refer to Anaximander's principle, the *apeiron*, or not. The Greek original has relative pronouns in the plural (here rendered by "whence" and "thence"), which makes it difficult to relate them to the *apeiron*. We can distinguish roughly two lines of interpretation, which may be labeled the "horizontal" and the "vertical" one. The horizontal interpretation holds that in the fragment nothing is said about the relation of the things to the *apeiron*, whereas the vertical interpretation maintains that the fragment describes the relationship of the things to the *apeiron*. The upholders of the horizontal interpretation usually do not deny that Anaximander taught that all things are generated from the *apeiron*, but they simply hold that this is not what is said in the fragment. They argue that the fragment describes the battle between the elements (or of things in general) that accounts for the origin and destruction of things. The most obvious difficulty, however, for this horizontal interpretation is that it implies two cycles of becoming and decay: one from and into the Boundless, and the other caused by the mutual give and take of the elements or things in general. In other words, in the horizontal interpretation, the *apeiron* is in fact superfluous. Aristotle already made this point when arguing that such a thing as an ἐνεϱγείᾳ ἄπειϱον σῶμα αἰσθητόν does not exist: "In order that coming to be should not fail, it is not necessary that there should be a sensible body that is actually unlimited. The passing away of one thing can be the coming to be of another, the all being limited" (*Physics* 208a8–11 = DK 12A14). As regards Anaximander's fragment, this is the strongest argument in favor of the vertical interpretation that holds that the fragment refers to the *apeiron*, notwithstanding the plural relative pronouns.

According to the "vertical" interpretation, then, the *apeiron* should be regarded not only as the ever-flowing fountain from which everything ultimately springs, but also as the yawning abyss (as some say, comparable with Hesiod's "Chaos") into which everything ultimately perishes. Recently, Graham has tried to defend the horizontal interpretation without making the *apeiron* superfluous. According to him "the boundless is the original matter out of which the world and its component stuffs come to be," whereas on the other hand "Anaximander did see cyclical changes of opposites as coming out of each other" (2006: 33 and 38). In Graham's interpretation, the *apeiron* has become a kind of *deus absconditus*: "In Anaximander the boundless is a divine, if inscrutable, being, the source (...) of the world," which enforces the law of reciprocity to which the elementary powers or bodies in our world are subject (2006: 44). After all, his attempt seems to end in an impasse: "What then is the ontological relationship between the boundless and the stuffs of our world? We simply do not know" (2006: 34).

The suggestion has been raised that Anaximander's formula in the first two lines of the fragment was the model for Aristotle's definition of the "principle" (ἀρχή) of all things in *Metaphysics* 983b8 (West 1971: 83 see also Ferber 1986: 554, and recently Graham 2006: 39). There is some sense in this suggestion. For what could be more natural for Aristotle than to borrow his definition of the notion of ἀρχή that he uses to indicate the principle of the first Presocratic philosophers, from Anaximander, the one who introduced the notion?

It is certainly important that we possess one text from Anaximander's book. On the other hand, we must acknowledge that we know hardly anything of its original context, as the rest of the book is lost. We do not know from which part of his book it is, nor whether it is a text the author himself thought crucial or just a line that caught one reader's attention as an example of Anaximander's poetic writing style. The danger is that we use this isolated text – beautiful and mysterious as it is – to produce all kinds of profound interpretations that are hard to ground. Perhaps a better way of understanding what Anaximander has to say is to study carefully the doxography that goes back to people like Aristotle and Theophrastus, who probably had Anaximander's book before their eyes and who tried to reformulate what they thought to be its central claims.

With these statements in mind, I suggest tentatively quite another interpretation, holding that Anaximander did not make τὸ ἄπειρον the origin but that he wanted to say something quite different.[5] Some scholars doubt whether or even deny that Anaximander could have used the noun τὸ ἄπειρον whereas they hold that he could have used it only as an adjective. For instance De Vogel (1957: 6), who quotes Aristotle: οἱ δὲ περὶ φύσεως ἅπαντες (ἀεὶ) ὑποτιθέασιν ἑτέραν τινὰ φύσιν τῷ ἀπείρῳ τῶν λεγομένων στοιχείων, which can be translated thus: "Those dealing with nature, on the other hand, all make the *apeiron* the attribute of some other nature, namely of one of the so-called elements" (Aristotle, *Physics* 203a17–18, not in DK).[6] De Vogel concludes from this: "Which means that according to Anaximander, being one of the περὶ φύσεως ἅπαντες, the ἄπειρον is not a subject, but a predicate" (1957: 7). This remark, however, does not bring us much farther, as the big question in the case of Anaximander remains what kind of so-called element this "some other nature" could be. Or, to be more precise, in the case of Anaximander, ἄπειρον cannot be taken as the predicate of some "element" because these are excluded by the arguments mentioned above. So it has to be predicated of something else.

The most fervent opponent of the idea that Anaximander could have used τὸ ἄπειρον as a noun is Havelock. In his analysis of the argument quoted above (*Physics* 203b6–10), he makes it into a kind of rendition of Aristotle's own cosmology with the prime mover, disguised in Presocratic terminology (1983: 77). However, for Aristotle, Havelock says, "the non-finite is to be incorporated into his own cosmology, although in the sense of 'matter'." And then, he concludes

[5] This interpretation was suggested in discussions with Radim Kočandrle. This does not imply, however, that I pretend to render his opinion here.

[6] Conche renders it freely as follows: "Tous les philosophes de la nature regarderaient l'infini comme l'attribut d'une substance différente de lui, et appartenant à la classe de ce qu'on appelle les στοιχεῖα, les 'éléments'" (1991: 91).

that Aristotle "has no choice in this part of his *Physics* but to identify 'the non-finite' (...) as itself a prime principle, a 'substance', loosely analogous (in this context only) to his own prime mover" (1983: 78). This sounds rather confused and not very convincing. A more natural reading of the quoted argument is the one given above, that according to Anaximander (who is mentioned by Aristotle himself at the very end of the argument) the origin (whatever this origin might be) must be *apeiron*.

I would draw the attention to the fact that both Aristotle and others, when speaking of Anaximander, use ἄπειρον as a predicate of φύσις ("nature"), as in the passage quoted by De Vogel. Another example is Simplicius (quoting Theophrastus): λέγει δ' αὐτὴν μήτε ὕδωρ μήτε ἄλλο τι τῶν καλουμένων εἶναι στοιχείων, ἀλλ' ἑτέραν τινὰ φύσιν ἄπειρον: "He (sc. Anaximander) says that it (sc. the origin) is neither water nor any other of the so-called elements but some other boundless nature" (*In Aristotelis Physicae commentaria* 24.13 = DK 12A9). And elsewhere: 'Αναξίμανδρος (...) ἄπειρον τινα φύσιν ἄλλην οὖσαν τῶν τεττάρων στοιχείων ἀρχὴν ἔθετο: "Anaximander (...) posed as the origin some infinite nature other than the four elements" (Simplicius, *In Aristotelis Physicae commentaria* 41.17, not in DK). In the same sense Hippolytus says: οὗτος ἀρχὴν ἔφη τῶν ὄντων φύσιν τινὰ τοῦ ἀπείρου: "He (sc. Anaximander) says that the origin of the things is some nature of the infinite" (DK 12A11). And Cicero: *is enim infinitatem naturae dixit esse, e qua omnia gignerentur*: "He (sc. Anaximander) says that it is the boundlessness of nature, from which all things have generated" (DK 12A13). Finally, in the *Turba Philosophorum* we read: *ait omnium initium esse naturam quandam*: "He (sc. Anaximander) says that the beginning of all things is a certain nature" (*Turba Philosophorum*, ed. Ruska: 109, not in DK).

These texts all seem to go back to the same text of Theophrastus that, as Kahn states, is most closely reflected in Simplicius' rendition (1994: 53). And, after all, Anaximander's book is said to have been entitled *On Nature* (περὶ φύσεως) (DK 12A2 and DK 12A7). Havelock maintains that the words "some other boundless nature" in Simplicius' text "represent the kind of language Anaximander must have used" and he suggests that Anaximander may have said something like "from the beginning, the nature of the all was, is, and ever shall be non-finite" (1983: 55 and 59. See also his semiserious reconstruction of Anaximander's imaginary hexameters on p. 81).

It is not quite clear to me what Havelock's last sentence could mean. But let us suppose that he was on the right track and that in these texts φύσις does not mean a colorless "something," as it is treated by many authors, but has a pregnant meaning. Then one could take these texts to mean that Anaximander never spoke of "the Boundless" as the origin of all things but that he postulated something like an eternal and all-embracing generative power that constitutes the cosmos and all that is in it. In that case, the words φύσιν ἄπειρον that were translated above tentatively with "boundless nature" may mean something like "the universal and inexhaustible power of becoming and growth that from the beginning eternally entails the very existence of all things." This "nature" should guarantee that becoming and growth will never stop, as is said in the second argument quoted above. It should be some countervailing power against the omnipresent powers of decay and perishing.

Anaximander's fragment could be understood, then, as the expression of these opposing powers as a kind of cosmic "justice" (δίκη), by means of which the things set boundaries to each other and even destroy each other.

The texts on cosmogony can also be understood in this sense: φύσις is the generating power of not just the living nature but the whole cosmos as well. Its eternal movement is said to have caused the origin of the heavens. Thus, Simplicius may say: ἄπειρον τινα φύσιν (...) ἧς τὴν ἀίδιον κίνησιν αἰτίαν εἶναι τῆς τῶν οὐρανῶν γενέσεως ἔλεγεν: "some unbounded nature (...) the eternal movement of which, as he says, is the cause of the origin of the heavens" (Simplicius, *In Aristotelis Physicae commentaria* 41.17–19, not in DK).

Hippolytus, obviously drawing from the same source, says that according to Anaximander "all the heavens and the worlds within them" have sprung from "some boundless nature" (DK 12A11). A part of this process is described in rather poetic language, full of images, which seems to be idiosyncratic for Anaximander: "a germ, pregnant with hot and cold, was separated [or: separated itself] off from the eternal, whereupon out of this germ a sphere of fire grew around the vapor that surrounds the earth, like a bark (φλοιός) round a tree" (DK 12A10). Subsequently, the sphere of fire is said to have fallen apart into several rings, and this event was the origin of sun, moon, and stars. Some authors have, quite anachronistically, seen here a kind of foreshadowing of the Kant–Laplace theory of the origin of the solar system. I would suggest that what we have here is a description of the working of the boundless power of φύσις in the development of the world.

Several sources even mention innumerable worlds (ἄπειροι κόσμοι), which looks like a plausible consequence of the "boundless nature" (DK 12A10, DK 12A14 and especially DK 12A17, and Simplicius, *In Aristotelis De caelo commentaria*, 202.14–16, not in DK). But this is presumably a later theory that was incorrectly read back into Anaximander. I return to this subject in the last chapter.

Finally, it is worth mentioning that the doxography tells us that according to Anaximander, life originated from the moisture that covered the earth before it was dried up by the sun. The first animals were a kind of fish with a thorny skin that they threw off when they entered the land (DK 12A10, DK 12A11, and DK 12A30). The Greek word used here for "skin" (φλοιός) is the same that was used for the metaphor "the bark of a tree" in Anaximander's cosmology. Very recently, Gregory has purported the interesting suggestion that Anaximander could have been inspired by the metamorphosis of the Caddis fly (Gregory 2011). As regards the origin of man, it is told that according to Anaximander men originally generated from fishes and were fed in the manner of a viviparous shark (DK 12A30). The reason for this is said to be that the human child needs long protection to survive. Some authors have, rather anachronistically, seen in these statements a protoevolutionist theory, but it is more probable that the idea of metamorphosis played a certain role in Anaximander's ideas concerning the development of animals and man.

Surveying all this, we must conclude that the tradition on Anaximander is to such an extent deficient that it is impossible to be reconstructed as a coherent whole. Our image of his work will always remain disrupted and incomplete just like the mutilated and decapitated statue that was found at the marketplace of Miletus and which bears his name (Fig. 7.2).[7]

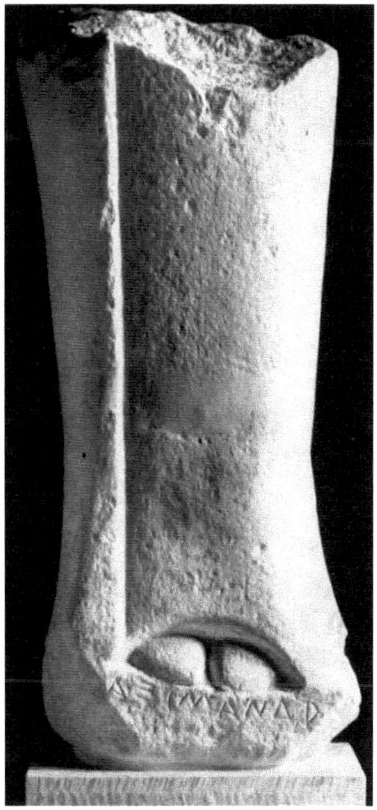

Fig. 7.2 A statue of Anaximander, found in the marketplace at Miletus (photograph from Darsow 1954: 103, Abb. 2)

We may, nevertheless, conclude on the basis of what we know that Anaximander has been one of the greatest thinkers that ever lived. By arguing on the "boundless nature," he was the first metaphysician, by making a map of the world, he was the first geographer, but above all, by his bold speculations about the universe, he broke with the ancient image of the celestial vault and thus originated the Western world picture. This will be the subject of the next three chapters.

[7] For an extensive discussion on this statue, and especially on the question whether it is a representation of Anaximander (on which question the author's answer is negative), see Darsow (1954: 101–117).

Chapter 8
The Discovery of Space: Anaximander's Cosmology

At first sight, Anaximander's cosmology looks like an eccentric vision sprung from a bizarre mind. Anaximander imagined the celestial bodies as huge rings, or more precisely, chariot wheels, consisting of opaque air-like (ἀεροειδής) stuff. Inside such a wheel (within its felloe), and invisible to us, fire is burning. The wheels have holes, through which we see the fire inside, and this is what we call the sun, the moon, or a star (DK 12A11, DK 12A18, DK 12A21, DK12A22, and (*Turba Philosophorum*, ed. Ruska: 109, not in DK)). Illustrative for the astonishment evoked by these images is, for instance, the desperate commentary of a French scholar: "Les idées d'Anaximandre sont tellement bizarres qu'on hésite à les reproduire" (Boquet 1925: 35). In a handbook on the history of astronomy, a Dutch author writes: "What he said about sun, moon, and stars (...) is rather obscure" (Pannekoek 1961: 98–99). Even an authoritative scholar like Charles Kahn doubts whether here authentic Anaximandrian images are at stake, and he suggests that they look like the style of a Hellenistic popularizer (1994: 87). And Dicks, the author of a standard work on early Greek astronomy, speaks about Anaximander's primitive astronomical ideas and peculiar notions (1959: 309, n. 1, see also 1970: 45–46).

In the next three chapters, I will defend the opposite position that Anaximander's cosmological ideas make full sense and that, properly considered, they generated a completely new world picture. Anaximander discovered space. These three words can summarize the new world picture introduced by him. I illustrate this thesis by means of an exposition of three of his fundamental insights that are well documented in the doxography:

1. The celestial bodies make full circles and pass underneath the earth as well.
2. The earth floats free and unsupported in the center of the universe.
3. The celestial bodies are not all at the same distance from the earth.

We tend to look upon these three as trivialities. I intend to show, however, that they are instances of creative imagination with which Anaximander blew up the archaic world picture. Or, as Cornford says: "He had very daringly broken with the old notion of a single *Ouranos*, the 'starry Heaven'" (1934: 10). Together, these three insights mark the origin of the Western world picture. As our world picture is the further development of the Greek conception of the universe, this statement can

D.L. Couprie, *Heaven and Earth in Ancient Greek Cosmology*, Astrophysics and Space Science Library 1, DOI 10.1007/978-1-4419-8116-5_8,

be considered as an extension of the credits given to Anaximander by Burkert: "The fundamental points of that specifically Greek conception of the world's structure are already to be found in Anaximander: the earth remains in place (. . .) in the center of the circling rings of fire which are the paths of the stars" (1972: 309).

The Celestial Bodies Make Full Circles Around the Earth

Anaximander imagined the celestial bodies as not stopping at the horizon but making full circles around the earth. This is stated in the doxography, where we can read that he conceived of the celestial bodies as rings or wheels, or to be more precisely, as the felloes of wheels (DK 12A11(4), DK 12A18, DK 12A21, and DK 12A22). What makes Anaximander's conception of the celestial bodies as wheels so important is that it implies that the celestial bodies make full circles around the earth and thus go underneath the earth as well.

The idea that the celestial bodies in their daily course do not stop at the horizon is so self-evident to us that it is hard to see how daring it originally was. Standing on the northern hemisphere, we can see some stars making full circles around the Polar star, and we can recognize as well that some other stars, being somewhat farther from the celestial pole, disappear regularly behind the horizon. We can argue, then, that those stars that we see not to describe full circles, in reality complete their circles below the horizon. As regards the sun and moon, we can see that the curves they describe daily in the sky are sometimes bigger and sometimes smaller, and we can predict where they will rise and set the next day. Therefore, it seems to be a simple conclusion to suppose that these two celestial bodies too will describe full circles around the earth. It is, however, of essential importance to notice that the idea that the celestial bodies make full circles and being sometimes under the earth is not something Anaximander could have *observed* but that it was a *conclusion* he must have drawn. In other words, Anaximander must have *imagined* the celestial bodies going underneath the earth. When Dicks writes: "The stars *are* seen to move in circular orbits across the sky, the sun *does* appear to go round the earth in a circle" (1970: 176), he uses a typical anachronism because we cannot see that, but only conclude it from what we see. For Anaximander, it was a very bold conclusion, precisely because it necessarily involved the idea of an earth floating free in the universe, as we see in the next section. How daring his conception was may become clear when we see that his successor Anaximenes taught that the celestial bodies do not go underneath the earth, but turn around it "like a felt hat turns around our head" (DK 13A7). Somewhat more on this strange image in Chap. 13.

The word "wheel" (τροχός) is marked by Diels/Kranz as authentic (see remark at DK 12B5: "auch τροχός (. . .) wohl echt"). With Burkert, I maintain that also the expression "wheel of a chariot" (ἁρμάτειος τροχός) must be considered as going back to Anaximander himself, and I would add the word "felloe" (ἁψίς) as well (1999: 180). The words "(chariot) wheels" are used in DK 12A21 and DK 12A22, and the word "felloe" in DK 12A22. I think the best way to understand

Anaximander's intentions is to take seriously the image of chariot wheels. Some doxographers, however, do not use the word "wheel," but "ring" (κύκλος) when they describe Anaximander's conception of the celestial bodies (DK12A11 and DK 12A18). This is clearly an anachronistic expression, used by the doxographers who tried to explain Anaximander's celestial wheels by means of the rings they were acquainted with in contemporary models, the so-called armillary spheres or armillaria. On the other hand, the word "ring" makes clear that in the case of Anaximander's celestial wheels, we do not have to envisage the complete wheel, but only its circumference or felloe.

The doxography also shows that later authors had some trouble in appreciating Anaximander's image of the celestial wheels. Achilles Tatius obviously does no longer understand it and thinks that the sun is the hub (πλήμνη) of a wheel, the spokes of which are the sun rays (DK 12A21). Sometimes modern authors are confused as well, as in the case of Dumont, who translates both πλήμνη and ἀψίς with "hub" (French: "moyeu") (1988: 34 and 35). Conche correctly distinguishes between hub ("moyeu") and felloe ("jante") (1991: 197). Modern scholars tend to think that Anaximander's celestial wheels look like bicycle tubes or, as Brumbaugh writes: "Rings of hollow pipe (a modern stove-pipe gives the right idea)" (1981: 21). Naddaf draws a cross-section of a celestial ring as a kind of hollow stovepipe, as can be seen in Fig. 8.1 (see Naddaf 1998: 10; see also Naddaf 2001: 16, Fig. 1.1, detail). Mugler and Krafft have tried to draw Anaximander's universe with such tube-like celestial wheels, as we see in Chap. 10. These authors have fallen prey to an apparent anachronism, as Naddaf seems to acknowledge, for he writes: "I don't mean to imply by this that 'chariot wheels' at the time were oval shaped" (1998: 15 n. 61).

Let us see what wheels looked like in Anaximander's time. There was nothing at

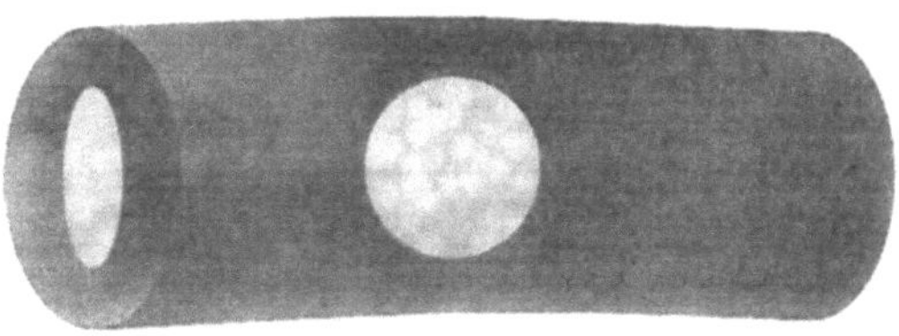

Fig. 8.1 Detail of a tube-shaped sun wheel with its aperture according to Naddaf (1998: 10, detail; by the courtesy of Gerard Naddaf)

all in Anaximander's environment that could have suggested him the idea of tube-like felloes, whereas chariot wheels will have been sufficiently present in the streets of Miletus. Chariot wheels looked very much like the wheels that are still nowadays used for coaches. In Fig. 8.2, we see wheels from the time of Anaximander, the cross-section of which (that is: of their felloes) is not a circle or an oval, as is the case with bicycle tubes or stovepipes, but a rectangle. Rescher is the only one who draws rectangular cross-sections of the celestial wheels. His rendition, though, is confusing, as he has a deviant opinion on the celestial wheels that is not supported by the doxographical texts: "stars, moon, and sun (...) are wheel-shaped masses of fire, separated from one another by wheel-shaped masses of air" (1958: 727, Figs. 9 and

10 = 1982: 24, Figs. 10 and 11).[1] The doxography has nothing to say about the shape of the air between the celestial wheels, but they are certainly not wheel-shaped.

Anaximander's wheels are as it were the materialization of the orbits of the

Fig. 8.2 Clay model of a four-wheeled vehicle, Athens 720 B.C. (drawing by Hans Exterkate)

celestial bodies around the earth. According to Anaximander, the celestial bodies not only *describe* full circles around the earth but they *are also* these full circles. We tend to look upon this as a curious and quixotic representation, used as we are to conceive of the celestial bodies as spheres, moving through space. Many authors have wondered what could have made Anaximander consider the celestial bodies as gigantic wheels turning around the earth. When we try to look at the heavens with Anaximander's eyes, however, the image is not as strange as it seems. Once he realized that the celestial bodies do not stop at the horizon but make full circles around the earth, Anaximander had to answer the question what the nature of these bodies could be that makes them move in circles. Most moving things he could observe moved in straight lines, for instance falling objects. Some made strange curves, such as thrown objects, but only circular objects like wheels were able to move naturally in a circle. It was a bold but reasonable association that made him conclude that the celestial bodies, too, must be wheels. One might perhaps say that without the invention of the wheel cosmology would never have developed. This is Bronowski's intriguing thesis, without, however, mentioning Anaximander. Bronowski argues that other cultures lacked a model of the heavens because they lacked the wheel. The Greeks, he says, built their model on the wheel: wheels within wheels, forever turning (1973: 74, 77, 194). Anaximander clearly was the first to make use of this image to elucidate his conception of the cosmos, in which the celestial bodies make full circles around the earth. This makes him the founding father of cosmology.

[1] For a critical exposition of Rescher's pictures, see Couprie (1995: 162–164).

Moreover, the function of Anaximander's celestial wheels will probably have been to prevent the celestial bodies from falling on the earth, as the danger would be when they were stones, as about a century later Anaxagoras thought them to be. The stone that fell from heaven at Aegospotamoi proved how dangerous that conception was. I return to this issue in Chap. 14. The doxography is silent on this point, but there can hardly be any doubt that the question why the celestial bodies do not fall upon the earth must have been as serious a problem for Anaximander as the question why the earth does not fall. This must have been one of the reasons why he imagined the celestial bodies as wheels: every point of such a wheel is at the same distance from the earth as any other (see also Bodnár 1988: 51; 1992: 339).[2] Aristotle's conception of the celestial bodies as connected to spheres moving around the earth has the same function of preventing them from falling. Aristotle explicitly says: "we are left with the conclusion that the circles move and the stars stay still and are carried along because they are *fixed* in the circles" (*On the Heavens* 289b31 ff., my italics). Aristotle has another conception of the celestial bodies than Anaximander, but apart from this these words could have been Anaximander's. The conception of celestial spheres carrying the celestial bodies was the main paradigm during the Middle Ages. According to Rossmann, a curious parallel with Anaximander's celestial wheels can still be found in Copernicus, whose concept of "orbis" is best understood not as "sphere," but as an "in sich rotierender Kreisring (...) an dem Erde, Mond, und Planeten (...) festhaften. Bei Kopernikus laufen nicht die Planeten auf ihren Bahnen um, wie die Vorstellung seit Kepler ist, sondern die Bahnkreise reißen die an ihnen hängenden Himmelskörper bei ihrer Umwälzung – revolutio – mit herum" (1986: 38). It is as if we read Aëtius' – admittedly anachronistic – words: "Anaximander says that (the celestial bodies) are set in motion by the circles and spheres on which each of them is attached" (DK 12A18).

Anaximander conceived of the celestial wheels as hollow: they consist of thick, condensed air, filled with fire (DK 12A 11(4), DK 12A18, DK 12A21, and DK 12A22). In a nicely illustrated number of pages of his latest book, Hahn shows that wheels with hollow felloes really existed in Anaximander's time and environment (Hahn 2010: 140–143).[3] When we look upon the heavens, however, we do not see the celestial wheels because they consist of (condensed) air. The fire inside is also invisible as it is within the cover of condensed air. There is, however, a hole in a celestial wheel, through which the fire shines, and this is what we call the sun, the moon, or a star (DK 12A11(4), DK 12 A11(5), DK 12A18, DK 12A21, and DK 12A22). According to Anaximander, the moon also has a light of itself, albeit much less intense than that of the sun (DK 12A22). Diogenes Laertius' report that according to Anaximander, the moon receives its light from the sun is certainly incorrect (DK 12A1(1)). The characteristic words used in the doxography to

[2] As is explained in Chap. 10, there is a second movement of the celestial wheels: up and down the celestial axis. But also during this movement the distance from any point of the wheel to the earth is equal to that of any other point of the same wheel at any given time.

[3] This does not mean, however, that I agree with the cosmological consequences about the earth axis, which Hahn draws from this (see Couprie 2010).

indicate the apertures in the celestial wheels are στόμιον ("mouth-like opening"), πόρος ("opening through which something can pass," "way out"), and ἐκπνοή ("outbreathing"). The image is clearly that of a mouth breathing out the fire that is inside the celestial wheel. In Chap. 11, we shall see how this image has been disturbed by quite another image – the nozzle of a bellows – which is the result of a wrong translation.

In the doxography, we read that sometimes the hole of the sun or moon wheel closes. Then, we speak of a solar or lunar eclipse. The regular phenomena of waning and waxing of the moon are explained as well by the gradual closing and opening of the hole in the moon wheel (DK 12A11, DK 12A21, and DK 12A22). How one has to understand this when an annular solar eclipse occurs, in which a ring of the sun remains visible, is not quite clear. Perhaps Anaximander was not acquainted with the possibility of annular eclipses. However this may be, we may conclude that Anaximander had no idea of the real cause of solar and lunar eclipses. There is, however, a text that suggests that he had a suspicion that the closing of the hole in the moon wheel was not all there was to say about lunar eclipses. Aëtius somewhere mentions that they have to do with the rising and descending (Greek: τροπαί) of the moon wheel (DK 12A22). This text is too incomplete to draw definite conclusions, but it seems that Anaximander realized that lunar eclipses can only occur when the openings in the wheels of the moon and the sun, as seen from the earth, stand exactly opposite one another. This would already have been one element of a correct explanation of a lunar eclipse.

The incompleteness of the tradition leaves us with all kinds of unanswered questions about Anaximander's conception of the celestial bodies as wheels of air and fire. For instance, how do we have to understand that the inner wheels (of the stars, which are nearer to the earth than sun and moon, see c below) hide the fire inside but let the fire of the sun and moon pass by? The same holds for the light of the sun with respect to the moon wheel. It has been suggested that the brighter light of the outer celestial bodies can penetrate easily the air of the inner wheels that can be thinner as their fire is less intense, but this sounds hardly satisfactory (see Bodnár 1988: 50. See also Von Fritz in a letter to Kahn 1994: 90 n. 3). Another question is how it can be explained in Anaximander's conception that the fully eclipsed moon stays vaguely visible (sometimes even with a beautiful orange-red glow).[4] Maybe we have to imagine that the condensed air that closes the opening of the moon wheel for some reason remains somewhat transparent.

The Earth Floats Free in Space

Anaximander's conception of the earth is as strange to us as that of the celestial bodies. According to the doxography he taught that the earth has the shape of a cylinder, much like a column drum, the height of which is one third its diameter.

[4] Dmitri Panchenko drew my attention to this problem.

The two drums marked A on Fig. 8.3 have approximately the dimensions 3:1 of Anaximander's drum-shaped earth. Some authors have wondered why Anaximander chose this strange shape for the earth. However, when we realize that Anaximander, just like his contemporaries, thought that the earth is flat and circular, as suggested by the horizon, the cylindrical shape is obvious. On the flat upper side of that cylinder, we live. This is the commonly accepted result of the interpretation of three rather mutilated texts. In two texts it is said that the earth is like a column of stone, and in the third it is said that the earth is cylindrical-shaped, its height being one-third of its diameter (DK 12A11, DK12A25, and DK 12A10). The ingenious suggestion by Dmitri Panchenko that a cylindrical earth, curved along the north–south axis, would explain at least some celestial phenomena better than when we are thought to live on a flat surface (2008: 193), is hardly believable and seems to conflict with DK 12A11(3), where two surfaces (ἐπίπεδοι) of the earth are mentioned, of which we live on the upper one. Diogenes Laertius' report that the earth according to Anaximander is spherical is usually taken to be false (DK 12A1(1)). In fact it is a plain anachronism. Fehling offers the interesting suggestion that Diogenes Laertius borrowed his report from his understanding of the symmetry argument which Aristotle ascribes to Anaximander and which presupposes a spherical earth (1994: 146).

Fig. 8.3 Columns and column drums at Priëne (photo used with permission of Matthew Recla)

In the doxography, the shape of the earth is called "curved" (γυϱόν), which is Röper's inevitable correction of the ὑγρόν (moist) of the manuscripts (DK 12A11; Röper 1852: 607. See also Diels 1879: 218–219 and Conche 1991: 215). Neuhaeuser's suggestion to read ὕπτιν ("flat") instead of ὑγϱόν has not been followed by other authors (1883: 349 n. 1). It is tempting to translate γυϱόν here with "(somewhat) concave" because we probably have to conceive of the surface of the earth as somewhat concave, as is the case with column drums, which is the result of a technique called ἀναθύρωσις, as Hahn has convincingly shown (2001: 169 ff. and 195–196).

Apart from the evidence of the making of a column drum, the same suggestion has been made already by Classen and Kahn, in conformity with the Greek conception of the οἰκουμένη as the lands around the basin of the Mediterranean Sea (Classen 1986: 64; Kahn 1994: 56 and 81 n. 3). This shape of a somewhat concave earth we still find in Archelaos and Democritus, as we had seen in Chap. 4. Why Guthrie says that it is "quite likely, that Anaximander thought of the disc of the earth as having a hole in the center," I do not understand (1962: 99 n. 3). Such a hole is hard to bring into agreement with Delphi as the center of the earth. Perhaps he means the realm of Hades situated inside the earth.

Anaximander's conception of the shape of the earth is clearly a relict of the archaic world picture (Fig. 8.4). But more important than the shape of the earth was his conception that the earth floats free and unsupported in the center of the cosmos (DK 12A1(1), DK 12A2(8), DK 12A26(1), and DK 12A26(6)).[5] This meant a complete revolution in the conception of the universe. The decisive step from the archaic to the new world picture did not consist in the transition from a flat to a spherical earth, as is commonly thought. As we will see at length in the Chaps. 17 and 18, the conception of the sphericity of the earth is a later development that requires a theory that explains why we do not fall off the spherical earth.

The doxography does not mention it, but we may surmise that Anaximander's bold idea of the earth floating free in space was the conclusion drawn from his conception of the celestial bodies as wheels or full circles that go underneath the earth as well. In other words, when Anaximander concluded from the daily rising and setting of the celestial bodies that they make full circles, completing their orbits under the earth, he must also have concluded that the earth is unsupported in the center of the cosmos. It is strange that hardly anybody in the literature on Anaximander has mentioned this simple link between the conceptions of the celestial wheels and the free-floating earth. Two exceptions are worth to be quoted. The first is Sticker: "Was veranlaßte wohl diese erstaunliche Hypothese (viz. of the free-floating earth)? Vielleicht die vernünftige Überlegung, daß nämlich von allen anderen denkbaren Möglichkeiten diese allein es zuläßt, sich vorzustellen, daß die Gestirne im Laufe von Tag und Nacht eine volle Kreisbahn – nicht nur immer wieder einen neuen Tagesbogen von Aufgang bis Untergang – um die in ihrer Mitte ruhend gedachte Erde vollführen können" (1967: 20). The other is Conche, quoting Flammarion: "Pour cela, afin de faire de l'hémisphère céleste une sphère complète, Anaximandre a dû prolonger le cours diurne des astres. Corrélativement, il a *isolé* la Terre, – peut-être la 'plus grande découverte de l'astronomie' –, l'a suspendue dans le 'vide', soutenue par rien, ne reposant sur rien" (Conche 1991: 37–38). Conche, however, speaks here less precisely of "sphère," although nowhere in the doxography is indicated that Anaximander taught a spherical universe.

How revolutionary Anaximander's conception of the free-floating earth was can be illustrated by the fact that his successor Anaximenes found it apparently too daring, so that he let the earth rest on the air like a lid (DK 13A6, DK 13A7, and

[5] In DK 12A26(6) I read κεῖται instead of κινεῖται. See also Conche (1991: 203 n. 23).

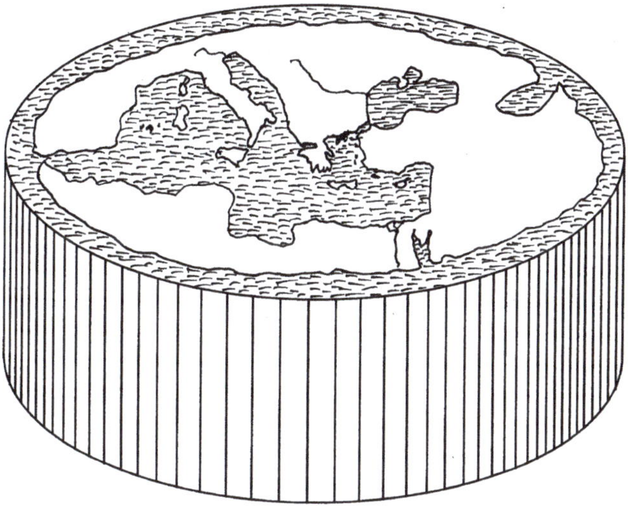

Fig. 8.4 This is how Anaximander must have imagined the earth

DK 13A20). In a halfhearted testimony, Simplicius ascribes to Anaximander the explanation of the earth's stability both by its symmetrical position and by the support of the air (Simplicius, *In Aristotelis De caelo commentaria* 532.14, not in DK). This must be wrong, as another source speaks of "a sphere of fire that grew around *the air that surrounds the earth*, like the bark around a tree" (DK 12A10, see also Kahn 1994: 55). It is again important to stress that Anaximander could not *see* that the earth floats free in space. We had to wait about 2,500 years to have astronauts really *see* the earth floating in space and therewith delivering the ultimate proof that Anaximander's conception was right (Fig. 8.5).

Fig. 8.5 The ultimate proof of Anaximander's conception of a free-floating earth (photograph Japan Aerospace Exploration Agency – JAXA)

We may suppose that Anaximander had to defend his bold theory against the question of his fellow citizens why we do not have to be afraid that the free-floating earth will fall down. His alleged answer to this question is handed down in two versions. The oldest version is that of Aristotle: "But there are some who say that it stays where it is because of symmetry (ὁμοιότης), such as among the ancients Anaximander. For that which is situated in the center and at equal distances (ὁμοίως) from the extremes has no inclination whatsoever to move up rather than down or sideways. And since it is impossible to move in opposite directions at the same time it necessarily stays where it is" (*On the Heavens* 295b10 ff. = DK 12A26). The second version, handed down by the church father Hippolytus, who obviously leans on Aristotle's text, sounds: "The earth is suspended, not supported by anything (literally: not dominated by anything), staying where it is by its equal distance from everything (διὰ τὴν ὁμοίαν πάντων ἀπόστασιν)" (DK 12A11(3)). The same argument is also ascribed to Parmenides and Democritus (DK 28A44). This reference, however, is doubtful, at least as regards Democritus, for he taught not only that the earth is flat (and somewhat concave) but also that it is supported by the air beneath it (Aristotle, *On the Heavens* 294b13 = DK 13A20, and Simplicius, *In Aristotelis De caelo commentaria* 520.28 = DK 59A88. See also Kahn 1994: 79 n. 4).

The argument ascribed to Anaximander by Aristotle and Hippolytus has been discussed by several authors. Usually, they stress the so-called mathematical character of the argument that, as Kahn says, "must, in substance, presuppose the standard definition of the circle as 'that which is in every way equidistant from the middle to the extremes'" (1994: 77). This kind of formulations is severely criticized by Dicks (1966: 35 and 37). Nevertheless, 16 years later Barnes uses an exaggerated phrase like this: "(the earth) is mathematically suspended by abstract reason" (1982: 27). Anaximander's argument has often been called the first instance of an argument based on the principle of sufficient reason (*ex principio sufficientis rationis*). This principle has been formulated by later Presocratics, and especially by Leucippus (DK 67B2). This text shows, by the way, that Heidegger was wrong when he said that it took 2,000 years since the dawn of Western philosophy, before the "Satz vom Grunde" was formulated (1957: 14–15). It is remarkable, however, that Aristotle, who handed down the argument, does not seem to be very deeply impressed by its strength. Actually he ridicules it, saying that by the same argument: "a hair that is subject to an even pulling power from all sides would not break, and that a man, being just as hungry as thirsty, placed in between food and drink, must necessarily remain where he is and starve to death" (Aristotle, *On the Heavens* 295b31ff.). Heath (1913: 25) calls this argument "amusing," but Panchenko does not recognize Aristotle's humor when he writes: "Aristotle makes it clear once again when illustrating the theory" (1994b: 35). In a similar version the argument is known since the Middle Ages as "Buridan's ass."

I tend to agree with Aristotle that, at least used as a cosmological argument, doubts about its soundness are justified. One main weakness of the argument is that it does not explain why it is precisely *the earth* that dwells in the center of the cosmos and not any other body, for instance fire, as Aristotle remarks. In other words, even granted that the argument suffices to explain why the earth remains in the center, it does not explain how the earth got there (Aristotle, *On the Heavens* 295b18–26). As we will see in Chap. 18, one of the attractive features of Aristotle's theory of falling is that it is able to explain this as well. Another problem of the argument is that it argues from the *nonexistence* of a sufficient reason. Absolute propositions about nonexistent things or facts (so-called negative existential propositions) are always in danger of becoming falsified upon closer investigation. Anaximander's case is a good example of this danger, as further investigation has taught us that the earth is *not* in the center of the universe, and thus the argument was falsified. For this reason, I disagree with those authors who call Anaximander's argument "clear and ingenious," or "überraschend und großartig," or "straordinario, e perfettamente corretto" (Barnes 1982: 25; Von Fritz 1971: 24; Rovelli 2009c: 53). How can the argument be perfectly correct when actually the earth does not dwell in the center of the universe?

As regards the authenticity of the argument, Aristotle's testimony seems to leave little room for doubt, as he mentions Anaximander explicitly. Moreover, elsewhere he mentions those who think that the earth is supported by air without counting Anaximander among them (Aristotle, *On the Heavens* 294b13). Notwithstanding this, there are reasons to doubt whether Anaximander was really the author of the argument as it sounds in Aristotle's text. As regards Aristotle's text, one may especially wonder whether the words "to move up rather than down or sideways" reflect Aristotle's own problems rather than Anaximander's. Anaximander had to answer the question whether the earth might fall, and not the sophisticated questions whether it might move into other directions, upward or sideways. Moreover, we may wonder whether Anaximander worried about the logical problem of an object moving into opposite directions at the same time, as is supposed in Aristotle's rejoinder. Some authors also doubt about the authenticity of the argument, for instance Robinson, who holds "that the view imputed to Anaximander by Aristotle not only *was* not but *could* not have been held by him" (1971: 116). In the same sense Fehling: "hier legt zunächst (...) die Vermutung nahe, daß er (sc. Aristoteles) auch das Symmetrie-Argument bei Anaximander nur 'erschlossen' hat" (1985b: 221).[6]

Simplicius already made the suggestion that Aristotle used Anaximander to oppose to an argument of Plato, obviously out of respect for his teacher (Simplicius, *In Aristotelis De caelo commentaria* 532).[7] In the *Phaedo*, Socrates on his deathbed uses an argument that sounds like that ascribed to Anaximander: "I am convinced that, if the earth is spherical and dwells in the center of the heavens, it needs neither the air nor any such force to keep it from falling, but the all-sided symmetry of the heavens and the equilibrium of the earth itself are sufficient to hold it in its place.

[6] For an extensive discussion with Fehling, see Couprie (2004a).

[7] Fehling has the same suggestion without, however, mentioning Simplicius (1985b: 222). And elsewhere: "Klar ist nur, daß Aristoteles ein Motiv hatte, auf einen alten und fast unbekannten Autor zurückzugreifen: er wollte die Polemik gegen Plato kaschieren" (1994: 144–145).

For whatever is in equilibrium and is set at the center of a homogeneous medium has no reason to incline in one direction rather than another, but being neutral it will remain immobile" (Plato, *Phaedo* 108e3–109a8). It is comprehensible that Aristotle did not want his teacher Plato to be overtly associated with an argument that he was about to deride and that he chose Anaximander, the first one who maintained that the earth floats in the center of the universe, as his target.

A comparison of Aristotle's version with that of Plato may clarify a misleading translation of the word ὁμοιότης. In the English translations of Aristotle's text by Guthrie, Burnet, and Stocks it is rendered with "indifference." This translation, however, is incorrect. The meaning of the phrase in question in *On the Heavens* that reads διὰ τὴν ὁμοιότητά becomes clear in the parallel text in the *Phaedo*, for there it reads more fully: τὴν ὁμοιότητά τοῦ οὐρανοῦ αὐτοῦ ἑαυτῷ πάντῃ, "the all-sided symmetry of the heavens."[8] Panchenko has argued that in the *Phaedo* passage ὁμοιότης "has more general and less definite meaning," whereas in Aristotle's account it should be translated as "equidistance" (1994b: 34). If we regard this part of Aristotle's text as a shortened version of Plato's, I do not see why we should not translate ὁμοιότης in both texts as "symmetry." The word ὁμοιότης, then, used in relation to the whole cosmos, hints at the conception of a spherical universe, whereas the word ὁμοίως, used of the earth, hints at its equidistance to the periphery. So I read the critical passage of Aristotle's text as follows: "(. . .) it (sc. the earth) stays where it is because of (the all-sided) symmetry (of the heavens) (ὁμοιότης) (. . .). For that (sc. the earth) which is situated in the center and at equal distances (ὁμοίως) from the extremes (. . .)." The idea is that the earth, being in the center of a symmetrical (read: spherical) universe is also symmetrical (read: equidistant) to the periphery of the cosmos. The same connotation is also present in Hippolytus' version.

The real problem is that a spherical universe is not Anaximander's but one that Aristotle is used to read into his predecessors, as Fehling says: "und so interpretierte er (sc. Aristotle) instinktiv die Himmelskugel überall hinein" (1985b: 224). This is the main reason why I hesitate to ascribe Aristotle's argument to Anaximander: it presupposes a spherical earth within a spherical cosmos.[9] Panchenko states without explanation: "I believe that Anaximander's cosmos was spherical too," and consequently: "the Aristotelian exposition of Anaximander's theory of the earth's stability appears yet as very reliable and even reflecting Anaximander's original wording" (1994b: 51 and 29). It is hard to see, though, how Anaximander's cosmological picture of celestial wheels, with the stars nearest to the earth, can be reconciled with a spherical cosmos. Moreover, Anaximander's conception of the earth was a cylinder, not a sphere.

---

[8] Fowler has: "the homogeneous nature of the heavens on all sides," and Tredennick: "the uniformity of the heavens."

[9] See Fehling: "Das Symmetrie-Argument setzt logisch und psychologisch die Erdkugel voraus" (1994: 143). Why "psychologisch" I do not fully understand. Instead of "logisch und psychologisch" I would say "mathematisch." Elsewhere Fehling also says that "Aristoteles' Angabe die Himmelskugel voraussetzt" (1994: 146).

I think a more natural explanation is that Anaximander's so-called argument is a no-longer-understood instruction for drawing a map of his universe as explained in Chap. 10. On such a map, the earth is equidistant (ὅμοιος) to the concentric rings of the celestial wheels. Aristotle, then, took the opportunity of a text saying that the earth has to be ὅμοιος to the concentric rings of the celestial wheels, to deride Anaximander instead of his teacher Plato.

Recently, and perhaps under the influence of the wrong translation of ὁμοιότης as "indifférence," Rovelli has interpreted the argument as meaning that according to Anaximander "things fall 'towards the Earth', but the Earth itself has no preferred direction (. . .) to fall. (. . .) There is no absolute 'up' and 'down'" (2009a: 54, see also 2009c: 61). He even calls this "l'intuition fondamentale d'Anaximandre" (2009c: 62, subscription of Fig. 13). To support his view, he quotes a text of the *Corpus Hippocraticum* that would show Ionian influences, and which says that for the antipodes "up" and "down" have changed places, and that the same is the case all around the earth. However, Mansfeld has argued persuasively that the treatise belongs to a much later date, quite late in the Hellenistic period, so that it cannot be used to illustrate an aspect of Anaximander's cosmology (see Mansfeld 1971). Rovelli obviously borrowed his quotation from Kahn, who in the second edition of his work unfortunately did not take the opportunity to revise his text according to Mansfeld's findings (1994: 84–85).

Moreover, Rovelli takes the argument to imply that according to Anaximander things do not fall perpendicular to the surface of the earth, in other words, that Anaximander already taught a centrifocal theory of falling. This makes the alleged argument that is meant to answer the question why the earth does not fall, into an argument why things fall to the center of the universe, which is the flat earth. Furley has already in 1987 articulated the absurdities to which a centrifocal theory of falling under the supposition of a flat earth leads: "It is difficult to combine a flat earth with centrifocal dynamics (. . ..). If lines of fall truly converge on the center from all directions, and the earth, being flat, lies in the center, it follows that falling bodies arrive at the earth's surface at all angles from horizontal to vertical. Even supposing the Greek world is the center of the earth, so that at Delphi all lines of fall might be thought of as theoretically vertical, at the extremes of the known world falling bodies should have been observed to fall at an angle. The contradiction of the theory with observable phenomena seems too obvious for the theory to be credible" (1987: 21). In contradistinction with Rovelli, Furley draws the conclusion that the argument cannot be Anaximander's. Furley's point can best be illustrated by Fig. 8.6 that shows that at the outskirts of Anaximander's earth – say, at the Pillars of Hercules – things (rain, stones, etc.) would fall at an unbelievable angle of about 20° in relation to the earth's surface, which means almost horizontally. Imagine you throw up a stone right above your head and see it fall down at a sharp angle.

Rovelli's illustrations, however, conceal rather than elucidate the problem of his interpretation. In the article in *Collapse* and in the *Internet*-version of the book, he draws a disk-shaped earth but changes without saying the dimensions of Anaximander's earth into 1:2 instead of 1:3 (see Fig. 8.7; Rovelli 2009a: 55, Fig. 4). The result is that the angle of falling at the outskirts of the flat earth looks less sharp.

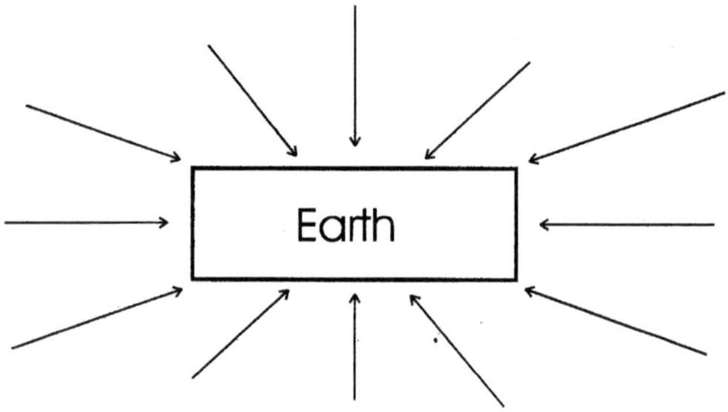

Fig. 8.6 Alleged centrifocal dynamics on Anaximander's earth

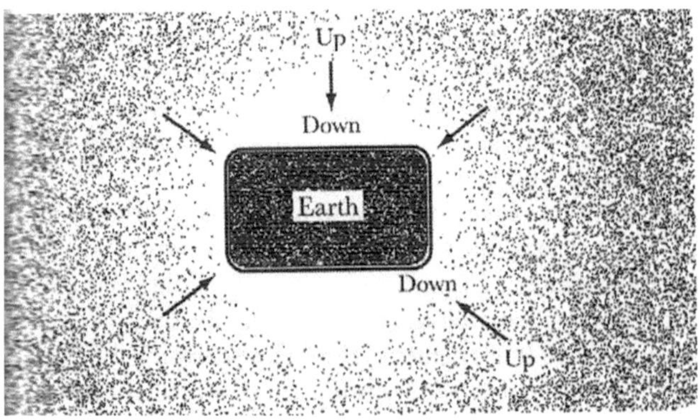

Fig. 8.7 Alleged centrifocal dynamics on Anaximander's earth, version Rovelli (2009a: 55, Figure 4, by the courtesy of Carlo Rovelli)

In the French edition of his book Rovelli even draws a curious representation of an ovoid earth, so that it looks as if everywhere the direction of falling on the earth is vertical or almost vertical (see Fig. 8.8; Rovelli 2009c: 62, Fig. 13, right). However, there is no indication at all in the doxography of such an ovoid shape of the earth, and Rovelli does not elucidate it himself either. I think it is a bad procedure to adjust your illustrations so that they fit your theory or hide the problems it involves.

As argued in the first two sections of Chap. 4, with the possible exception of Parmenides, all Presocratics we know of preserved the concept of a flat earth. And, as far as we know, none of them combined this conception with a centrifocal dynamics, in which the notions "up" and "down" are relative to the earth, which means that the concept of "beneath the earth" would not make sense. On the contrary, according to Aristotle not only Anaximander's successor Anaximenes but even thinkers as late as Anaxagoras and Democritus, who also believed that the earth is flat, did not know of the relativity of up and down. They taught, as

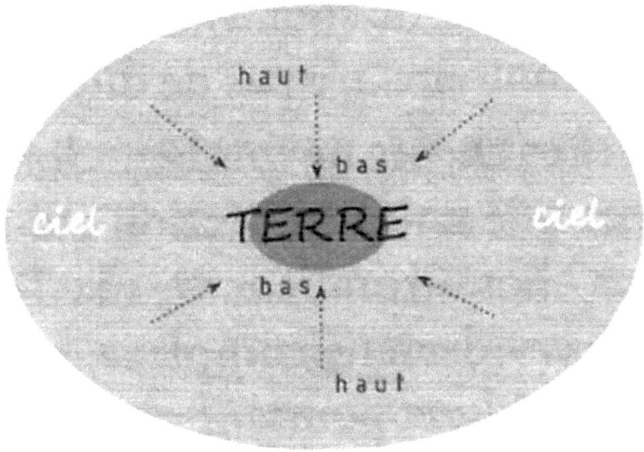

Fig. 8.8 Alleged centrifocal dynamics on Anaximander's earth, version Rovelli (2009c: 62, Figure 13, right, by the courtesy of Carlo Rovelli)

Aristotle says, that "its flatness causes the earth to stay where it is, so that it does not cleave the air *beneath* it, but rests on it like a lid, as is usual for flat objects" (*On the Heavens* 294b16 = DK 13A20, my italics). According to these Presocratics "the flat earth *is upheld by* the air *beneath*" (Simplicius, *In Aristotelis De caelo commentaria* 520.28 = DK 59A88, my italics). And perhaps most clearly in Simplicius' description of the answer Empedocles and Anaxagoras gave to the question why the earth does not fall: "Some people say that a physical mechanism keeps the sky from falling – namely the action of a vortex which holds it up – since *the downward pull on the heavens* is less than the force exerted by the vortex. Empedocles and Anaxagoras say this" (*In Aristotelis De caelo commentaria* 374.32, not in DK, my italics).[10]

It would be strange and historically incomprehensible when Anaximander, the first of them all, was the only exception, teaching a centrifocal theory of falling.

Actually, Rovelli's interpretation of Anaximander's alleged argument comes very close to Aristotle's theory of falling. What we call "falling" is, according to Aristotle, nothing but the natural movement of that which is heavy toward its natural place. Precisely because the heavy element, earth, falls toward the center of the spherical-shaped cosmos, the earth stays there immovably. The spherical shape of the earth is also the consequence of Aristotle's theory of falling, as the sphere is the stereometric body in which all parts of its circumference have the

[10] Perhaps one could maintain that Simplicius, in *In Aristotelis De caelo commentaria* 374.32 tries to say that the sky is kept from falling downward by the force of the cosmic vortex and that in this text "downward" means "toward the central earth," implying a centrifocal theory of falling. However, in *In Aristotelis De caelo commentaria* 375.25, not in DK, also commenting on Aristotle's *On the Heavens* 284a14 ff., Simplicius speaks of the downward tendency of the heavens *and of the earth*, which implies a noncentrifocal falling.

shortest possible distance to its center (see Chap. 18). It would be strange if Aristotle would deride an argument that is virtually the same as his own. All things taken together, it is improbable that Anaximander himself was the originator of the argument why the earth floats free in the center of the cosmos. But even if he used it, its interpretation cannot have been the one Rovelli proposes.

The conception of the free-floating earth was obviously the consequence Anaximander drew from his fundamental insight that the celestial bodies do not stop at the horizon, but make full circles around the earth. These circles he visualized as huge celestial wheels. My guess is that Anaximander was so convinced by the evidence of his idea of the celestial bodies making full circles that he also took the consequence of a free-floating earth, even though he was not able to deliver a conclusive proof of how this was possible. Moreover, granted that Anaximander did not use a convincing argument to explain why the earth does not fall, it also becomes easier to understand why his successor Anaximenes let the earth rest on the air like a lid to prevent it from falling (DK 13A6, DK 13A7, and DK 13A20). Or perhaps Anaximander himself already thought that the earth is supported by the air below, as is Simplicius' already quoted guess (Simplicius, *In Aristotelis De caelo commentaria* 532.14, not in DK). However this may be, it does not mean, as Rovelli maintains, that without his own interpretation "les explications d'Anaximandre sur la centralité de la Terre deviennent absurde" (2009c: 62 n.). Anaximander's central earth is simply the consequence of his conception of the celestial bodies making full circles around the earth. We may compare Anaximander's situation with that of Copernicus and his followers (Galilei, Kepler), who maintained that the sun, and not the earth is in the center of the universe, although they were not yet able to deliver a convincing theory to replace Aristotle's theory of falling that had convinced people for two millennia that it must be the earth that is in the center of the universe. For such a new theory, mankind had to wait until Newton formulated his laws of gravitation.

The Celestial Bodies Lie Behind One Another

Anaximander's idea about the order of the celestial bodies is, in our eyes, as strange as those about the earth and the celestial bodies themselves. According to the doxography, Anaximander placed the celestial bodies in the wrong order: the sun is farthest away from the earth, the stars (and planets) are nearest, and the moon is in between (DK 12A11 and DK 12A18). As Anaximander's order of moon and sun is correct, we could ask how he acquired that knowledge. One is tempted to think that he must have inferred it from the phenomenon of a solar eclipse because then the moon shoves before the sun, for instance, at the famous eclipse of 28 May 585 B.C. that was said to have been predicted by Thales. Anaximander was at the age of 25 at that time. His theory about the origin of solar eclipses, however, prevents such an explanation. As we have seen, both solar and lunar eclipses originate from the closing of the hole in their celestial wheels, not by the relative positions of sun, moon, and earth. This means that Anaximander can never have concluded from the phenomenon of a solar eclipse that the sun is farther away than the moon. Nevertheless, Anaximander must have had some criterion for his strange order of the celestial bodies that obviously has to do with their difference in brightness. For some reason, he thought that the celestial bodies had to be placed in the reversed order of brightness so that the brightest one is farthest away from the earth. Let us first point out that Anaximander was not the only one who made this kind of mistake. It is said that Metrodorus of Chios, a pupil of Democritus, as well as a certain Crates held the same strange opinion, whereas Leucippus defended the order moon, stars, and sun (DK 12A18 and DK 67A1(33)). Apparently, something we tend to look upon as completely self-evident was not so normal in the eyes of these early cosmologists. In many modern languages, it is still idiomatic to use the expression "sun, moon, and stars," keeping Anaximander's order, like an archaic fossil in our language. This point was already made by Burkert: "The natural and unsophisticated way of grouping the heavenly bodies has been – to the present day – sun, moon, and stars" (1972: 310 n. 62).[11]

Maybe we have in a text of the doxography on Leucippus an indication that may explain Anaximander's order of the celestial bodies. According to Leucippus, the celestial bodies set on fire as a result of the velocity of their orbits around the earth (DK 67A1(33)). Apparently, Leucippus thought, in contradistinction with Anaximander, that the stars were brighter than the moon, the moon being bigger but fainter. We may suppose that the celestial bodies get hotter and brighter the faster they move. As they all turn around the earth in about 1 day, the outermost wheels (to speak with Anaximander) necessarily must move the fastest. If we take as distances for the sun, moon, and stars the numbers that are discussed in the next section, and take the diameter of Anaximander's earth to be 5,000 km, and calculate with $\pi = 3$, then the velocity with which the sun wheel turns around the earth equals $2 \times 3 \times (27 \times 5{,}000)\colon 24 = 33{,}750$ km/h.[12] For the ancient Greeks, this must have been an enormous speed. That they were aware of these velocities is attested in the doxography on Anaxagoras, where it is said that objects in heaven whirl around at a velocity that cannot be compared with anything on earth (DK 59B9). At a speed like this, it is conceivable that the celestial bodies will glow because of the friction with the air, and that the celestial body that is farthest away will show the most intense glow. Maybe this kind of considerations has led to Anaximander's order of the celestial bodies, with the sun farthest away from the earth. This is also Guthrie's guess: "The reason for the position of the sun may have been (...) that as the brightest and hottest body it must be where the speed of the revolution is greatest" (1965: 420).

[11] Krafft, who points out this peculiar parallel with the German idiom: "Sonne, Mond und Sterne," probably took it from the German edition of Burkert's book (Krafft 1971a: 106).

[12] The Babylonians used the assumption $\pi = 3$, according to Dicks (1959: 307 n. 3). We may suppose that this still held for Anaximander as well. See also Needham: "Although there is evidence that the ancient Egyptians and Old Babylonians had values such as 3.1604 and 3.125, the commonest practice in ancient civilisations was to take the ratio simply as 3" (1959: 99).

Several authors have wondered how Anaximander could have been so stupid as to hold the wrong order of the celestial bodies, as he should have been able to observe star occultations by the moon (So e.g., Boll 1950: 257; Dreyer 1953: 14; Dicks 1970: 226 n. 51). But let us try to put ourselves in Anaximander's position. In the first place, a star occultation is not so easy to observe. How many of us have ever consciously witnessed one? More important is that nowadays we *know* that the stars are far away, and thus we speak of a star occultation when we see a star disappear behind the moon. From his point of view, however, Anaximander had no reason to speak of the occultation of a star when he saw a star disappear when the moon was at the same place in the sky. Maybe he saw stars disappear and appear again, but he did not observe this as the occultation of the star, as this interpretation did not fit his paradigm of the order of celestial bodies from faint to bright. The simplest way to describe his interpretation of the phenomenon – if he ever observed it – is to say that he must have thought that the brighter light of the moon outshined for a while the fainter light of the star. And when the moon is still young, showing only a small sickle, and the dark part of the moon covers a star, it must be, according to an analogous argumentation, the bright light of the sun that outshines that of the star, for when the moon is crescent it stands near the sun. A similar argumentation is still used by the sophist Antiphon who held – just like Anaximander – that the moon has a light of itself, and that its invisible parts are outshined by the brighter light of the sun. The same holds, he says, for the stars (DK 87B27). Therefore it is a *petitio principii* to say that for Anaximander star occultations were easy to observe. In other words, those who think that Anaximander has to be blamed for not acknowledging the true nature of star occultations are victims of another case of the anachronistic fallacy. Also, the fact that the stars look smaller than the sun and the moon does not automatically lead to the conclusion that they must be farther away.

Taking for granted that Anaximander did not yet discern the planets from the stars, we might think of the possibility that he observed, with the help of a bowl filled with oil as described in Chap. 2, the transition of a planet before the sun. If so, he might have considered such an event as a verification of his ideas about the order of the celestial bodies. Unfortunately, however, during his lifetime, no transition of Venus took place (that of 22 November 549 B.C. was not visible at Miletus), whereas transitions of Mercury, of which there were four during his lifetime, on 16 April 589 B.C., 17 October 585 B.C., 14 April 556 B.C., and 14 October 553 B.C., are not visible with the naked eye.[13] Perhaps one may even surmise that he observed sunspots moving over the sun's surface and took them to be stars passing before the sun, but nothing is known of ancient Greek observations of sunspots.

Some have tried to show that Anaximander's conception of the universe was not original, but that he has borrowed it from elsewhere, and especially from Iranian sources. In the *Avesta*, some texts mention stars, moon, and sun – in the same order

---

[13] See http://transit.savage-garden.org/VenusCatalog.html for Venus, and http://transit.savage-garden.org/sspt.html?inferior=1&superior=3 for Mercury.

as Anaximander's – as stadia in the journey of the soul to the heaven, which is to the throne of Ahura Mazda (Eisler 1910: 90–91 n. 3; Burkert 1963: 106 and 110–112; West 1971: 89–90). Diels even saw a parallel with shamanistic rituals concerning the journey of the soul to heaven (1897: 233). Others, however, have pointed to the fact that it concerns rather late texts, and they wonder whether the influence was not the other way round. According to Bousset, these texts stem from the time of the dynasty of the Achaemenides, which is between 550 and 330 B.C. (1960: 26 n. 2). This is later than Anaximander, who died in 547 B.C. Schmitz' conclusion is straightforward: "Der umgekehrte Weg, auf dem prägende Motive Anaximanders in den Iran gelangt sein können, ist viel besser zu verfolgen" (1988: 77–78). Kahn, too, surmises that "perhaps the most plausible explanation of this curious parallel would lie in the influence of sixth-century Greek ideas upon the religious cosmology of the Achaemenid period" (1994: 90 n. 1. See also Duchesne-Guillemin 1966: 425). However this may be, there is a real and significant difference between these texts, embedded as they are within a religious and mythical context, and Anaximander's conception of the universe, in which such a context is completely absent.

Other authors think of Assyrian or Babylonian influences as regards Anaximander's order of the celestial bodies. However, of the numerous astronomical clay tablets from Mesopotamia, only two are possibly concerned with depth in the universe. The first is a tablet dating from ±1150 B.C. that can be rendered thus: "19 from the moon to the Pleiades; 17 from the Pleiades to Orion, 14 from Orion to Sirius, 11 from Sirius to δ canis majoris, 9 from δ canis majoris to Arcturus, 7 from Arcturus to Scorpio, and 4 from Scorpio to AN.TA.GUB (an unidentified star)," and ends with the question: "how far is one god (i.e., star) distant from another god?" (Hilprecht-text HS229. See Neugebauer 1952: 94, and Rochberg-Halton 1983: 212). Scholars differ on the question whether this text is about radial or transversal distances (i.e., distances in depth or along the celestial vault). The last mentioned, angular distances between different constellations, seems to be the most probable, although in both interpretations the numbers are evidently wrong. Experts tend to think that the numbers "are not astronomically significant but purely the outcome of a mathematical problem as it is set up in the text," perhaps as an exercise for pupils (Rochberg-Halton 1983: 216. See also Van der Waerden 1974: 62–63).

The other clay tablet can be translated thus: "The upper heaven of *luludanitu* stone is Anu's. He (i.e., Marduk) settled the 300 *Igigi* gods there. The middle heaven of *saggilmut* stone is of the *Igigi* gods. *Bel* sits there in a high temple on a dais of lapis lazuli and has made a lamp of electrum shine there. The lower heaven of jasper is of the stars. He drew the constellations of the gods on it" [tablet KAR 307, lines 30–38. see Ebeling (1931: 30 and 33), and Lambert (1975: 58)]. In this text, only the lowest heaven is associated with celestial bodies (stars) that are painted on it, not unlike as is handed down of Anaximenes, who considered the stars (or the constellations) as fiery leaves painted on the firmament (DK 13A14). The other two heavens apparently have a more religious meaning. We may conclude that in the neighboring cultures there exists hardly any evidence, before and during Anaximander's time, of a conception of the heaven that can be compared with his.

What is important is not that Anaximander placed the celestial bodies in the wrong order, but simply that he placed them *behind each other*. That the celestial bodies are at *different* distances from the earth is an important insight that has been overlooked by most authors. One exception is Burkert, who writes in this context: "The arrangement of the celestial bodies (. . .) is wrong, in his account, but still he 'discovered' a fundamentally important fact" Burkert (1972: 310). When Anaximander looked at the skies he saw, for the first time in history, *space*. In this sense, we may say that Anaximander was the discoverer of space. Anaximander's conception of the celestial bodies being behind each other implies that there is *depth* in the universe. This is also unambiguously expressed in the doxography: "Anaximander was the first who gained insight into these measurements and *distances* (of the celestial bodies), as Eudemus reports" (DK 12A19, my italics).

Although to us it looks like a simple and obvious statement, in reality the discovery that the celestial bodies are not all at the same distance from us but lie behind each other was amazing because it cannot be based on observation. It is a complete misunderstanding when Naddaf counts me with those who hold "that the progression of numbers could very well be the result of observational astronomy (however rough)" (2005: 80). I wrote: "Our conclusion must be that Anaximander neither could have observed the spatial relations between the celestial bodies, nor that he could have *inferred* those relations in the way we are used to" (Couprie 2001a: 33). And again on p. 38: "It is important to realize that Anaximander's numbers cannot be based on observation, not only because they are apparently wrong, but particularly for the simple reason that with the naked eye one does not *see* distances in the heavens. What is more, he couldn't have measured those distances" [see also Couprie (2003: 168, 201, 208, 210)]. Against Kahn, who thinks that "the celestial dimensions cannot have been based upon any kind of *accurate* observation," I argued that: "*Anaximander's numbers are not based on observation at all*" (Kahn 1994: 96, my italics; Couprie 2003: 214, my italics).[14] Generally speaking, my contention was (and still is) that depth in the universe cannot be observed and that, consequently, Anaximander's numbers are not the result of observation but that we can draw a representation of Anaximander's cosmos *that does not conflict with* simple observational data. This is the subject of Chap. 10.

Since Anaximander, we know that there is depth in the universe, but we do not *see* that the celestial bodies are behind each other, as explained already in the "Introduction." To quote Upgren again: "(. . .) no one can see any depth in the sky; everything appears as if at the same distance" (2002: 64). The most natural way of expressing how we *see* the universe is the image of the celestial vault, a roof a cupola, onto which the celestial bodies are fixed, all at the same distance from us. This is also, in one kind or another, the archaic conception of the universe. How revolutionary Anaximander's conception of depth in the universe was, becomes

[14] The same misunderstanding already in Naddaf 2001: 12–13. My discussion of the angular diameter of the sun, which Naddaf brings forward as a proof that I let Anaximander take observational data into account, has nothing to do with the distances of the celestial bodies and Anaximander's numbers.

clear when we compare it with the ideas of his successor Anaximenes, who apparently found Anaximander's conception too daring, and reintroduced the idea of the celestial vault by imagining the stars as nails driven into the crystalline heaven, or as flowers painted on the celestial roof (DK 13A14). Incidentally one meets in the literature a hint on the epoch-making character of Anaximander's new conception of the universe. Burnet writes that Anaximander "had shaken himself free of the old idea that the heavens are a solid vault" (1930: 69). Conche expresses it in this way: "Anaximander, au contraire, a dissous le 'bol solide' du ciel, et conçu le ciel comme 'ouvert', contredisant ainsi la vue traditionelle" (1991: 72), although I would say "dôme solide" rather than "bol solide." Heidel gives the most extensive account: "the solid earth, still regarded as the disk suggested by the horizon, was thought to stand still, while the outer bands, composed of mist and fire, continue to revolve about it. This was a very bold hypothesis. While it saved the appearances in respect to the earth, it did away with a stroke with the notion of the 'inverted bowl' or hemisphere of the sky" (1937: 7).

In the next chapter, I discuss Anaximander's numbers that are associated with his celestial wheels. One obvious interpretation of these numbers is that he used them to explain to his cocitizens his idea of depth in the universe. The numbers are, according to Tannery's interpretation, multiples of 3 (plus 1), viz., 27 and 28 for the sun wheel, 18 and 19 for the moon wheel, and 9 and 10 for the wheel(s) of the stars. More on Tannery's suggestion in the next chapter. In the Greek counting system the number $9 = (3 \times 3)$ meant "very big," "very long," whereas $10 = (3 \times 3 + 1)$ meant its completion. Thus, Troy was conquered in the 10th year after withstanding the siege for 9 years, and Odysseus scoured the seas for 9 years before reaching his homeland Ithaca in the 10th year. A God who inadvertently drinks from the river Styx is said to be exiled for 9 full years from the everlasting gods and to be allowed to return in the 10th year (Hesiod, *Theogony*: 775–806). Or even more appropriate: it takes a brazen anvil 9 days to fall from heaven to earth, until it arrives on the 10th day (Hesiod, *Theogony*: 722–725).[15] Accordingly, when Anaximander's numbers are taken to refer to the distances to the front and backside of his celestial wheels, they may be taken to mean: the stars are far away (9 and 10), the moon is even farther away (18 and 19), and the sun is farthest away (27 and 28). See also Fig. 9.5 in the next chapter.

Maybe it is not too daring to read in a text on Anaximander's cosmogony that he was fully aware of the fact that he blew up the archaic conception of the universe. This text reads: "He says that, at the genesis of the cosmos, out of the Eternal a germ of the warm and the cold separated itself off and that out of it a sphere of fire grew around the air which is around the earth, like the bark around a tree. When this (sphere of fire) was torn open and had fallen apart into a number of circles, sun, moon, and stars originated" (DK 12A10). When we consider the tree as the usual metaphor of the celestial tree, which is identical with the celestial axis, then the bark around that tree can be taken as an image of the celestial vault, which explodes in a

[15] More examples in Hahn (2010: 84).

kind of celestial firework, resulting in the system of concentric wheels of the celestial bodies.

Anaximander's cosmos is no longer closed by a firmament. There is no textual base whatsoever for Rescher's suggestion that the original sphere of fire (which he identifies with the *apeiron*) that grows around the earth like the bark around a tree still resides at the periphery of Anaximander's universe (1958: 724, Fig. 5 = 1982: 16, Fig. 6). The same holds for West's suggestion that we have to add the number 36 to Anaximander's numbers as an indication of the diameter of the outermost heaven (1971: 92). On the contrary, on the basis of texts that say that the *apeiron* embraces everything, or even explicitly embraces "all heavens," it is tempting to suppose that Anaximander's universe of concentric wheels round a central earth has to be thought of as surrounded by a boundless space (DK 12A11 and DK 12A15). Kahn is rather apodictic on this point: "The ἄπειρον of Anaximander is then primarily a huge, inexhaustible mass, stretching away endlessly in every direction (. . .). *The Boundless is in fact what we call infinite space*" (1994: 233, my italics). According to him "the conception of the ἄπειρον as a universal body or mass surrounding the world is a permanent feature of Ionian cosmology" (1994: 234). Kahn is here perhaps thinking of Anaximenes, who stated that air surrounds and supports the earth and is also indicated as ἄπειρος (DK 13A1, DK 13A5, DK 13A6, DK 13A7, DK 13A9, DK 13A10, and DK 13B3). We have to note, however, that on the other hand Anaximenes is reported to have conceived of the cosmos as finite, with a crystalline vault, into which the stars are driven like nails and on which the constellations float like leaves. After an analysis of the different meanings of the word *apeiron* Graham concludes: "Thus we can take *to apeiron* as a spatially unlimited stuff," although he maintains that we are not allowed to identify it with the technical term "infinite" (2006: 31). The same could be said of the technical meaning of "space." In other words, when we take Anaximander's *apeiron* as "boundless space," we may not understand this as our abstract concept of space, but as an unlimited and undetermined stuff that is not identical with one of the known "elements" (earth, water, air, and fire). Burnet brings the two notions of the collapse of the firmament and the *apeiron* as an endless space together in one statement: "Anaximander has shaken himself free of the old idea that the heavens are a solid vault. There is nothing to prevent us from seeing right into the Boundless" (1930: 69). However, these interpretations must be abandoned if my suggestion at the end of the previous chapter is right that Anaximander did not speak at all about "the Boundless," but had in mind some eternal and all-embracing generative power (φύσις).

Chapter 9
Anaximander's Numbers: The Dimensions of the Universe

In this chapter, I deal with the problems that arise when we try to draw a map of Anaximander's universe, and suggest ways to solve them. In discussing several authors who have studied the subject, I will point out some bothersome inconsistencies and mistakes in their renditions. One problem, however, which has not been noticed before, will prove to be insolvable within the context of the doxographical evidence. I argue that the only way to meet this problem is to look upon it as circumstantial evidence for the supposition that Anaximander never made a three-dimensional model whatsoever of his conception of the universe.

Diogenes Laertius and Pliny report that Anaximander made a globe (σφαῖρα), (DK 12A1 and Pliny, *Naturalis historia* VII, 203, not in DK, but see *Nachtrag zum ersten Band*: 487). The meaning of the word σφαῖρα cannot be "earth globe," as this presupposes the conception of a spherical earth, whereas Anaximander, as we have seen, conceived of the earth as a cylinder in the shape of a column drum. Moreover, the first earth globe dates from many centuries later and was made by Martin Behaim in Germany in the year 1492 A.D. The word σφαῖρα cannot mean "celestial globe" either, as this presupposes not only the conception of a spherical universe but also knowledge of the stars and constellations in the southern hemisphere, which Anaximander could not possess. The meaning of σφαῖρα must be, then, a kind of armillary sphere as they were fabricated in Diogenes Laertius' and Pliny's time. But even so understood, these statements must be due to an anachronistic misunderstanding because as argued in this chapter, it is not probable that Anaximander made a three-dimensional model of his conception of the cosmos. Generally speaking, Anaximander's universe of a column-drum-like earth at the center of the concentric wheels of the celestial bodies can hardly be called spherical, as a σφαῖρα is supposed to be. The conception of a spherical universe presupposes that the stars are at the periphery of the cosmos, making up the outermost sphere, whereas for Anaximander the stars are nearest to the earth.

It must be called an anachronistic misunderstanding, then, when we read, as already quoted in the preceding chapter, that Panchenko believes "that Anaximander's cosmos was spherical too" (1994b: 51). The background to this remark is a note on another page, which says: "doxographic tradition informs us that Anaximander had such a theory" (sc. of the celestial sphere) (1994b: 32 n. 12). Here obviously the above-mentioned testimonies of Diogenes Laertius and Pliny are

D.L. Couprie, *Heaven and Earth in Ancient Greek Cosmology*, Astrophysics and Space Science Library 1, DOI 10.1007/978-1-4419-8116-5_9,

meant. These texts, as we saw, do not talk about the celestial sphere – the heavens, but about an armillary sphere – an instrument. Diogenes Laertius says explicitly that Anaximander *made* a σφαῖρα. There is no text in the doxography that credits Anaximander with the idea of the celestial sphere. The three-dimensional reconstruction of Anaximander's cosmos in the next chapter will show how far his ideas are from that of a celestial sphere (see Figs. 10.6 and 10.7).

Nevertheless, we may assume that Anaximander, who drew a map of the earth, offered some representation of his cosmological ideas as well. We may readily imagine that he drew a two-dimensional picture, a kind of ground plan or map of his conception of the cosmos. The doxographers would erroneously have taken an account about this map as standing for a celestial globe or armillary sphere such as they knew from their own experience.

The doxography offers several hints about the dimensions of Anaximander's universe. The relevant texts are:

1. The earth (. . .) has a height which is one-third of its width (DK 12A10).
2. Anaximander says that the sun is as big as the earth (DK 12A21).
3. The sun is not smaller than the earth [DK 12A1(1)].
4. Anaximander says that there is a circle which is 28 times as big as the earth (. . .), and this is the sun (DK 12A21).
5. Anaximander says (...) that the circle (of the sun), from where the outbreathing happens and through which it is moved around, is 27 times as big as the earth (DK12A21).
6. The circle of the sun is 27 times as big as <the earth, that of> the moon <18 times> (DK 12A11).
7. Anaximander says that the circle of the moon is 19 times as big as the earth (DK 12A22).

The additions in (6) between angle brackets, adopted by Kranz (*Nachtrag* in DK, Bd. I: 487), were first put forward by Tannery who in a brilliant and generally accepted reconstruction argued that in these texts the diameters of the wheels of the celestial bodies are meant to be ratios of the earth's diameter. The measurements are, according to Tannery, multiples of 3 for the inner perimeters of the celestial wheels, and the resulting numbers + 1 for their outer perimeters. As the stars, according to Anaximander, stand nearest to the earth, Tannery extrapolated their numbers from those of the sun and the moon. The result is: 27 and 28 for the sun, 18 and 19 for the moon, and 9 and 10 for the stars (1887: 91–92 and 119). Perhaps this interpretation was also suggested by Plato's use of one of the series of geometrical proportions (1, 3, 9, 27) that plays a role in his description of the composition of the stuff out of which the world soul is made (Plato, *Timaeus* 35a–36b).

The most natural reading of Anaximander's numbers is to treat them as instructions for drawing a map of the universe. This is, by the way, why I think that Anaximander himself drew such a map. The little history of the attempts to reconstruct this map in the next sections shows how difficult it is to follow even these easy-looking instructions. On the basis of Tannery's reconstruction of Anaximander's numbers, Diels made the map of Anaximander's cosmos reproduced in Fig. 9.1

(1897: 236; 1923b: 21). I call the interpretation of Anaximander's numbers as presented by Tannery and Diels the "standard interpretation." The orthodox variant of this standard interpretation, which is that of Tannery and Diels, and accepted by others, is that in which the numbers stand for the *diameters* of the celestial wheels as multiples of the earth's diameter. For Diels' and for all similar drawings holds, by the way, that they are no real "plan views," or "cross-sections" as they are sometimes called. As the celestial wheels do not lie in the same horizontal plane as the earth, but lie aslant, making an angle of about 52° with the plan of the horizon, as we will see in Chap. 10, on a real plan or cross-section they would not even appear. If the wheels were in the same plane as the earth, the sun and the moon would always be at the horizon.

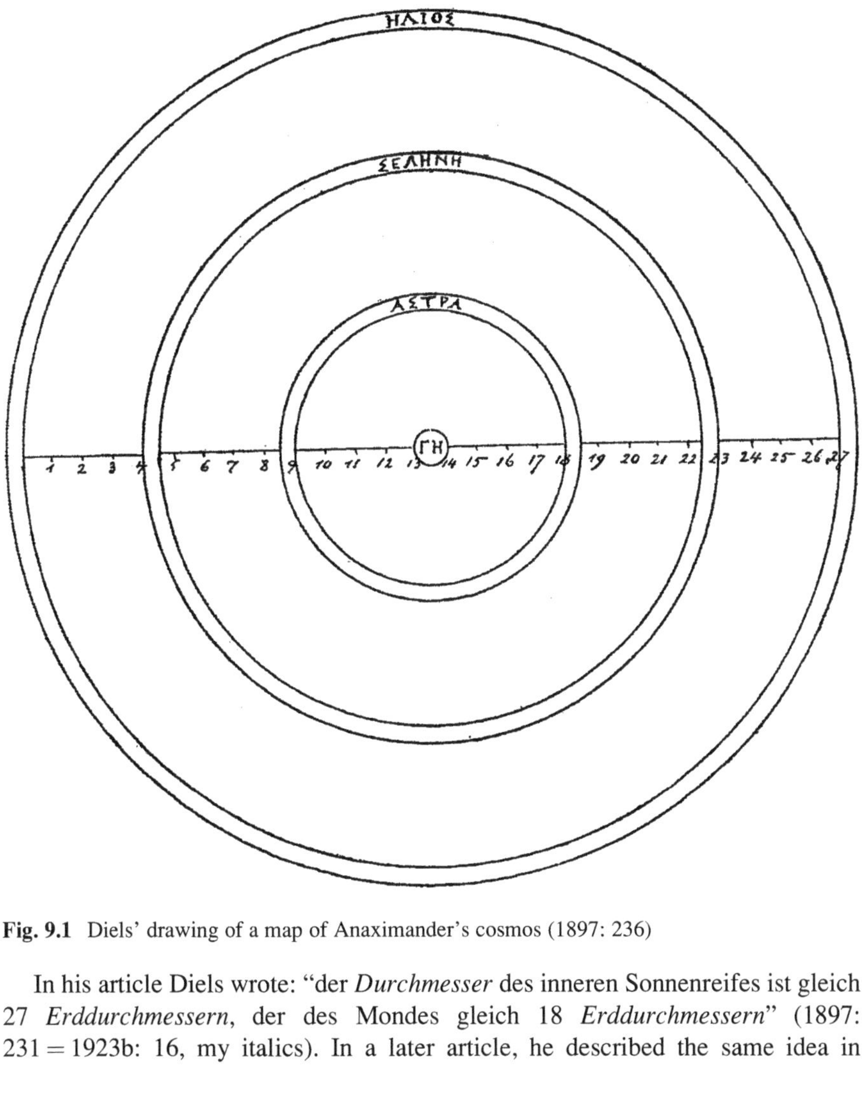

Fig. 9.1 Diels' drawing of a map of Anaximander's cosmos (1897: 236)

In his article Diels wrote: "der *Durchmesser* des inneren Sonnenreifes ist gleich 27 *Erddurchmessern*, der des Mondes gleich 18 *Erddurchmessern*" (1897: 231 = 1923b: 16, my italics). In a later article, he described the same idea in terms of radii instead of diameters: "Der Mondkreis stand 18, der Sonnenkreis endlich 27 *Erdhalbmesser von der Erde entfernt*" (1923a: 72 = 1923b: 8, my italics). That this comes to the same thing can easily be checked in Fig. 9.1. In the first quotation Diels compares the *diameters* of the celestial wheels with the *diameter* of the earth, and in the second quotation he compares the *radii* of the celestial wheels with the *radius* of the earth. Conche suggests that in the later article Diels changed his mind, but he obviously does not fully understand that in both articles Diels expresses the same idea in different words (1991: 209 n. 35). For, as O'Brien rightly states: "whether we think in terms of radius, diameter, or circumference makes no difference in itself. For the figures are relative, not absolute. The *proportions* of the universe will remain the same, *provided we compare like with like*" (1967: 425, second italics mine).

All the same, there still is a problem with this interpretation, namely, the mention in the doxography that the sun, according to Anaximander, is as big as the earth [see (2) and (3) above]. In the orthodox version of the standard interpretation, the cross-section (or width) of a celestial wheel (or ring, or circle) from inner to outer perimeter is only one half of earth's diameter, as can be clearly seen on Diels' drawing. Tannery was well aware of this consequence, as he wrote: "La double épaisseur du cerceau est ainsi égale au diamètre de la Terre" (1887: 94). And Diels saw it as well: "so ist die Breite dieser Ringe auf einen Erdradius zu veranschlagen" (1897: 232 = 1923b: 17). The consequence of this state of affairs is that the sun, being as big as the earth, does not fit into its own wheel, as Kirk has pointed out (2009: 136 n. 1). I will call this for short "Kirk's problem." Mark, that when here "the sun" is mentioned, the aperture in the sun wheel, through which we see the fire inside, is meant, so that Kirk's problem actually should be read as: the aperture of the sun wheel cannot be bigger than the diameter of the sun wheel itself.

Kirk himself thinks the problem can be solved by supposing that "the larger figure might represent the diameter from outer edge to outer edge, the smaller one that from points half-way the outer and inner edges of the actual felloe" (loc. cit.). Kirk does not offer a picture, but Fig. 9.2 (by Naddaf, see below) shows how in his proposal the diameter of the sun wheel from outer edge to outer edge measures 28 earth diameters, and the diameter of the sun wheel from halfway its width on one side to halfway its width on the other side measures 27 earth diameters. The same holds, *mutatis mutandis*, for the wheels of the moon and the stars. The result is that the celestial wheels are one earth diameter wide, which means that Kirk's problem is solved. However, there remains something unsatisfying in measuring both from edge to edge of a celestial wheel and from halfway to halfway its width. When we consider Anaximander's numbers as instructions for making a drawing, Kirk's solution needs three extra numbers. These missing numbers for the diameters of the celestial wheels from inner perimeter to inner perimeter are: 26 for that of the sun wheel, 17 for that of the moon wheel, and 8 for that of the star wheel. The doxography, however, does not mention such numbers. So, after all Kirk's solution is not satisfactory.

The best picture of Kirk's idea is given by Naddaf (2001: 18–19, Fig. 1.3, reproduced here as Fig. 9.2; see also 2005: 77–78). To read this picture well, one has to understand that at both ends of the row of earth diameters a half-disk is drawn so that the sum total amounts to 28. As it is a perfect illustration of Kirk's idea, this means that the same objections apply to it as well: the lines that should indicate the diameters of the inner perimeters (26, 17, and 8, respectively) are not drawn. Naddaf mentions this but obviously does not regard it as a problem (2005: 76).

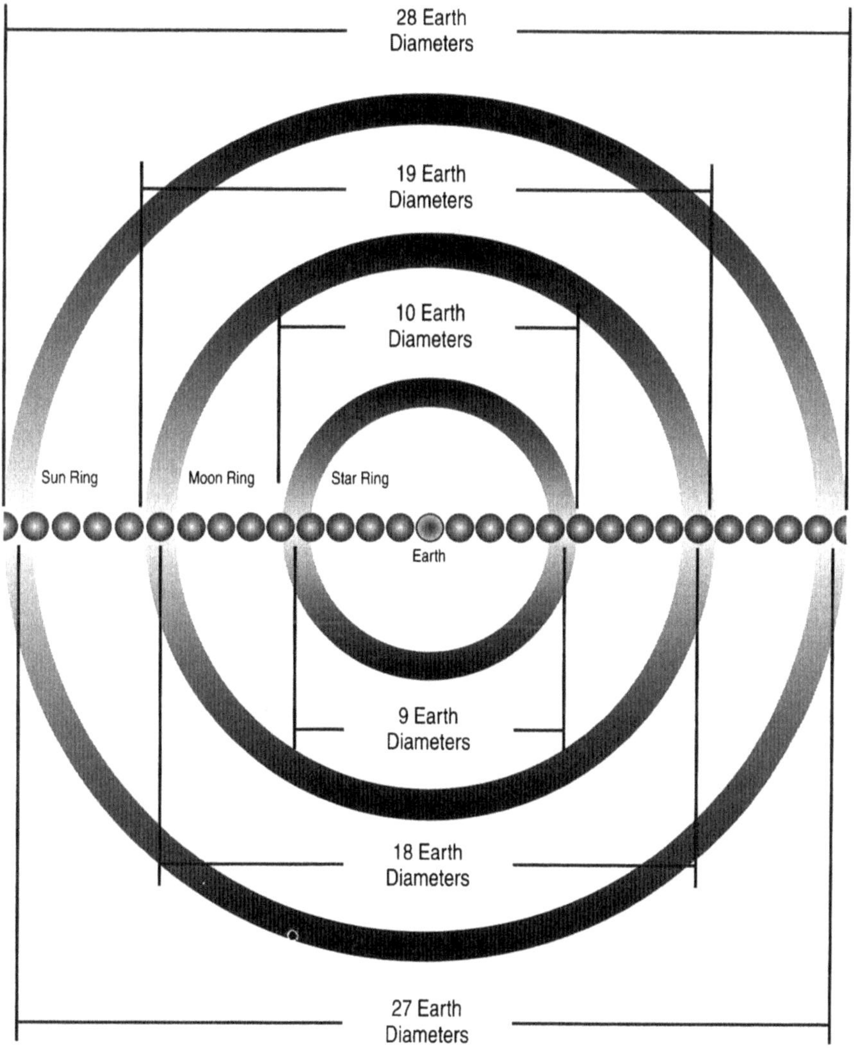

Fig. 9.2 Naddaf's version of a map of Anaximander's cosmos, revealing Kirk's problem (2001: 18, Fig. 1.3, by the courtesy of Gerard Naddaf)

When we bear in mind that it makes no difference whether we think in terms of radius, diameter, or circumference, provided we compare like with like, Rescher's solution is a variation on the same theme: when we take the numbers as indicating

Table 9.1 Anaximander's numbers (after Rescher)

Object	Radius	Circumference	Ratio of circumference to earth's circumference
Earth	1	$2\pi \times 1$	1
Star circle (inner perimeter)	9	$2\pi \times 9$	9
Star circle (outer perimeter)	10	$2\pi \times 10$	10
Moon circle (inner perimeter)	18	$2\pi \times 18$	18
Moon circle (outer perimeter)	19	$2\pi \times 19$	19
Sun circle (inner perimeter)	27	$2\pi \times 27$	27
Sun circle (outer perimeter)	28	$2\pi \times 28$	28

the *circumferences* of the celestial rings as ratios of the earth's circumference, then between the *radii* of the celestial wheels (Rescher prefers to speak of "circles"), taken as multiples of the earth's radius, the same relations hold as between the circumferences (1958: 727 = 1982: 22).[1] Rescher shows this in a table which is reproduced here (and to which I have added the numbers for the stars, Table 9.1).

In Rescher's system, Kirk's problem remains unsolved as well, as the cross-section of a celestial wheel is again one half of the earth's diameter. For if the earth's radius = 1, as can be read in the second column of the table, then its diameter = 2, whereas the cross-section of the wheels, which is the difference between their outer and inner perimeters, remains 1. This means that in his version again the sun, which is according to the doxography as big as the earth, does not fit into its own wheel, which was Kirk's problem. Rescher's drawing is somewhat difficult to read, but the intention is that the number 1 next to the earth indicates the earth's *radius*. The arrows marked "9," "18," and "27" are supposed to start *in the center* of the earth. The little stripe next to the number 1 that seems to be but is *not* the extension of the arrow marked "9" intends to point to the earth's radius, which is the little line that seems to be but is *not* the extension of the arrow marked "27." The earth itself should have drawn about twice as big (see Rescher 1958: 727, Fig. 8 = 1982: 23, Fig. 9; here reproduced as Fig. 9.3).[2]

O'Brien gives another variation of the same theme. He holds that the width of the rings should be compared with the radius of the earth as an "unexploded wheel" and maintains that "there is no reason why we should not think of the thickness or width of the sun wheel as equal to one half of the earth's diameter" (1967: 423–424). O'Brien is well aware of Kirk's problem, and he offers a solution that makes the sun fit into its own wheel. He remarks rightly that "the opening in the sun wheel, through which the visible sun appears, is presumably on the inner face of the

[1] As stated already in Chap. 8, Rescher distinguishes, rather strangely, between "rings" of fire and "wheels" indicating the airy spaces between the rings, whereas I use both words, as is usually done, as indicating the same phenomenon of Anaximander's materialized orbits of the celestial bodies. A similar account of the several dimensions, although not in a table, in Naddaf (2001: 19, and 2005: 77 and 78).

[2] I explain this rather extensively because I myself was misled by these features of Rescher's drawing (see Couprie 2009a: 172 and 175). This does, however, not impede my criticism of Rescher's solution.

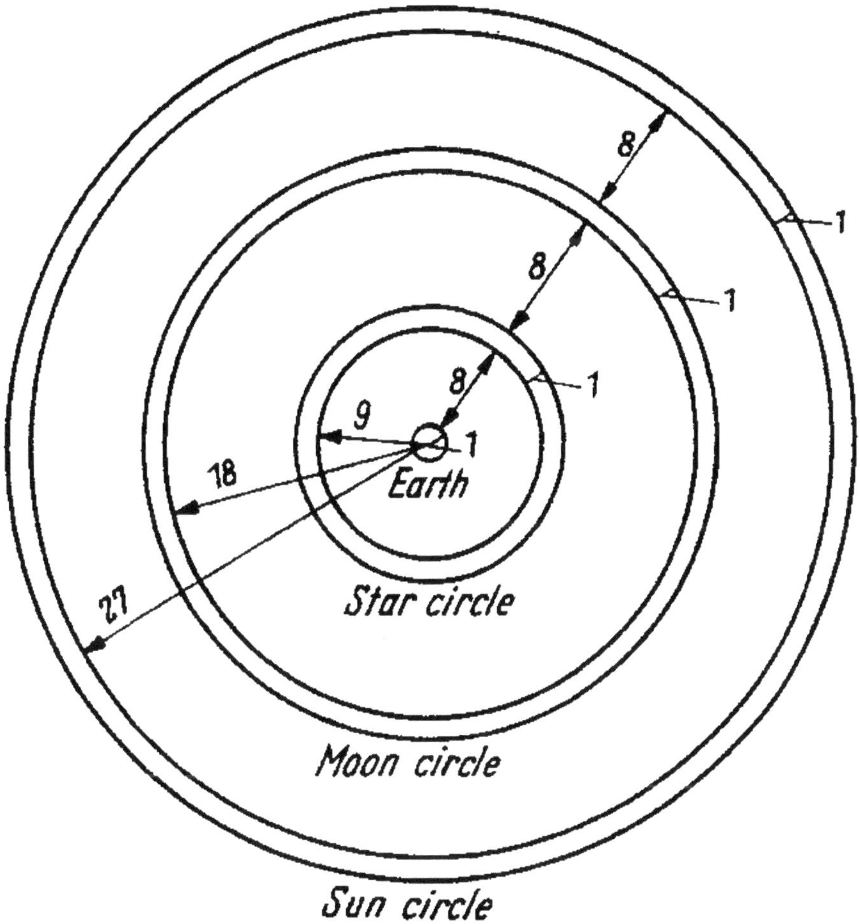

Fig. 9.3 Rescher's drawing of a map Anaximander's cosmos (1982: 23, Fig. 9: , by the courtesy of Nicholas Rescher)

sun wheel, on what we may perhaps call the *height* of the sun wheel" (1967: 424). Actually, in the plan drawings shown here, only the *width* of the celestial wheels is pictured, and not their *height*, which is the side of the wheel that is turned toward the earth. O'Brien suggests that the dimensions of the celestial wheels are the same as those of the earth, a suggestion which Rescher already called a "defensible conjecture" (1958: 727 = 1982: 24). O'Brien takes the dimensions of Anaximander's earth to be "a height three times the size of the diameter," instead of the usual translation "the height of the earth is one third of its diameter" (1967: 424–425). Here, he follows Martin's translation that was later also adopted by Dumont (Martin 1879: 66; Dumont 1988: 28). Transferred to the celestial wheels this would mean: "the height of the wheel is three times its width," from which it follows that the sun (the aperture in the wheel), being equal to the diameter of the earth, fits into the height of the wheel, which equals three earth diameters. However, as Conche remarks, this translation is not right, as τὸ τκρίτον does not mean "three times," but τὸ τρίτον μέρος, "the third part of a whole" (1991: 193 n. 5). Moreover, the height of column drums can vary between relatively wide ranges but seldom measures three times its diameter, whereas the ratio of a height of one third its diameter is more common (especially for the upper part of a column). After all, O'Brien's solution for Kirk's problem is not satisfactory either.

Rescher's drawing (Fig. 9.3) may suggest another solution to Kirk's problem. When we take the numbers as indicating the distances from the celestial bodies to the earth, which is to say as the *radii* of the celestial rings or wheels, and in addition take the *diameter of the earth* as our standard unit (= 1), instead of its radius, as Rescher did, the result will be that the width of the celestial rings is equal to the earth's diameter. The result is also that the interval from earth to stars, stars to moon, moon to sun is eight *earth-diameters*, as can be seen in Figs. 9.4 and 9.5 [and not "nine diameters," as Cornford abusively writes (1934: 12, my italics)]. We may call this the unorthodox variant of the standard interpretation of Anaximander's numbers. This interpretation fails, however, says Kirk, because the Greek text excludes it (2009: 136 n. 1). Notwithstanding this verdict other authors think that Kirk's problem can be solved in this way (Cornford 1934: 12 and 15; Burch 1949–1950: 155 n. 41; Conche 1991: 209–210; Hahn 2001: 191; Couprie 2003: 213). O'Brien would object that this solution conflicts with the idea of comparing like with like (the diameters of the celestial wheels with the diameter of the earth, the radii of the celestial wheels with the earth's radius). However, it depends on what one wants to consider as "like." In the unorthodox version of the standard interpretation, the width of the celestial wheels is "like" or equal to the diameter of the earth. This is in accordance with the texts that say that according to Anaximander, the sun is equal in size to the earth. The advantage of the unorthodox variant of the standard interpretation as compared to the orthodox variant is that the sun fits into its own wheel.

Conche expresses this idea when he explicitly writes that "le *demi-diamètre* – le rayon – du κύκλος solaire" is 27 times "le *diamètre* de la surface circulaire de la Terre" (1991: 209–210, second italics mine). Notwithstanding this, he draws the diameter of the earth about three times too big, so that the sun, being as big as the earth, does not fit into its own wheel. It is easy to see that in Conche's drawing, the half diameter of the ring of the stars is not nine times the diameter of the earth, as he claims, and the same holds, mutatis mutandis, for the rings of the moon and the sun (1991: 210, here redrawn as Fig. 9.4; Conche prefers to speak of "anneau").

I prefer to treat Hahn's version of the problem of the numbers in Anaximander's map of the cosmos in plan view, which contains a curious calculation error, in Chap. 12 in connection with his overall interpretation of Anaximander's cosmology as being inspired by the work of contemporary temple architects (see also Couprie 2010).

In defense of the unorthodox interpretation, I would underline that I think the numbers of the celestial bodies must be meant as radii because radii indicate the *distances* of the celestial bodies to the earth. As Simplicius explicitly confirms, Anaximander was especially interested in the distances of the celestial bodies: "Anaximander was the first who gained insight into these measurements and

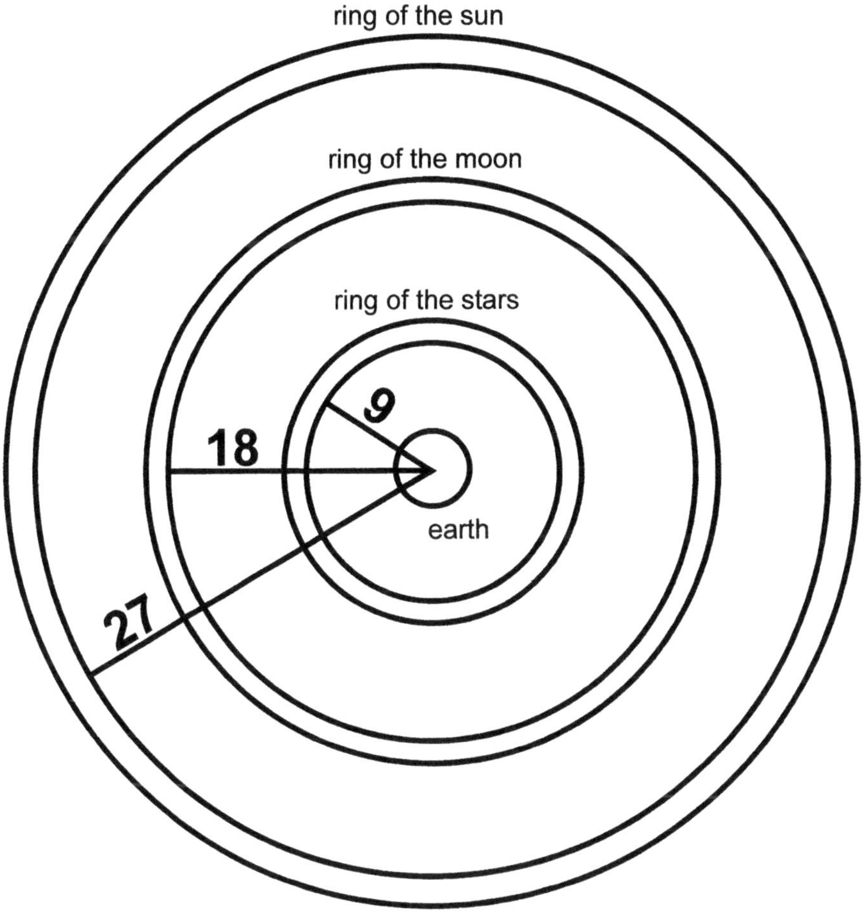

Fig. 9.4 Author's copy of Conche's version of a map Anaximander's cosmos (1991: 210)

distances, as Eudemus reports" (DK 12A19). A similar conclusion was reached recently by White: "the figures for Anaximander's rings are more naturally interpreted as radial. (. . .) Eudemus included them in his account of Anaximander as the first to specify *distances*" (2008: 108, my italics). Moreover, when one looks upon Anaximander's numbers as instructions for drawing a map of the cosmos, their interpretation as radii is the most plausible one. When you draw the little circle of the earth and leave one leg of a pair of compasses in its center, you may circle the different celestial wheels by measuring off their radii. I follow Conche in taking the earth's *diameter* as the main unit (=1) instead of its radius so that the distances of the celestial wheels to the earth are measured in earth diameters. This results in celestial wheels, the width of which equals the earth's diameter ("comparing like with like") so that Kirk's problem is solved. For clarity's sake in Fig. 9.5, I show a drawing of my own interpretation as well.

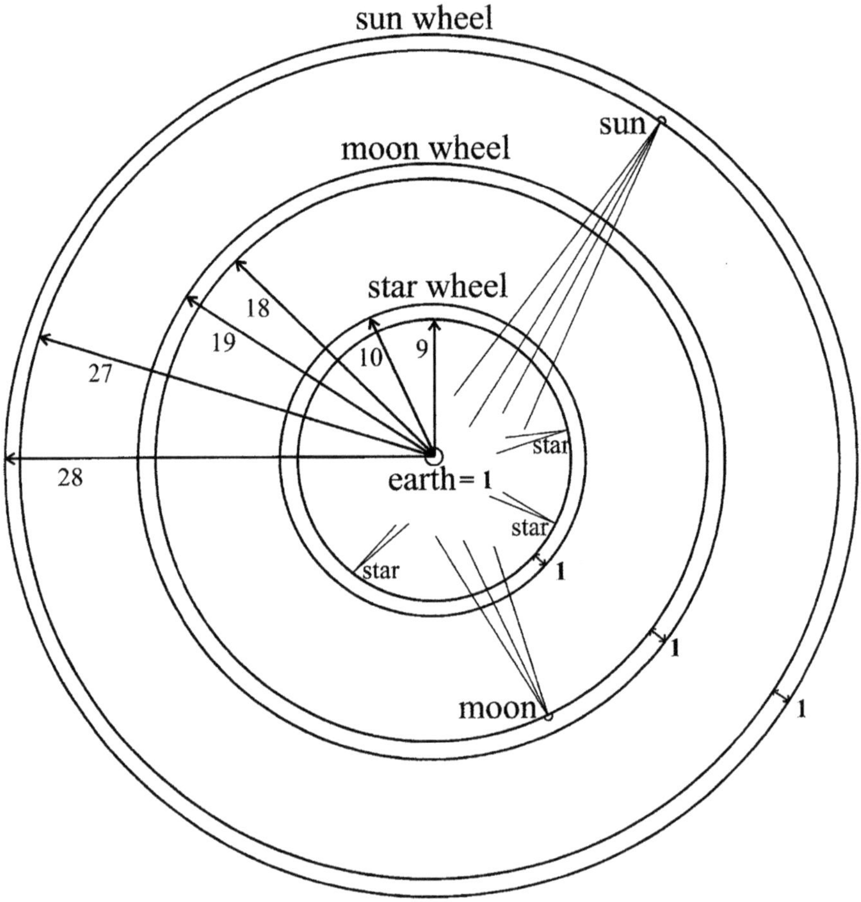

Fig. 9.5 My drawing of a map of Anaximander's cosmos

Nevertheless, also in this solution one problem remains. The distances between the wheels of sun and moon, and between the wheels of moon and stars equal: 8 earth diameters, but the distance between the earth and the star wheel equals 9 earth diameters. I would like to stress that nowhere in the doxography is said that these distances must be equal. Nevertheless, it seems to be at least an aesthetic flaw when they appear to be not all equal. After all, I must confess that Engmann is right when she remarks that "the figures do not yield equal distances" (1991: 22; see also Couprie 2001a: 38, note 47, where I said the opposite). Whether Engmann is hinting at the figures just mentioned is not clear, as her remark is no more than a stray note.

In Kirk's solution, this problem does not seem to exist, as can easily be seen in Naddaf's drawing (Fig. 9.2), where all the distances between the rings as well as between the earth and the star ring add up to 3½ earth diameters. But even in this interpretation, the distance from the earth to the stars is 3½ earth diameters only if one calculates it from the earth's rim, as if the celestial wheels lie in the same horizontal plane as the earth. As Anaximander's earth is flat and the celestial wheels are tilted in relation to the earth's surface, this figure has to be somewhat higher.

According to O'Brien, the equal distances between the rings are one of the most important factors that determined Anaximander's choice for his numbers (1967: 427). He also advances the idea that possibly the equal distances between the celestial bodies has been the source of the report on "unlimited worlds" at equal distances of one another (1967: 427 n. 4). And also Cornford, quoted with approval by Kahn, already wrote: "The statement probably involves the (...) confusion of coexistent worlds (...) with Anaximander's (...) heavenly bodies (...), spaced at equal distances from one another" (1934: 12. See also Kahn 1994: 46–53, especially 50). I agreed to these statements in the above mentioned publication (2001a: 38), and in a sense I can stay with that opinion, as also in the unorthodox interpretation of Anaximander's numbers (viz. as relating to radii), the distances *between the celestial bodies* are equal. It is only the distance between the earth and the stars that is different. Notwithstanding this flaw, I think the arguments for the unorthodox interpretation are so convincing and the objections to the orthodox interpretation are so strong that I stay with the opinion that Anaximander's numbers must be read as indicating the radii of the celestial wheels, which are the distances to the celestial bodies.

Maybe the subtle differences between the respective versions treated above do not look worth the reader's while. The second problem, however, is more serious. It is the problem of the sun's angular diameter. All authors discussed so far start from the supposition that the sun, which is the hole in the sun wheel, is equal in size to the earth, as the doxography says [see (2) and (3) above]. The commonest suggestion is to suppose that the aperture is exactly as big as the width of the sun wheel. This defines Kirk's problem: the sun must fit into its own wheel. Conche expresses this thought in a rather complicated way: "Cela étant, le soleil a même diamètre que la Terre, ce qui signifie, si l'on admet que l'ouverture solaire occupe toute la largeur de l'anneau, que le soleil est aussi large que la Terre" (1991: 209). We have to bear in mind, however, that the aperture in the sun wheel is situated in the side that is turned toward the earth that is not visible on a map. For most authors, this does not make any difference, as they suppose the height of the sun wheel to be the same as its width. The problem that remains, then, is that of the diameter of the aperture in the sun wheel as compared with the circumference of the wheel (which equals the sun's orbit around the earth), or, in other words, the problem the angular diameter of the sun. In reality, the angular diameter of the sun is about 0.5° (see Fig. 2.8).

Let us take, for example, the situation in the unorthodox version of the standard interpretation. If the sun hole (which is the sun as we see it) is taken to be as big as the earth (and therefore as big as the width of the sun wheel), and the sun wheel has a radius of 27 earth diameters, then $2\pi \times 27 = 162$ suns in a row fit into the circumference of the sun wheel (taking $\pi = 3$, as was usual in Anaximander's time). This would mean that the angular diameter of the sun equals $360: 162 = 2°15'$ instead of 0.5°. Figure 9.6 illustrates this. This means that the sun would be seen more than four times as big in diameter as it is in reality. In reality, however, as the angular diameter of the sun is 0.5° and the circle of the sun wheel is 360°, 720 suns in a row would make up a full circle.

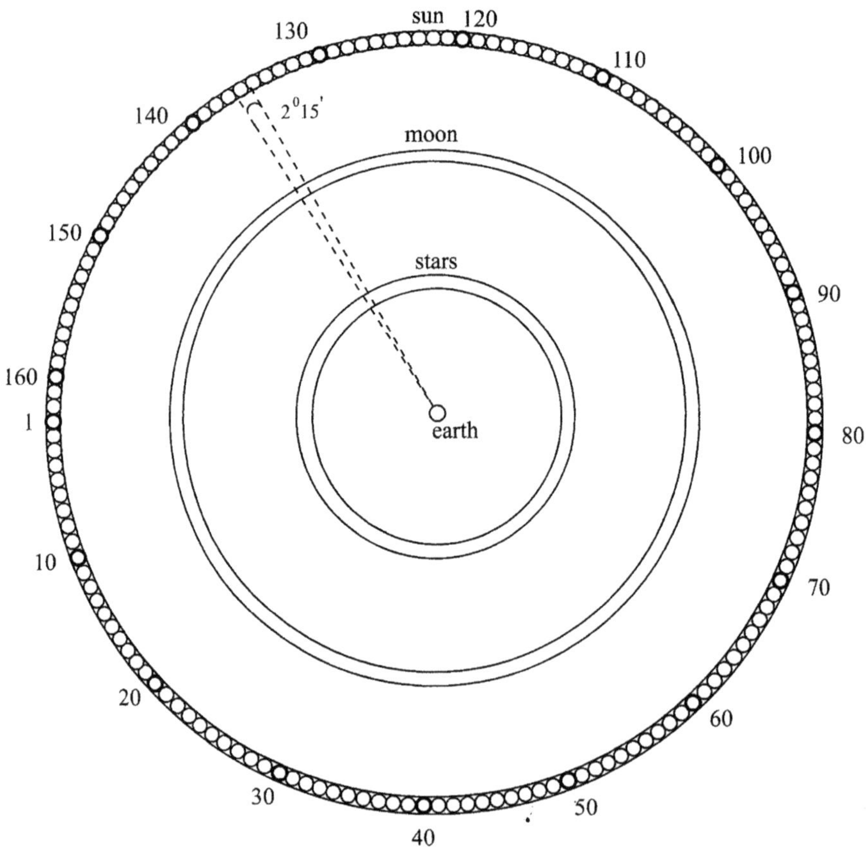

Fig. 9.6 The problem of the angular diameter of the sun

If we take Anaximander's numbers to stand for diameters, as in the orthodox version of the standard interpretation, the angular diameter of the sun would be even bigger, as no more than $2\pi \times 13.5 = 81$ suns fill up the circle of the sun wheel. Accordingly, Heath calculates an angular diameter of "a little over 4°," so that the sun would appear eight times as big as it actually does (1913: 32). In O'Brien's option, discussed above, where the aperture of the sun wheel is rightly situated in the side that is turned toward the earth, this would make no difference, as he also takes the aperture to be as big as the earth so that no more than 81 suns make up the full circle of the sun wheel. This is also the case in Rescher's text version, although he, strangely enough, calculates an angular diameter of "somewhat *under* 4°" (1958: 727 n. 21a = 1982: 24 n. 24, my italics). Burch and O'Brien mention, not quite precise, that the angular diameter of the sun would be about 4° (Burch 1949–1950: 155; O'Brien 1967: 426). This results, then, in a major discrepancy between the angular diameter of the sun thus calculated and its real angular diameter as it is seen in the sky. As regards this problem, Burch thinks that "there seems to be no way of explaining this discrepancy between fact and theory" (1949–1950: 155). One might think that this way of calculating surpassed Anaximander's capabilities, or, with Burch, that he never gave it a thought. However, according to the doxography, Thales already calculated that 720 suns in a row make up the circumference of the sun's orbit around the earth [DK 11A1(24) and 11A19]. A possible method of calculating this number was treated in Chap. 2. It makes more sense to ascribe this discovery not to Thales but to Anaximander because it was he who taught that the orbit of the sun (in his terms, the sun wheel) passes under the earth as well, making a full circle.

The only way I see to solve the problem of the angular diameter of the sun is to look more closely at the meaning of the text that says that the sun is equal in size to the earth. After all, the word "sun" is ambiguous in Anaximander's universe, as it can indicate both the sun wheel and the aperture in the wheel through which the fire shines. I understand the report that according to Anaximander the sun is as big as the earth as a drawing instruction for the width of the sun wheel (= one earth diameter) that has erroneously been taken by later doxographers as a measure for the size of the sun itself. Thus interpreted, the text says nothing about the size of the aperture in the sun wheel, which means that this aperture can be any size so as to fit the real angular diameter of the sun. In other words, the size (or diameter) of the aperture in the sun wheel, which is the sun as we see it, may be much smaller, so that 720 suns in a row fit into the circumference of the sun wheel. This is especially crucial when we take the usual interpretation of the dimensions of the earth as having a height which is one-third of its diameter and suppose that the celestial wheels have the same dimensions. Then, the aperture has to fit within the height of the sun wheel, which is one third of its width (and not three times its width, as O'Brien thought). Accordingly, in Fig. 9.5, the aperture in the sun wheel (and also that in the moon wheel) is drawn much smaller than the width of the wheel. We may conclude that, when we take Anaximander's numbers to stand for the distances of the celestial wheels to the earth, when we take the report that the sun equals the earth in size to have bearing on the width of the sun wheel (as well as that of the other celestial wheels), and take the numbers as drawing instructions, both Kirk's problem and that of the angular diameter of the sun can be solved.

Recently, White has tried to show that there is an observational basis for Anaximander's figures. He claims that "an alternative interpretation of the relevant testimony raises an intriguing possibility," which especially entails that "the figures for the outer ring (sc. that of the sun, D.C.) would have a straightforward observational basis" (2008: 107 and 108). In a sense, White presents as a solution to what I called above one of the problems with Anaximander's numbers: the far too big angular diameter of the sun. To reach the desired result, White needs to make two conjectures. The first is that Anaximander calculated with $\pi = 3\frac{1}{3}$, for which there is, however, no historical evidence. The second conjecture is that Anaximander estimated the angular diameter of the sun as 2°. White calls the diameter of the earth Anaximander's "astronomical unit (AU)," and he concludes: "then the entire circle is 180 AU, its diameter is 54 AU (three-tenths of 180), and its radius is 27 AU" (2008: 108). In this way, the number 27 is said to be derived from observation. As stated above to take the angular diameter of the sun to be 2° is the kind of mistake that would not have stayed unnoticed. Moreover, and as already quoted in Chap. 2, White himself argues that Anaximander could have measured the angular diameter of the sun by comparing the size of a disk that covers the sun with the circle of the horizon. The alleged value of 2° would mean a measuring error of about 300%.

Perhaps this is the reason why White stresses that the results of the measurement at stake "are bound to be wildly inaccurate" (2008: 109). But the opposite is the case: as was argued in Chap. 2, the results are rather accurate and remain within an acceptable range of about 5%. In defense of his conjecture that Anaximander took the angular diameter of the sun to be 2° White refers to Aristarchus' puzzling hypothesis that "the moon subtends one-fifteenth of a zodiacal sign," which is to say that it has an angular diameter of 2° (2008: 109). Aristarchus' hypothesis has worried many scholars. Perhaps the best explanation is that Aristarchus, being a mathematician, was not so much interested in the exact values of the astronomical data, but rather in the ingenuity of solving a geometrical problem. The same holds for other astronomical data in Aristarchus' book (Aristarchus, hypothesis 6, in Heath 1913: 353. See also Pannekoek 1961: 120). It makes little sense, however, to burden Anaximander with a bizarre mistake made by Aristarchus. Moreover, one cannot both bestow Anaximander with a method for measuring the sun's angular diameter, and at the same time accuse him of so big a mistake as to make the figure for the sun's angular diameter roughly four times too large.

There remains one final but serious problem that turns up as a consequence of the standard interpretation of Anaximander's numbers, both in its orthodox and in its unorthodox form. This problem is not immediately clear from visualization in a two-dimensional ground-plan or map, as drawn above, but needs a three-dimensional model or at least visualization in a front view or elevation. The problem that has never before been recognized as such is that there is no place on Anaximander's earth where the sun at any moment of the year stands in the zenith. In reality, however, this is the case in places of which we would say that they lie on or between the tropics of our spherical earth, for instance, Syene in Egypt. A simple calculation and a drawing may illustrate this. As Miletus lies at 37.5° N, the greatest height of the sun there, during the summer solstice, is 76° (52.5° + the obliquity of the ecliptic, which is 23.5°). When we render this in a drawing, we see immediately that the sun will never stand in the zenith on Anaximander's flat earth. The drawing of Fig. 9.7 is made according to the orthodox version of the standard interpretation (that of Tannery, Diels, Kirk, and others). In the unorthodox version, the effect would even be twice as big.

At the time of Anaximander, there was a vivid commercial traffic between Miletus and Egypt, to say nothing of the fact that it is told that his cocitizen and teacher Thales has been in Egypt. It is hard to believe that Anaximander would not have been acquainted with the fact that at Syene during the summer solstice a gnomon does not throw any shadow at all, since the sun stands in the zenith at that time. Naddaf has tried to argue that, given the Milesian relations with Egypt, Anaximander must have been there. He even writes: "it would appear strange if he had not visited the great country" (2005: 100), but this remains speculation. However, generally speaking, Anaximander, who introduced the gnomon in Greece, must have noticed that the shadow of a gnomon gets shorter as one goes to the south, which easily should have led him to the conclusion that at some place the gnomon would not throw any shadow at all. As far as I can see, there is no solution to this problem within the context of the standard interpretation of Anaximander's numbers.

Fig. 9.7 On Anaximander's earth the sun never stands in the zenith

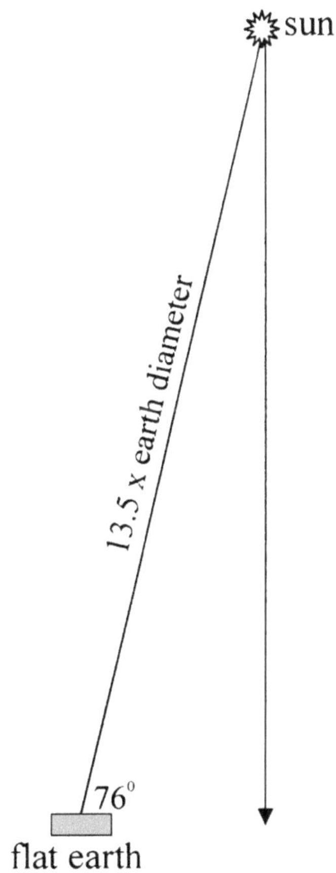

One possible and far-reaching conclusion would be that the standard interpretation of Anaximander's numbers is false, or in other words that Tannery's original suggestion was not so bright after all. It is very hard, however, if not impossible, to think of any interpretation of Anaximander's numbers that would make it possible for the sun to be in the zenith. If we would argue that Anaximander's numbers do not indicate real distances but are mainly meant symbolically as "far," "farther," "farthest," which I believe is the real meaning of Anaximander's numbers, it would not help either. For, on the contrary, on a flat earth like Anaximander's, the sun has to be nearby and consequently rather small, to make it possible to stand in the zenith, as is shown in Fig. 9.8.

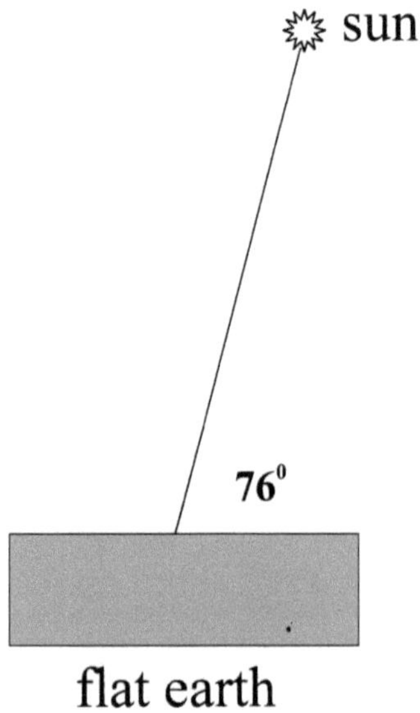

Fig. 9.8 On a flat earth the sun must be nearby and rather small

I tend to draw another conclusion from this state of affairs. I think it shows that Anaximander, who probably drew a map of his conception of the universe, certainly did not make a three-dimensional representation of it. For if he had, he would have noticed immediately that something was wrong because nowhere on his earth the sun stood in the zenith. It must have been precisely this consequence that made Anaxagoras conclude that, when the earth is flat, the sun must be rather nearby and that accordingly the size of the sun must be in the order of that of the Peloponnesus, as will be discussed extensively in Chap. 16.

Chapter 10
The Visualization of Anaximander's World Picture

In Chap. 9, we drew a map, a two-dimensional representation, of Anaximander's universe (see Fig. 9.5). One of the conclusions was that Anaximander probably did not fabricate a three-dimensional model of his cosmos. This does not preclude, however, our attempt to deliver a three-dimensional representation for our own understanding. In Chap. 8, we saw that the comparison of the celestial bodies with chariot wheels has to be taken seriously. In Chap. 9, it was argued that the width of such a celestial wheel, which is to say of its felloe, equals one earth diameter. The height of the earth, according to Anaximander, is one third its diameter. It is a reasonable guess to suppose that the height of the celestial wheels (which is the side that is turned toward the earth, where the hole is through which the fire shines) is also one third the width of the wheel. The drawing in Fig. 10.1 illustrates this.

In trying to make a spatial representation of Anaximander's universe, we will have to choose what his celestial wheels represent. As regards the sun and the moon, there are two options. The first is that they represent the ecliptic (the annual path of the sun along the constellations of the Zodiac), and the monthly path of the moon. The second option is that they represent the daily orbit of these celestial bodies, just as is the case with the star wheels. To make a choice, we will first have to answer the question whether Anaximander was acquainted with the (inclination of the) ecliptic. The notion of the ecliptic, the annual orbit of the sun along the starry sky, does not fit very well into Anaximander's astronomy because it presupposes the concept of the sphericity of both the heaven and the earth. The obliquity of the ecliptic is about 23.5° as regards the celestial equator, which is the projection of the earth equator on the celestial sphere. This means that the concept of the ecliptic is rather senseless without an earth and a heaven conceived as spherical. Anaximander, however, thinks of the earth as a flat drum, and the stars are, according to him, the innermost ring and not the outermost, as is the case with those who teach the sphericity of the cosmos. The first mention in Babylonian astronomy of the ecliptic is in a clay tablet dating from about 410 B.C., thus a considerable time after Anaximander (see Thurston 1994: 60). It is reported that according to Anaxagoras, Democritus, and Cleanthes, who still thought that the earth is flat, all the celestial bodies turn around the earth from east to west (DK 59A78). This seems to be a triviality, but that is exactly why this remark

D.L. Couprie, *Heaven and Earth in Ancient Greek Cosmology*,
Astrophysics and Space Science Library 1, DOI 10.1007/978-1-4419-8116-5_10,

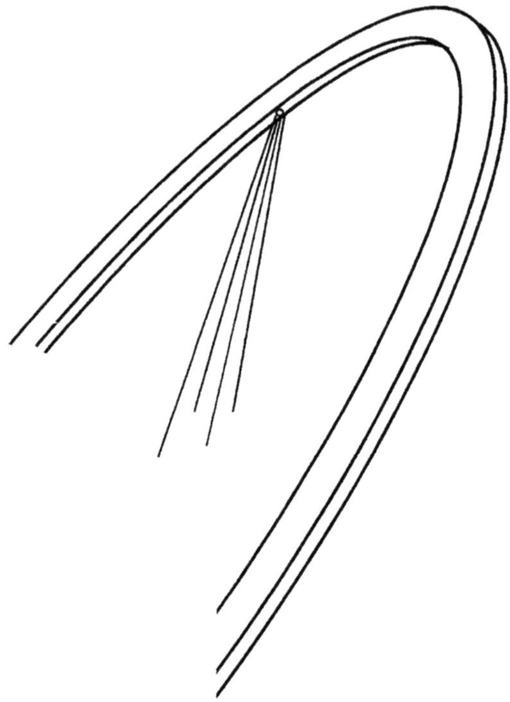

Fig. 10.1 Detail of Anaximander's wheel of a celestial body

is so important. It looks as if the author wants to underline that Anaxagoras, Democritus, and Cleanthes did not yet describe the movements of sun, moon, and planets as a retrograde movement from west to east along the ecliptic (or along the signs of the Zodiac), but as their daily orbit from east to west, with different velocities. Lucretius says explicitly that according to Democritus the sun moved slower than the stars, and the moon slower than the sun (DK 68A88), which also implies that they were talking about the daily movements of the sun and the moon.

All this leads to the conclusion that Anaximander's sun wheel cannot be identified with the ecliptic. Mutatis mutandis, the same holds for the moon. Having discarded the interpretation of the sun wheel as the ecliptic, the second option remains: the celestial wheels of the sun and the moon will simply have to be identified with their daily orbits around the earth from east to west, just like the wheels of the stars. We may describe the movements of the sun and the moon (and the planets as well) without making use of the concept of the ecliptic, as follows: sun and moon turn around the earth in the same direction as the stars, albeit somewhat slower, which makes them stand regularly in another constellation. This is also the way in which Dreyer describes the movements of the celestial bodies: "the planets (. . .), as the Ionians and other early philosophers thought, travel from east to west, only somewhat more slowly than the fixed stars" (1953: 39).

"Planets" here also refer to the sun and the moon, as will be explained in Chap. 13. It is not sure whether Anaximander already discerned what we call the planets from the stars. In the same sense, West writes: "Without qualification, however, 'lying aslant' naturally refers to the more obvious fact that the circles lie oblique to the plane of the earth's surface (...). For the sun's circle in his system does not correspond to the ecliptic but to the path of its daily revolution" (1971: 85).

This way of explaining the movements of the planets has been called the Ionian theory, which was replaced by the theory of the contrary movement, which some authors call the Pythagorean theory (see Burkert 1972: 333). This is what Kahn, following a suggestion of Howard Stein, calls "the weaker interpretation (...) which does not ascribe so much theoretical insight to our Milesian astronomer: Suppose that the circle or wheel of sun (or moon) is designed to explain not its annual (or monthly) path among the stars but simply its apparent daily motion. Then 'aslant' will mean that the circles of sun and moon lie *aslant the earth*, i.e., inclined to the plane of the horizon, just as the daily motion of the stars itself is 'tilted' with respect to the visible surface of the earth" (1970: 102).

The Greek words used for "lying aslant" are κείμενον λοξόν (DK 12A22). Kahn rightly remarks that the Greek word λοξός is usually used for the ecliptic and not for the inclination of the heavens (see also Chap. 5). His conclusion is that this favors what he calls "the stronger interpretation" (1970: 102). The obvious answer, however, is that here the same anachronism is at stake that was already discussed in Chap. 5: Aëtius, talking about the *circle* of the sun, automatically associated it with the ecliptic and used the word that belongs to that circle. Moreover, he says that the circle of the moon lies aslant, *just like* that of the sun (DK 12A22). Strictly speaking, this can only be said of the daily circles of these celestial bodies, for the moon's monthly orbit is tilted as regards the ecliptic.

Pliny's report that Anaximander knew the obliquity of the ecliptic has to be considered, then, as incorrect (DK 12A5); (see also Panchenko 1999: 36–37). Nevertheless, Mugler and Krafft have tried to draw a picture of Anaximander's universe, apparently taking the sun wheel as the ecliptic (Mugler 1953: 14; Krafft 1971b: 297). In addition to the fact that these attempts are anachronistic, they both make the same two mistakes. The first is that they draw the tube-like celestial wheels that we discussed in Chap. 8. The second, and more important, is that they make the angle between the ecliptic (the sun wheel in their conception) and the moon wheel more (and in the Krafft's drawing even much more) than the 5° it should be. Mugler's drawing, here reproduced as Fig. 10.2, is especially disturbing, as the arrows on the rings of sun and moon point into the same direction as the arrow of the stars. If he really wanted to take the sun wheel as the ecliptic and the moon wheel as the monthly path of the moon, then the arrows of their wheels in Fig. 10.2 must point into the other direction, opposite to that of the stars. Moreover, the ring of the sun in his drawing lies in the same plane as that of the stars, which has to be at an angle of 23.5° if it were meant as the ecliptic. And if he intended to draw the daily path of the sun and the moon, as the arrows and the plane of the sun ring seem to indicate, then the moon ring has to be not at an angle but parallel to the sun ring, as will be explained hereafter. In short, Mugler's rendition of

Anaximander's universe is completely misleading. Krafft's version gives less controversial information for its interpretation but is also mistaken in much the same way as Mugler's. A typical difference between these two representations is that Mugler draws the earth tilted, as if here were drawing Democritus' alleged tilt of the earth, whereas Krafft draws the earth in the horizontal plane. Moreover, Mugler lets the tilting of the flat earth in relation to the celestial axis be 23.5°, whereas it should be 51.5°, as explained in Chap. 6 (see Fig. 5.6). Both Mugler and Krafft draw the totality of the star wheels as a sphere. This solution is discussed below.

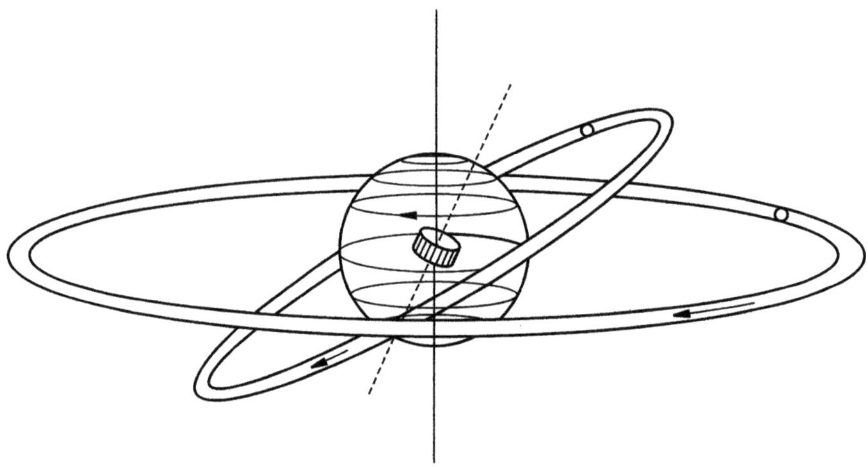

Fig. 10.2 Author's copy of Mugler's drawing of Anaximander's universe (1953: 14)

The sun and the moon do not always stand at the same height in the sky. The height of the sun, for instance, varies through the seasons. Nowadays, we are used to describe this phenomenon with the help of the ecliptic, but in Anaximander's world picture, it has to be described otherwise: additionally to their movement from east to west, which is slower than that of the stars (and that of the moon slower than that of the sun), the sun and the moon have a movement of their own from south to north and vice versa. This means that their wheels, turning around the celestial axis, shove also up and down parallel to the celestial axis. The same idea has already been worded by Van der Waerden: "Instead of resolving the motion of the sun in a diurnal motion, along with the fixed stars, and an annual motion, in opposite direction, along the ecliptic, it is resolved into two components, one parallel and the other perpendicular to the equator (1974: 11)."[1]

[1] Sartorius has tried to render the same idea of a double movement of the sun wheel without using the concept of the ecliptic, by drawing two circles: one at the celestial equator for the daily movement, and a smaller one, a kind of epicycle, perpendicularly attached to it, for its yearly movement (1883: 215–223; see also Heath 1913: 33–35). This idea is discussed extensively in Couprie (1995: 166–168); see also Panchenko (1999: 39).

The ancients called these second movements of the sun and the moon their "turnings" or "tropics" (τϱοπαί ἡλίου καί σελήνης). These tropics make the sun stand higher in the sky in summer than in winter. The distance between two turnings of the sun is two times the inclination of the ecliptic, although Anaximander would not have called it that way. For the moon, the same holds, mutatis mutandis, although its north–south movement goes to both sides 5° further than that of the sun. (We would express this by saying that the plane of the moon makes an angle of 5° with the plane of the ecliptic).

As we saw in Chap. 5, the celestial axis is inclined with regard to the surface of the earth, which is flat according to Anaximander. This means that the celestial wheels, visualizations of the daily orbits of the celestial bodies, are perpendicular to the celestial axis and thus make an angle with the surface of the earth that is complementary to the inclination of the celestial axis. Accordingly, as this inclination is about 38° at Miletus (or at Delphi), the celestial wheels make an angle of about 52° with the plane of the earth. As a consequence of the tilted position of the celestial wheels, the sun and the moon and the noncircumpolar stars have their daily risings and settings. The "lying aslant" of the wheels of the celestial bodies is the above-mentioned κείμενον λοξόν of the doxography.

It is time to elucidate this conception with the help of some drawings. In Fig. 10.3, we see the sun wheel, turning around the celestial axis, which is tilted with regard to the surface of the flat and drum-shaped earth.

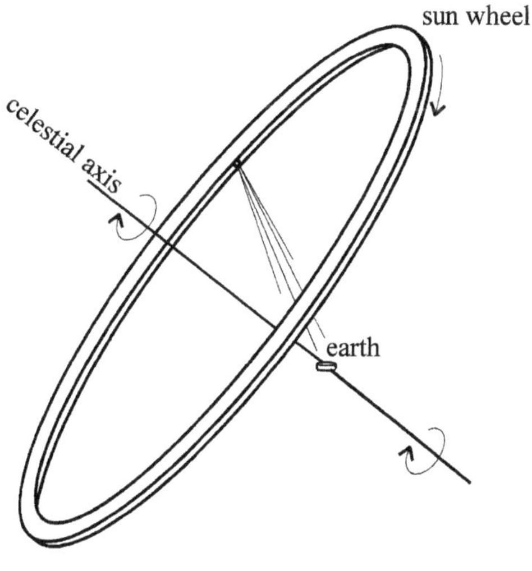

Fig. 10.3 The sun wheel

We may observe that the sun in its daily orbit describes a bigger curve in summer than in winter. During the summer, the sun rises farther to the north and also sets farther to the north than in the other seasons. During the spring and autumn equinoxes, the sun rises exactly in the east and sets in the west. During the winter, the sun rises and sets farther to the south. On a flat earth, this looks like as depicted in Fig. 10.4 (see also Fig. 2.2).

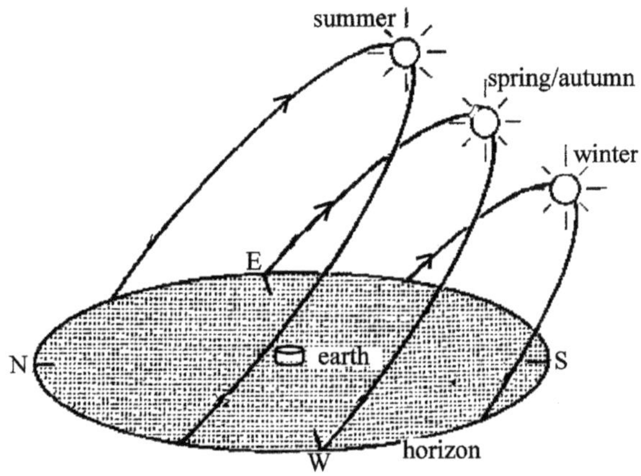

Fig. 10.4 The daily track of the sun during the seasons as visible above the horizon

When we draw the full circles of the daily track of the sun, which Anaximander would call the sun wheel, it looks like Fig. 10.5.[2]

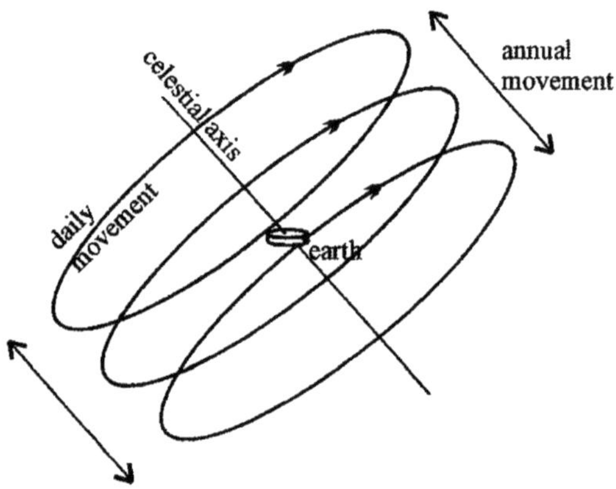

Fig. 10.5 The full daily orbits of the sun during the seasons

Now, we can draw a three-dimensional picture of Anaximander's universe. In Fig. 10.6, the full orbits of the sun and the moon are drawn as Anaximander's celestial wheels. Depicted is a situation in summer: the sun stands high, which means that the sun wheel has slid upward along the celestial axis, and it is day: the aperture in the sun wheel shines from above the earth. The moon wheel happens to be at the other end of its up-and-down movement along the celestial axis. The moon

[2] A similar drawing, but much more beautiful, can be found in Cellarius, albeit with the sun as a globe instead of a wheel, which results in a spiral (2006: plate 17).

(the aperture in the moon wheel) is not visible on the surface where we are thought to live, for it shines from beneath. The totality of the star wheels, which are according to Anaximander nearer to the earth than sun and moon, is rendered as a sphere, which seems to be the best solution, given the lack of information on this point in the doxography.

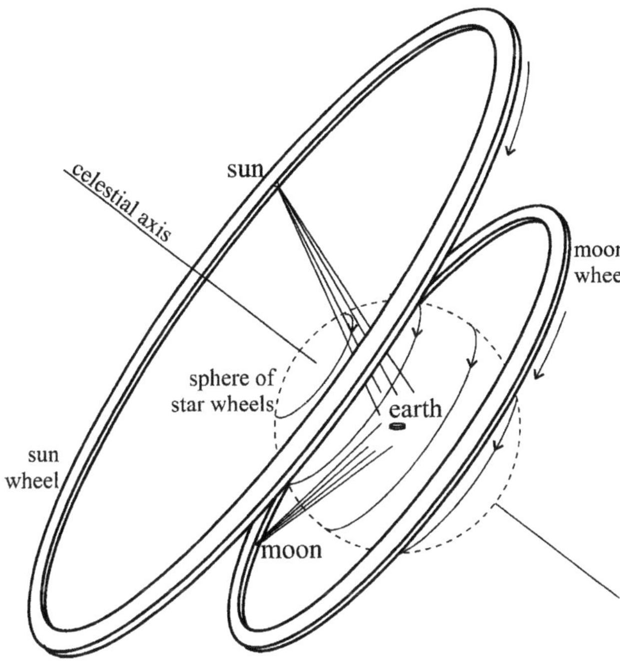

Fig. 10.6 Anaximander's universe on a summer morning

In Fig. 10.7, it is winter (the sun stands low in the sky because the sun wheel has slid downward along the celestial axis), and it is night (the sun shines at the lower side of the drum-shaped earth). The moon wheel happens to be at the upper side of its movement along the celestial axis.

In former publications, I used to elucidate the up-and-down movements of the wheels of sun and moon by drawing an imaginary or virtual cylinder as high as twice the inclination of the ecliptic for the sun wheel, and another imaginary cylinder for the moon. Unfortunately, this has caused a misunderstanding, as though I meant that these imaginary or virtual cylinders should be identified with the celestial wheels (see, e.g., Graham 2006: 7 n. 17; Hendrix 2004: 6; Hahn 2001: 216 and the drawing on 217). This is why I leave out such drawings here. Maybe the misunderstanding is caused by a former suggestion of Tannery, that the height of the sun wheel, as seen from the earth, is twice the inclination of the ecliptic, whereas the aperture in the sun wheel moves over the surface of the wheel during the year so as to cause the seasons (and the same idea

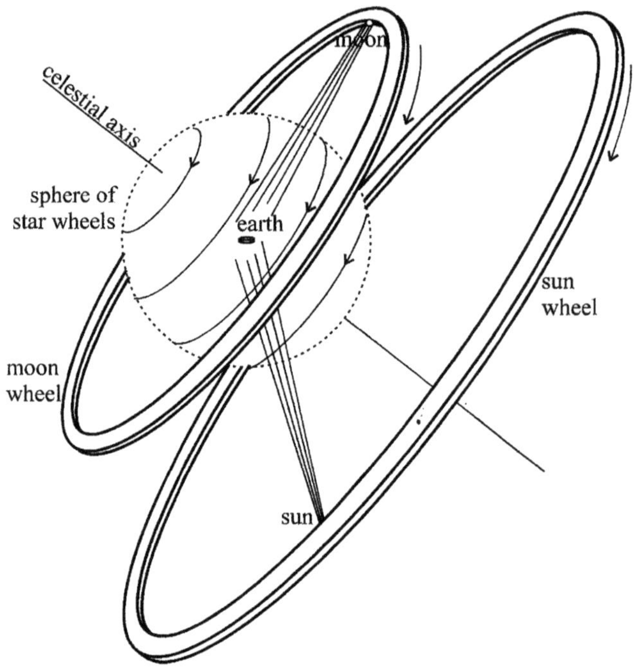

Fig. 10.7 Anaximander's universe on a night in winter

mutatis mutandis for the moon wheel) (1887: 97). However, as Heath rightly remarks: "there is nothing in the texts to support this" (1913: 31); see also (Couprie 2003: 251 n. 195).[3]

[3] Unfortunately, in Hahn's drawings of Anaximander's cosmos in perspective that are ascribed to me, another two flaws have crept in as well: next to the earth it reads "virtual earth," which has to be simply "earth," and the line marked "57°" has to be somewhat longer, as it indicates the height of the virtual cylinder that is supposed to render the up-and-down movements of the moon wheel (Hahn 2001: 217 and 218, Figs. 4.20 and 4.21). The same mistakes, unfortunately, in Hahn (2010: 29, Fig. 1.16, right). However, in Fig. 3.6, right, on p. 74, the wrong "virtual earth" at least has been replaced by "cylindrical earth."

Chapter 11
Bellows or Lightning? A Curious Terminology Explained

Anaximander's image of celestial wheels is in itself clear: it visualizes the circular orbits of the celestial bodies, and it explains why these bodies turn in circles around the earth, as well as why they do not fall on earth, as we have seen in Chap. 8. In the same chapter, we saw that the openings in the celestial wheels, through which we see the fire inside, are designated by the words στόμιον ("mouth-like opening"), πόρος ("opening through which something can pass," "way out"), and ἐκπνοή ("outbreathing"). These fire-breathing, mouth-like openings are what we see as the celestial bodies. The combination of these two images (wheels and mouths) in itself is already rather surprising and, one might say, daring. All the same, next to the image of a mouth breathing out the fire that is inside the celestial wheel, in the doxography on Anaximander, yet another image is used twice that apparently has the intention to explain the same phenomenon of how we see the light of the celestial bodies. This image has aroused much discussion and, as I show, much confusion as well. It concerns a rather technical question, namely, the translation of a curious expression: πρηστῆρος αὐλός. In this chapter, I explain why its usual translation is wrong, and defend another translation. My argument is that it does not so much concern an image to elucidate how we see the light of the celestial bodies, as a characterization of this light itself on the analogy of a meteorological phenomenon. At the end of this chapter, it will also be possible to elucidate why Anaximander could have chosen the image of outbreathing mouth-like openings.

According to Aëtius, Anaximander says that we see the light of the sun ὥσπερ διὰ πρηστῆρος αὐλοῦ, "as through the αὐλός of a πρηστήρ," and that of the moon οἷον πρηστῆρος αὐλόν, "as the αὐλός of a πρηστήρ" (DK 12A21 and DK 12A22, cf. DK 12B4). Presumably, the same kind of mechanism accounts for the light of the stars as well. The question is, of course, what is the meaning of the words πρηστήρ and αὐλός in this context, and consequently, what is the meaning of the quoted expressions. The first clause usually is translated thus: "as through the nozzle (or the mouthpiece) of a bellows," and the second "as the nozzle of a bellows." This translation goes back to a suggestion of Hermann Diels in *his Doxographi Graeci*: "immo πρηστήρ est *follis fabrorum*" ("a πρηστήρ is the bellows of blacksmiths"). Elsewhere, he translates πρηστῆρος αὐλός as "das Mundstück eines Blasebalges" (1879: 26–27 and 1897: 229). In Diels' *Die Fragmente der Vorsokratiker* his translation is more or less tucked away. In the second

D.L. Couprie, *Heaven and Earth in Ancient Greek Cosmology*, Astrophysics and Space Science Library 1, DOI 10.1007/978-1-4419-8116-5_11,

edition, in a *Nachtrag zum ganzen Werk*, after having stipulated that the word πρηστῆρ should be printed wide (1910, *Nachtrag VI*: "πρηστῆρος zu sperren als Wort des Anaximandros"), indicating that he considers it as authentic, he says that such a "Blasebalgröhre" is mentioned by Hippocrates. We will discuss this *locus* further on in this chapter. In later editions of *Die Fragmente der Vorsokratiker*, this reference appears in a footnote at DK 12A21. And in the *Wortindex* of the 1910 and 1922 editions, made by Walther Kranz, in the lemma πρηστήρ the translation "Blasebalg" is listed.

However, the image of the nozzle of a bellows, somehow connected to a celestial wheel, tends to complicate rather than elucidate the meaning of the text. Why should there, in addition to the image of a mouth breathing out fire, be need of another image, that of the nozzle of a bellows? The combination of the image of a bellows and its nozzle with the images of wheels and mouths makes everything unclear. Actually, the idea of looking again at the translation of πρηστῆρος αὐλός came to me during a visit Robert Hahn paid me in May 2000. He had gathered all kinds of pictures of ancient Greek bellows, such as can be seen now in Hahn (2010, Figs. 4.1–4.14). When we discussed these pictures, it struck me that the whole idea of celestial bellows is awkward. If we were to understand that every celestial body has such a contraption, the result would be hundreds of nozzles, extending from the celestial wheels toward the earth. And all this should have the intention of explaining how we see the light of the celestial bodies! Moreover, a regular bellows is supposed to blow air into the fire and not the other way round – fire into the air – as must be the case with the would-be celestial bellows.

Since Homer, the usual word for bellows is not πρηστῆρ, but φῦσα. The only evidence Diels produces in *Doxographi Graeci* for his translation of πρηστῆρ as "bellows" is one *locus* in Apollonius of Rhodes' *Argonautica*.[1] This case, however, is not very strong. It reads as follows: "In the second place, she (sc. Hera) went to Hephaestus and caused him immediately to stay his iron hammers: the sooty πρηστῆρες withheld their breath" (*Argonautica*, lines 775–777). In his annotations of this text, another German scholar, Hermann Fränkel, denies that the meaning of πρηστῆρες here is "bellows": "Nicht der 'Blasebalg' (...), sondern der 'Glutwind' der oben aus dem Felsen aufstieg" (1961: 532 note at line 777; see also Livrea 1973: 223 note at line 777; Vian 1981: 105, translates: "souffles"). The background of Fränkel's argument, although he does not say this in so many words, is that the word πρηστῆρ is never used for the instrument or contraption from which something blows, but always for the blown-out stream itself, as we will see in due time. If Fränkel is right, the only evidence for Diels' translation collapses.

But let us play the devil's advocate and suppose that Diels was right and that πρηστήρ in this text of Apollonius means "bellows." In that case, Apollonius uses the word πρηστήρ within a context – Hephaestus' forge – which would have made it evident to every Greek at the time that "bellows" could be at stake. This would have given Apollonius the opportunity to exaggerate to stress that his story is not

[1] The same *locus* is mentioned by Neuhaeuser (1883: 367).

about an ordinary forge with normal bellows (φῦσα), but about the workshop of a god with its huge and impressive bellows (πρηστήρ), emitting a thunderstorm's blast. On the contrary, in the context of Anaximander's description of the universe, the context is definitely not that of a forge, and therefore it is not immediately evident that bellows, let alone extraordinarily huge bellows, would play any role in the celestial mechanics. *If* Anaximander had meant to compare the light of the heavenly bodies with nozzles of bellows, then he would have used the ordinary word φῦσα, and not the word πρηστήρ that every Greek at the time would have understood in this context of heavenly phenomena to indicate a violent weather event, as we will see further on. Hahn has recently argued that "Couprie's argument has it exactly backwards.(...). If Appollonius is 'exaggerating' to highlight the more formidable forge of the god over and against the blacksmith's to show the enormity of its power, Anaximander's metaphor is more powerful still, the cosmic fire itself. Had he used *phusa*, only an ordinary forge would be conjured" (2010: 92). Hahn's argument presupposes what has to be proven. In Apollonius' story, the context is that of a forge. This context is completely absent in the case of Anaximander's celestial wheels. His readers could only have understood πρηστήρ as "bellows," if they were already acquainted with this meaning, which is not the case.

Strangely enough, in *Die Fragmente der Vorsokratiker* Diels does not mention Apollonius of Rhodes any longer, but instead he refers to two *loci* in Hippocrates' *De articiculatione*, viz. sections 47 and 77 (ed. Kuehlewein 1902: 181 and 235. See Diels 1910, *Nachtrag VI*; see also Diels 1922: note at 2.21, and note at DK 12A21). In these texts, Hippocrates describes the use of a device made from a leather wineskin (ἀσκός) with a brass pipe (αὐλός ἐκ χαλκείου) connected to it, which is used in an unsuccessful experiment to cure a hump on the spine (section 47) as well as in the (also not very effective) treatment of dislocation of the hip joint (section 77). The idea is that the empty wineskin is placed under the patient and is gradually filled with air blown into it through a pipe attached to one of the loose feet of the skin. There is, however, a significant difference between Hippocrates' wineskin and Anaximander's alleged bellows. In Hippocrates' apparatus, the air has to be blown through the pipe *into* the wineskin (somehow like the blowing up of a child's balloon), whereas the important characteristic of the alleged celestial bellows is that it is supposed to blow the fire *out* through the pipe. More important is that Hippocrates does not use the word πρηστήρ. And the word φῦσα that appears in both texts is not used here in the meaning of bellows but indicates the wind or air blown into the wineskin, and not the sack itself. For these reasons, it is hard to see how these texts, in which the very word πρηστήρ does not even appear, could be put forward as an evidence for the translation of πρηστήρ as bellows.

Apparently, Diels was so convinced that his suggestion was right, that he even changed the reading of another source: where the manuscripts render Hippolytus' text as τόπους τινὰς ἀερώδεις ("some air-like places"), Diels reads πόρους τινὰς αὐλώδεις ("some pipe-shaped openings") (note at DK 12A11(4) and Diels 1879: 560 n.). Here obviously the wish was father to the thought. Conche remarks rightly: "la correction de Diels (...) est inutile" (1991: 192). There is no need for tube-like gadgets to understand this text. What Hippolytus presumably wanted to say is that the inner fire of the celestial wheels is emitted from their envelopes of air at certain points. After Diels' death, Kranz apparently became suspicious of Diels' suggestion, since in the fifth and later editions of the *Vorsokratiker* which he edited, πρηστῆρος αὐλός is translated as "Glutwindröhre" and not as "Blasebalgröhre" (DK 12B4).

Perhaps the source of Diels' misunderstanding is Achilles Tatius, who in the same context speaks of a trumpet. The text reads: "(...)[2] as in the case of a trumpet (σάλπιγξ), it (viz. the sun) emits the light through a narrow hole like πρηστῆρας" (DK 12A21). The word σάλπιγξ (a kind of trumpet) here is obviously an alternative for αὐλός that may not only mean "pipe" or "tube" but also "flute." Now, Achilles Tatius is not a very reliable ally, for in the same text he is confused about Anaximander's image of the celestial wheels. Kahn and Guthrie omit him from their considerations because his text is "a mere distortion" of Aëtius' and "an unintelligently garbled version of what is described more clearly by Aëtius" (1994: 25; Guthrie 1962: 93 n. 1). But even here, there is no need to think that Achilles Tatius meant pipes emerging from celestial wheels. All he seems to be saying is that the celestial wheels have apertures just like a trumpet, through which the light (τὸ φῶς) flows in the case of the celestial wheels, and air in the case of a trumpet. It is important to notice that the word πρηστήρ here clearly means the stream that is blown out of the narrow opening (ἐκ κοίλου τόπου καὶ στενοῦ) in the celestial wheel, and not something out of which the stream is blown. Conche translates: "(...) comme par une trompette, le soleil, d'un lieu creux et étroit, renvoie la lumière *comme des souffles émanées*" (1991: 197 n. 13, my italics). And Dumont: "(...) qu'il envoie la lumière *comme un souffle* qui jaillirait de la cavité étroite d'une trompette." It is, again, forcing the text in conformity with his own preconceived idea, when Diels proposes the emendation: "(ἀπὸ) πρηστῆρος" instead of πρηστῆρας (note at DK 12A21).

Sometimes even stranger translations of πρηστῆρος αὐλός have been tried. Mansfeld's overtly anachronistic translation, for which he does not offer any explanation, is "Lötrohr," which means something like "soldering-pipe" (1987: 77). A similar suggestion has been made already by Heath: "the idea [is] that the stars are like gas-jets, as it were" (1913: 31). The idea is also repeated by Guthrie: "there is a hole, through which the fire streams through a leak in its pipe" (1962: 94). It is hardly believable, though, that Anaximander had in mind gas jets and leaking gas pipes, or soldering pipes that were nonexistent in his time. Kratzert, following Riedel, sees in the word αὐλός a reminiscence of the Dionysian cult because a surname of Dionysus is φλοιός, another word used in Anaximander's cosmogony (Kratzert 1998: 39; Riedel 1987: 11). It is hard to see, however, what the flutes that accompanied the dithyrambs of a Dionysian trance should have to do with the way the light of the celestial bodies reaches us. Another translation of ὥσπερ διὰ πρηστῆρος αὐλοῦ: "as through the funnel of a tornado," has been

[2] The text between brackets says "Others say." Here Diels rightly remarks: "vielmehr derselbe A (naximandros)."

proposed by Hall, and earlier by Teichmüller (Hall 1969: 57–59; Teichmüller 1874: 13 n.). Neuhaeuser discusses Teichmüller's suggestion but rejects it (1883: 365–367). The translation "funnel of a tornado" presents difficulties similar to those encountered with the "bellows" translation: the funnel-shaped cloud of a tornado is difficult to relate to fire or light. Plass points to the incidental cases of lightning within a funnel of a tornado, but although his contribution is perhaps instructive, it is not very convincing (1972: 179–180). Moreover, the image of hundreds of these funnels emerging from the holes in the celestial wheels is decidedly odd. However, this last translation has at least the advantage of employing a meteorological image for a celestial phenomenon.

Let us investigate somewhat closer into the meaning of the word πρηστήρ. Etymologically speaking, a πρηστήρ is a "blower" or a "burner." Elsewhere in the doxography on Anaximander, and also in the doxography on Heraclitus, Anaxagoras, Democritus, and Metrodorus of Chios, πρηστήρ is a meteorological phenomenon (DK 22A14, DK 59A84, DK 68A93, DK 70A15).[3] Aristotle treats the πρηστήρ in relation with thunder and lightning (and whirlwinds): "when a wind that is drawn down (which is when it has become weaker) catches fire, it is called a πρηστήρ" (*Meteorologica* 369a11 and 371a16). According to Metrodorus of Chios, when a bolt of lightning is weakening it becomes a πρηστήρ (DK 70A15). And more important, the doxography says explicitly that for Anaximander himself the πρηστήρ is a meteorological phenomenon, originating from wind (πνεῦμα), just like thunder, lightning, and windstorm or whirlwind (DK 12A23). The meaning of πρηστήρ can perhaps be rendered best by something like "scorching wind" and always seems to be associated in some way with fire. As Gilbert expresses it: "es ist der als (....) Wind sich fühlbar machende Glutodem des Feuers" (1967: 454 n. 1). And as a meteorological phenomenon it is associated with lightning. As Gilbert writes: "Der πρηστήρ tritt uns zuerst bei Hesiod entgegen und erscheint hier in durchaus charakteristischer Weise als ein Gluthauch des brennenden Feuers, welches namentlich im Gewitter sich fühlbar und sichtbar macht" (1967: 454, ref. to Hesiod, *Theogony* 844 ff.). Elsewhere, Gilbert almost identifies πρηστήρ with lightning (κεραυνός): "der πρηστήρ also wesentlich nicht verschieden vom κεραυνός, nur geringeren Feuergehaltes und danach auch geringerer Wirkung" (1967: 625 n. 1, see also 453–455 and 619–631). Here, again, it is important to remark that a πρηστήρ is always the emitted storm or glow itself, but never indicates an instrument, contraption, device, or mechanism whatsoever, from which it is blown.

Classen notes that the Arabs (tenth century A.D.) translated the phrase ὥσπερ διὰ πρηστῆρος αὐλοῦ as "like the lightning," or more fully: "the fire comes out of them through a mouth-piece (in the same way) as the flashes of lightning appear" (Daiber 1980: 155, my translation; Classen 1986: 87 n. 114). My guess is that the Arabs were on the right track. Lloyd has given a casual hint that points in the same direction: "It seems better to retain a meteorological sense for πρηστήρ,

[3] The meaning of πρηστήρ in Heraclitus, DK 22B31, is uncertain (DK translate "Gluthauch").

then, e.g. (...) lightning" (1966: 314 n. 1, see also 313 n. 1). So let us see what the doxography has to tell about Anaximander's opinion on the phenomenon of lightning. Hippolytus reports that according to Anaximander bolts of lightning (ἀστραπαί) generate when an invading wind (ἄνεμος) breaks up the clouds [DK 12A11(7)]. More generally, Aëtius reports that, according to Anaximander, not only πρηστήρ but also other meteorological phenomena like βροντή (thunder), ἀστραπή (bolt of lightning), κεραυνός (lightning), and τυφών (whirlwind), result from wind that is enclosed within a thick cloud (νέφει παχεῖ), from where it escapes. The resulting rupture, by contrast with the dark cloud, looks like a flash of light (διαυγασμός). [DK 12A23, cf. DK 12A11(7)].[4] Mark that the word αὐγή may be used both of a beam of lightning and a sunbeam.

And now, let us compare this with what the reports say about Anaximander's opinion on the light of the celestial bodies. Anaximander's celestial wheels are made of compressed air that hides the fire within (πιλήματα ἀέρος τροχοειδῆ, πυρὸς ἔμπλεα) (DK 12A18). The fire of the sun shines permanently through the hole in the sun's wheel. The resemblance with the above-mentioned wind, enclosed in a thick cloud that generates a πρηστήρ, is obvious. "It is impossible to mistake the parallel between this meteorological fire and that of the celestial rings," Kahn rightly remarks (1994: 102). Maybe some commentators have been led on the wrong track, thinking that the only translations of the word αὐλός are "pipe," "tube," or "flute." The meaning "stream," "jet," or "squirt," however, that fits better to meteorological phenomena is attested in Homer, who somewhere uses it in the sense of a jet, squirt, or stream (of blood): "and immediately a thick stream (αὐλός) of human blood gushed from his nostrils" (*Odyssey* XXII, 18). Remarkably, also the word πρηστήρ can be used in a similar context meaning "squirt" or "stream": "and from his eyes two streams (πρηστῆρε) of blood floated down" (Euripides, fragment 384).

The comparison Anaximander must have been thinking of is obvious: whereas an ordinary flash of lightning is a *momentary* flash of fire appearing in a rupture in a thick airy substance (a cloud), the light of the sun (and of the other celestial bodies as well) is like a *permanent* jet, or stream (αὐλός) of lightning fire (πρηστήρ) emanating from a hole in a compressed airy substance (a celestial wheel). Tannery already made the same suggestion: "un astre est donc comme un éclair qui durerait toujours" (1887: 92). Perhaps he borrows this interpretation from Teichmüller, who writes "dass die Erzeugung der Gestirnflammen nach der Analogie des Blitzes von Anaximander erklärt wurde" (1874: 31). Elsewhere, Teichmüller yet prefers the already quoted analogy with a whirlwind, as he translates ὥσπερ διὰ πρηστῆρος αὐλοῦ with "wie durch einem aus einen Blasinstrumente herausfahrenden

[4] Seneca's testimony of Anaximander on lightning (DK 12A23) is difficult to understand: "fulguratio" is a violent movement of the air tearing apart and imploding, which unveils a lazy (?) fire, incapable of escaping ("languidum ignem nec exiturum aperiens"). A "fulmen," on the other hand, is the course of a stronger and tighter wind ("spiritus"). According to Bicknell, Seneca's account is more trustworthy than that of Aëtius (1968: 184). But neither he nor anyone else, as far as I know, can make sense of it.

Wirbelwind" (1874: 14). Burnet sees the same parallel: "lightning is explained in much the same way as the heavenly bodies. It, too, was fire breaking through condensed air, in this case storm clouds. It seems probable that this was really the origin of the theory, and that Anaximander explained the heavenly bodies on the analogy of lightning" (1930: 68). Nevertheless, and rather surprisingly, without any further explanation, Burnet accepts the translation "as through the nozzle of a pair of bellows."

The expression οἷον πρηστῆρος αὐλόν that is used to describe in DK 12A22 the light of the moon seems to convey this idea of an analogy with the phenomenon of lightning more clearly than the words ὥσπερ διὰ πρηστῆρος αὐλοῦ that are used to explain the light of the sun in DK12A21. The translation of οἷον πρηστῆρος αὐλόν by "like a stream of lightning," evokes the image of a beam of light thrown through the hole of the moon wheel. This expression easily goes with what was stressed already several times, namely that a πρηστήρ is always that what streams but never that out of which the stream flows. My suggestion is that Aëtius no longer fully understood what Anaximander had meant with οἷον πρηστῆρος αὐλόν and, perhaps thinking that Anaximander was speaking about some sort of pipe, squirting fire, tried to elucidate his words by means of the circumscription ὥσπερ διὰ πρηστῆρος αὐλοῦ. In other words, I do not think that οἷον πρηστῆρος αὐλόν is a kind of shortened version of ὥσπερ διὰ πρηστῆρος αὐλοῦ, as most commentators seem to think, but that the first phrase expresses Anaximander's thought better than the second one.

Now, we are also able to understand why Anaximander used and even needed the image of a mouth, breathing out the fire that is inside the celestial wheel. Once we have understood that we see this fire like a stream or squirt of lightning fire (οἷον πρηστῆρος αὐλόν), we may readily conceive of the apertures in the celestial wheels as a kind of fire-spitting mouths. Perhaps the combination of such heterogeneous things as a (celestial) wheel and a mouth-like opening looks less strange when we read that according to Workman, the words ἐκπνοή ("outbreathing") and στόμιον ("mouth-opening") are technical terms of bronze founding, indicating the airholes in the casting mold of a bronze foundry (1953: 46). Here, too, two heterogeneous things (casting mould and mouth) are combined to describe how a stream of hot air escapes from an object. Moreover, a mouth is easily conceived of as opening and closing, and this is the way in which Anaximander is said to have explained both solar and lunar eclipses, and the waning and waxing of the moon in DK 12A11(4), DK 12A21, DK12A22, and DK 12A11(5).

As regards the translation of the words πρηστῆρος αὐλός, we may say that generations of scholars have been misled by the imagination of the great Diels. The words πρηστῆρος αὐλός are not meant as an image (viz., of a bellows and its nozzle), but as an attempt to explain the light of the celestial bodies on the analogy of the phenomenon of lightning.[5]

[5] The translation defended in this chapter was recently accepted in Graham 2010: 59 and 68.

Chapter 12
Critique of an Alleged Cosmic Architecture

According to Anaximander, the earth has the shape of a cylinder and looks like a column drum, the height of which is one third its width. This datum has induced Robert Hahn to outline in several studies what he sees as the importance of the contemporary temple architecture for Anaximander's cosmology. In the introduction of his *Anaximander and the Architects*, he writes: "Anaximander's conception of the shape and size of the earth as a 3×1 column-drum is the point of departure in chapter 2" (2001: 5). That chapter is titled *The Ionian Philosophers and Architects*. And in his latest book, *Archeology and the Origins of Philosophy*, the starting point of Hahn's studies is again Anaximander's identification of the earth with a column drum, which "was no throwaway at all, as scholars must have assumed by their silence, but rather the so-called tip of the iceberg" (2010: 15). Hahn argues that Anaximander, just like his contemporaries, must have been impressed by the first big temples with stone columns that were built, as he expresses it, "in his backyard." The architects, such as Theodorus, Rhoikos, Chersiphron, and Metagenes, and the early philosophers, such as Thales and Anaximander, were part of the intellectual elite of their time, and they certainly will have taken cognizance of their mutual works. Moreover, they were all occupied with what Hahn calls "applied geometry": the architects in designing their temples based on certain measuring units or modules, Thales in measuring the height of a pyramid or the distance of a ship at sea, and Anaximander in sketching a συγγραφή (a description in words and numbers) of the universe.

What Hahn tries to show is rather modestly formulated. He repeatedly assures us that "Anaximander did not simply copy the architects," "but rather that his philosophical imagination drew upon architectural techniques," and he emphasizes that "the argument is not that Anaximander copied the architects" (2001: 10, 148). What he intends to do is to place Anaximander within the historical and cultural context of his time and show that he discussed with and learned from the architects that built the impressive Ionian temples. Hahn was not the first to point to architectural influences on Anaximander's cosmology. In 1986, Lebedev published an article called *Der geometrische Stil und die Kosmologie Anaximanders*, the third section of which is called *Architektonische Metaphorik* (1986: 41–43). In the previous section, he already wrote: "Der Kosmos Anaximanders ist ... ein architektonisches Gebilde" (1986: 40).

D.L. Couprie, *Heaven and Earth in Ancient Greek Cosmology*, Astrophysics and Space Science Library 1, DOI 10.1007/978-1-4419-8116-5_12,

We may readily admit that the column drum is an architectural building block, and even that it was an architectural novelty in Anaximander's days. Comparing the earth with a column drum, then, was using an architectural image. Hahn's most important contribution to the study of Anaximander is, according to me, that he has extensively shown how a column drum, by means of a technique called ἀναθύρωσις, was made to fit exactly to other drums (see, e.g., 2001: 150 ff.; 2010: 42). This technique consists of carving smooth the edge of a drum surface, whereas the interior surface is chiselled somewhat concave. This yields a clear confirmation of the rightness of the translation "(somewhat) concave" of the word γυρόν (literally: "curved"), handed down in the doxography as an indication of Anaximander's drum-shaped earth [DK 12A11(3)].

In Chap. 9, I stressed that Anaximander's numbers may be looked upon as instructions for making a drawing, or more precise, a plan view of his universe, made with the help of compasses. Drawing a plan view, as for instance in making a map of the earth, is as such not necessarily an architectural technique, but the use of compasses was an architectural skill. One may say that in taking the diameter of the earth as a module for the distances of the celestial bodies Anaximander was using an architectural technique. And finally, the architects used simple ratios for the dimensions of their temples and so did Anaximander for the dimensions of his cosmos, although his ratios were different from theirs. Together, in a general sense, Hahn is certainly right in saying that Anaximander was indebted to the architects of his time.

However, as we will see, Hahn's claims eventually go much further and tend to be more specific than this, and this is where my question marks begin. Anaximander chose the ratio 3:1 (meaning: the diameter is three times the height) for his drum-shaped earth not accidentally, Hahn says: "The column-drums at the archaic Heraion, Artemision, and Didymaion were roughly 3 × 1" (2001: 188; Hahn uses sometimes "3:1," and sometimes "3 × 1," indicating the same dimensions). The examples, however, that he gives of archaic column drums have the ratios 3.4:1 and 3.9:1 (2001: 156 and 158). And elsewhere, he writes that column drums occurred with ratios varying from 3:1 to 5:1 (2001: 147; 2010: 5). Drums with a ratio bigger than 5:1 or smaller than 3:1, he says, obviously posed a building problem (2001: 147–148). These figures can hardly count as a support to Hahn's aforementioned thesis that the ratio of archaic column drums was about 3:1.

It is not very helpful that little is left from the temples of Anaximander's time. Moreover, in the literature on temple building, ancient as well as recent, much can be found on the lengths and diameters of columns, but relatively little about the heights of column drums. Nevertheless, the available data show quite other figures than the 3:1 ratio presented by Hahn. Two remaining column drums of the archaic Didymaion show ratios of 3.4:1 and 2:1 (see Schneider 1996: 78–83). The only remaining drum of the temple of Pherae measures approximately 2:1 (see Østby 1992: Abb.16.). Still upright standing columns of the temple of Apollo in Priëne (fourth century B.C.), the height of which is about 13.5 times the diameter of the lowest drum, count nine or eight drums, which results in an average ratio of about 1:1.5. The measures of the only remaining column of the Heraion IV (fourth century B.C.) also do not comply with Hahn's thesis. They differ from 1.6:1

(the lowermost drum) to 2.5:1 (the uppermost minus one), as is shown in Fig. 12.1. The upper part of this column is missing, but we may suppose that the ratios of the missing drums went up to 3.5:1. As the relatively flatter drums weighed less, they were probably used for the upper part of the column, where they had to be lifted up to 10 m and higher. The column drums brought forward by Hahn, then, are presumably from the upper part of a column, and not somewhere in the middle of the column, as Hahn suggests in his latest book (2010: 50, Fig. 2.9; besides, in this figure a pointer marked "3 × 1 Column Drum" points to a drum with the dimensions 2 × 1).

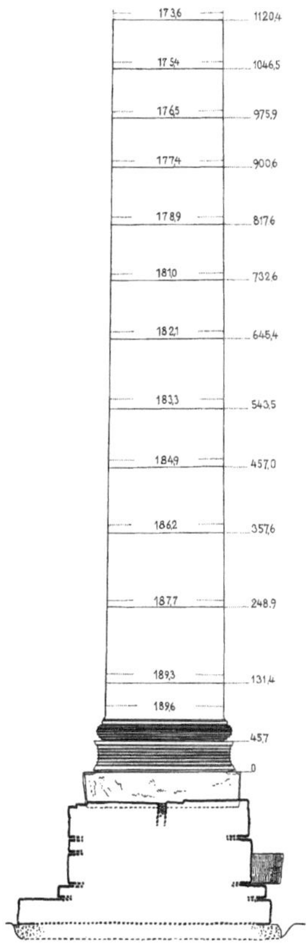

Fig. 12.1 Measures of column drums of the temple of Hera at Samos (Reuther 1957: 13)

Finally, it is known that according to the restoration plan for the naval arsenal at Piraeus, written by Philo of Eleusis (340 B.C.), the lower diameter of the columns measures 2 ft and 3 span, every column containing one drum of 5 ft high and six of 4 ft high. Taking a span to be approximately two-thirds of a foot, this results in ratios of about 1:1.25 and 1:1 (see Martin and Stierlin 1966: 42). It is strange that Hahn, who quotes from this plan (2001: 110), does not quote the above cited part on the columns and column drums. In virtue of this documentation,

Hahn's assertion that "the architects made use of column drums roughly 3 × 1 in size" (2007: section C) must be qualified as an inapt simplification. My conclusion is that Anaximander did not derive his ratio of 3:1 for the measures of the drum-shaped earth from the mean or most common column drum.

Hahn's next thesis is that Anaximander, in imitation of the architects, used the number 3 as a measuring unit or module for the measurements of his universe: "the earthly module of '3'," and again, "the module number, '3'" (2003: 89 and 146). To be precise, however, not the number 3 but the diameter of the earth is the module. There is some confusion here between the notion of the earth's diameter taken as a module and the earth's diameter being three times its height. The distances of the celestial bodies are measured as multiples of the diameter of the earth, namely, 9 earth diameters as the distance to the stars, 18 to the moon, and 27 to the sun. If the module were 3, the numbers for the distances of the celestial bodies would have been 27, 54, and 87 for the inner perimeters and 30, 57, and 90 for the outer perimeters.

Hahn quotes Vitruvius (*De architectone* III, 3.7) to support his view that the *diameter* of the column-drum-shaped earth is Anaximander's module. Lebedev, however, points to another text in Vitruvius as well, in which is indicated that the front of the Doric temple is 27 times the module, which is the lower *radius* of the column (Lebedev 1986: 41, Vitruvius, *De architectone* IV, 3.3–4). This text is, as far as I can see, overlooked by Hahn. Lebedev, who is obviously an adherent of what I called in Chap. 9 "the orthodox variant of the standard-interpretation," concludes that Anaximander used both "modules": "Bei Anaximander ist (a) der *Eddurchmesser Modul* für die Kosmoskonstruktion, [und] erreicht (b) der ganze Kosmosdurchmesser 27 *Erdhalbdurchmessern* (*Moduln*)" (1986: 41, my italics). However this may be, it illustrates once more the difficulties we met when we try to determine Anaximander's "module" as an architectural feature to construct a celestial temple.

Several times, Hahn offers a drawing of Anaximander's universe in plan view to elucidate his view (Hahn 2001: 189, Fig. 4.5 (4), 191, Fig. 4.6, and 218, Fig. 4.21 on top; Hahn 2003: 84, Fig. 2.4, 145, Fig. 2.22, and 147, Fig. 2.23; Hahn 2007; section C; and in the first edition of Hahn 2010: 29, Fig. 1.16 left, 17, Fig. 1.2, and 73, Fig. 3.5). The fact that Hahn uses this picture so many times obviously indicates that it is meant as an essential illustration of his ideas. There are slight differences between the several versions, but not as regards the issue to discuss here. The picture reproduced here as Fig. 12.2a is that of Hahn (2003: 145). In these pictures, Hahn visualizes both the earth's diameter as a celestial module and the formula 9 + 1 as the recurring measure between the celestial wheels. In his own words: "This is the pattern of organic growth, the *radii* of the wheels grow in size by increments of '3 × 3' or 9 units" (2003: 146). However, when we count the little circles to the right of the earth we have 2 × (9 + 1) = 20 earth diameters instead of 19 to the moon wheel and 3 × (9 + 1) = 30 earth diameters instead of 28 to the sun wheel. In other words, in order to get the right sum total of 19 and 28 there will have to be eight little circles instead of nine between the wheels of the stars and the moon, as well as between the wheels of the moon and the sun.

I have been informed that this strange error that for almost 10 years remained unnoticed will be rectified in the second edition of *Archaeology and the Origins of Philosophy* and replaced by Fig. 12.2b. In this drawing, we get nine little circles (modules) between the earth and the stars, but eight between the stars and the moon as well as between the moon and the sun. At first sight, then, this adjusted drawing no longer suits Hahn's idea that "the cosmic numbers appear as iterations of the 9 + 1 formula," also expressed as "the appearance of the formula 9 + 1 in Anaximander's map of the cosmos" (2003: 84). I guess, then, that textual adjustments in Hahn's book will be needed as well, but the revised text was not available when I finished the manuscript of my book so that I have to suspend my final judgment.

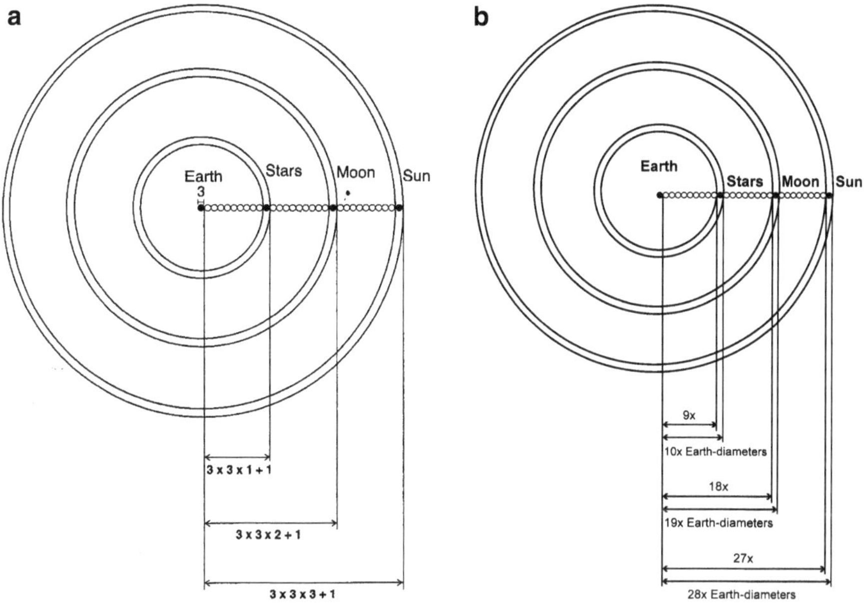

Fig. 12.2 (**a**) The former, miscalculated version (*left*, Hahn 2003: 145) and (**b**) the adjusted version of Anaximander's architectural formula for the cosmos according to Hahn (*right*, by the courtesy of Robert Hahn)

The first number of the series of cosmic distances, 9, also has an architectural basis, Hahn says, as it is related to the dimensions of the Ionian column: "the fundamental architectural module was the lower column diameter," and "the specific rule for column height (...) was nine or ten times the lower diameter" (2001: 147, repeated in 2007: section C). As there are no complete columns dating from that time left, this is a mere conjecture. In the literature quoted by Hahn himself, this relation is estimated by several authors at 6.4:1, 8:1, 8.2:1, 10:1, 12.7:1, and even 13.3:1 (2001: 145–146). Vitruvius mentions in his handbook a ratio of 9.5:1 for Ionic columns (*De architectone* III, 3.7; Hahn 2001: 146). The height of the Doric column, including the capital, is 14 times its half-diameter, according to Vitruvius (*De architectone* IV, 3.4). This measure is strange, for as Lebedev remarks, in reality the height of the column in Doric temples varies between 8½ and $9\frac{2}{3}$ times the lower diameter (1986: 41). However this may be, some authors admit that actually no conclusion can be drawn as to the length of the columns of archaic temples (Schaber 1982: 76; Gruben 1963: 153). One of the conclusions of a recent study is that looking for "round" ratios between diameter and length of a column does not correspond to the practice of the architectural design (De Jong 1994: thesis 6). Against all this evidence, Hahn's explicit phrase "Anaximander's module is *the very same one* that the architects in archaic Ionia identified when they set out to make their cosmic temples" cannot be maintained (2007: section C, my italics). In short, when Hahn appeals to a standard column with a ratio of, roughly, 9:1, it looks like pressing the data within the frame of a preconceived idea.

Hahn suggests that Anaximander chose the shape of the column drum for his earth because the temple column divides (or connects) heaven and earth, just like the celestial axis (2001: 188). The axis of the heavens has to be thought of, according to Hahn, as a big column at the center of the world, around which the firmament revolves and of which the earth is one of the drums (2010: 142 and 49–50, Fig. 2.9). A difficulty Hahn pays insufficient attention to is that the axis of the heavens is tilted, whereas a column is meant to stand right up. Elsewhere, Hahn takes much pain to show how column drums are made to fit exactly upon each other with the help of the technique of ἀναθύρωσις and a device called ἐμπόλιον, in order to make together a column (2001: 154–157; 2010: 42–44 and Fig. 2.5). The ἐμπόλιον is a kind of pin put into holes at the centers of two drums, assuring that they are placed exactly on top of one another. Here, one could object that the celestial axis is tilted in relation to the flat surface of Anaximander's earth. This tilting of the celestial axis, which is about 38° at Miletus, was a main subject of discussion among the Presocratics, as explained in Chap. 5. For Anaximander, this would have meant that his column-drum-shaped earth does not fit into the tilted celestial axis, thought of as a temple column.

Although they also can be seen as instructions for drawing a map of the universe, cosmologically speaking, Anaximander's numbers are best understood as the expression of his revolutionary insight that the celestial bodies are not all glued at the firmament, but at different distances from the earth so that these numbers can be taken to mean far ($1 \times 3 \times 3$), farther ($2 \times 3 \times 3$), farthest ($3 \times 3 \times 3$), as explained in Chap. 8. This fundamental cosmological understanding of Anaximander's numbers is independent of any possible architectural practices or influences.

Column drums make up a column, and columns make up a temple. According to Hahn, just like the architects who used simple proportions for their temples, Anaximander chose a simple proportion for the measurements of his cosmic temple. On closer inspection, however, the proportions used by the architects differ from those used by Anaximander. Hahn shows how the plan of temples is made according to a ratios of 1:2:4 (height:width:length) (2001: 78, Fig. 2.7). On the other hand, the ratios used by Anaximander are 1:2:3 (the distances of the celestial bodies to the earth, respectively 9, 18, and 27 earth diameters). Moreover, the architects compared different things (height, width, and length), whereas Anaximander compared similar things (viz., distances to the earth).

But Hahn goes even further. The columns of a temple, he says, have a cosmic meaning, as they connect heaven and earth (or divide heaven and earth, depending on how you look at it) (2001: 87, 88, and 188). From this, he concludes that "Anaximander imagined the cosmos to be a kind of temple, the cosmic house, along the analogy of the cosmic meaning of the column" (2001: 188). In a later study, Hahn writes about "Anaximander's cosmology, a vision of cosmic architecture" (2007: section C). In his latest work, Hahn seems to avoid using the word "temple" for Anaximander's cosmos. Instead, he uses expressions like "the house that is the cosmos" (2010: 51, 120), of which Anaximander sought to explain the structure and stages of its construction. These words are a long way from the modest goal of showing how Anaximander discussed with and borrowed ideas from the architects that Hahn earlier said to pursue. Let us examine somewhat closer the image of the heavens as a temple. The temples of the architects were rectangular, with a triangular roof. Anaximander's alleged cosmic temple, on the contrary, consists of rings or wheels. A bigger contrast is hardly conceivable.

More important, however, is a metaphysical argument that can be brought forward against Hahn's interpretation of Anaximander's cosmos as a temple. This can be elucidated by his discussion of the cosmic meaning of the temple roof. There is an ancient tradition of the cosmic symbolism of (temple) roofs. The ceilings of several Egyptian burial chambers, tombs, and temples, as well as the inner side of many coffin lids, for instance, were decorated with heavenly themes. The roof of the Greek temple must also have had "cosmic and symbolic significance"; "the 'Heaven' in that architectural analogy was represented by (. . .) the roof" (2003: 135 and 137). Hahn shows the reconstructed akroterion that stood on top of the roof of the archaic Heraion in Olympia and which is, according to Yalouris, a solar symbol (1972: 92–94). The view of this and similar akroteria could have inspired Anaximander to make his model of the cosmos, Hahn maintains (2003, 140). This is, of course, a speculation. Moreover, the likeness of this akroterion with a map of Anaximander's universe is only superficial (Fig. 12.3).

Fig. 12.3 Reconstruction of the akroterion from the Heraion in Olympia, about 600 B.C. (photograph by Jelle Abbenes)

Let us repeat Hahn's conclusion: "*The point is that the original meaning of Anaximander's cosmology is cosmic architecture*," which is to say that Anaximander "imagined the cosmos to be a kind of temple, the cosmic house" (2007: section C; 2001: 188; 2003: 148–150). Even admitted that we are looking with awe to the Greek temples, their ultimate goal is to transfer a feeling of safety. Just like the human house offers a safe haven within a hostile world, the temple, the house of the gods, provided a safe place to mankind. Transposed to the universe in a metaphysical sense we may say that the archaic world of Homer, with the celestial dome vaulting over the flat earth can be seen as a cosmic refuge that protects people against the surrounding Chaos. But it was precisely Anaximander who broke with this idea of the universe as a safe and closed space. As we have seen in Chap. 8, it was Anaximander who blew up the celestial dome when he imagined the earth hanging free in space and the celestial bodies behind each other. Anaximander's universe was not a safe refuge, which shows from the fact that he had to answer the anxious question why the unsupported earth does not fall. Anaximander's cosmos was the opposite of a cosmic temple with a solid ceiling.

As far as I am concerned, the lesson of all this is that one must be aware of making too much of one of the images that Anaximander used. According to Theophrastus, Anaximander wrote a kind of rather poetic prose. This implies that he obviously was fond of all kinds of images from different fields: architecture (the earth compared with a column drum), war equipment (the celestial bodies conceived of as chariot wheels), the human body (the apertures in the celestial wheel as blowing mouths), flora (the bark of a tree as an image for the original fire that surrounds the air around the earth), meteorology (the light of the celestial bodies explained on the analogy of lightning fire), and jurisdiction (in the remaining fragment DK 12B1). We may wonder whether a better understanding of Anaximander is fostered by absolutizing one of these images. In a similar way, one who is impressed by Anaximander's image of the celestial bodies as chariot wheels could be tempted to imagine his universe as a celestial chariot (like that by which Apollo transports the sun), or perhaps as a celestial machinery like that in Fig. 1.1. Or someone who thinks (wrongly, see previous chapter) that Anaximander imagined the light of the celestial bodies coming to us through a kind of cosmic bellows could look upon Anaximander's universe as a kind of cosmic forge, like that of Hephaestus in the Hades, or perhaps as a breathing and living organism, according to the image of fire-blowing mouths. Or even someone, having in mind the text of fragment DK 12B2, could think that Anaximander conceived of the cosmos as a celestial *agora* where the celestial bodies achieved justice in a state of equilibrium.

I even leave open the possibility that there may be some truth in some of these ways of looking at Anaximander's cosmology. My point is that they do not help us very much to get sight of the historical importance of Anaximander's cosmological ideas: his earth floating unsupported in the center of the concentric wheels of the celestial bodies, which meant his definitive break with the archaic concept of the celestial vault, and thus the creation of a new world picture that is, essentially, still ours. What ultimately counts is not so much what could possibly have inspired Anaximander, but the significance and impact of his cosmological ideas.

Part III
The Completion of the New World-Picture and the Debate on the Shape of the Earth

Chapter 13
A Survey from Anaximander to Aristarchus

With his three fundamental insights, that the celestial bodies make full circles around the earth, that the earth dwells unsupported in the center of the universe, and that the celestial stars are behind each other, Anaximander made himself the founding father of cosmology. It is not the intention of this book to give a complete history of Greek cosmology. There exist standard works that fulfill this task.[1] Yet, some remarks on developments after Anaximander are needed, as they may throw additional light on the unique project of his new world picture. Some developments deserve special attention in separate chapters, the most important of them being the replacement of the image of a flat earth by that of a spherical earth (see Chaps. 17 and 18). Anaxagoras, who thought of the celestial bodies as fiery stones, proposed a proof that the earth is flat and tried to measure the size of the sun, gets three chapters (Chaps. 14, 15, and 16). The final chapter discusses how the ancient firmament that was shattered by Anaximander returned via the back door in later authors, with the one exception of Heraclides Ponticus. In this connection, the issue of the "infinite worlds" will also be treated.

The introduction of Anaximander's new world picture has been accomplished relatively quickly. But some elements of the archaic world picture offered a tough resistance, or even made a comeback after a while. Not only Anaximander still conceived of the earth as a flat disk, like a column drum, but also most Presocratics shared the idea of a flat earth, as we saw in Chap. 4. Anaximander's successor Anaximenes seems to have returned in some points to the archaic world picture. He conceived, as is told, of the stars as fiery nails in the celestial vault, or their constellations as painted upon it like leafs [(*Meteorologica* 354A28–33) = DK 13A14]. Moreover, Anaximenes obviously was not able to believe that the celestial bodies continue their orbits underneath the earth, for it is said that he taught that they – in a never fully clarified image – go around the earth "like a felt hat around the head" [DK 13A7(6)]. According to the church father Hippolytus, who has handed down this text that is usually treated as authentic, he meant that the sun after sunset is hiding behind a mountain in the northern regions, to appear from

[1] See for instance Heath (1913), Dicks (1970), O'Neil (1986), North (2008: 67–133), and, rather popularizing, Wright (1995).

D.L. Couprie, *Heaven and Earth in Ancient Greek Cosmology*,
Astrophysics and Space Science Library 1, DOI 10.1007/978-1-4419-8116-5_13,

behind it the next morning. If we may believe Aristotle, Anaximenes was only one of many of the ancient cosmologists who held this view. Still much later, Severianus (born in the second half of the fourth century A.D.) and Cosmas Indicopleustes (±550 A.D.) adopted a similar idea (see Dreyer 1953: 217). I must confess that I am not able to imagine what could have led these people to such a peculiar conception (Fig. 13.1).

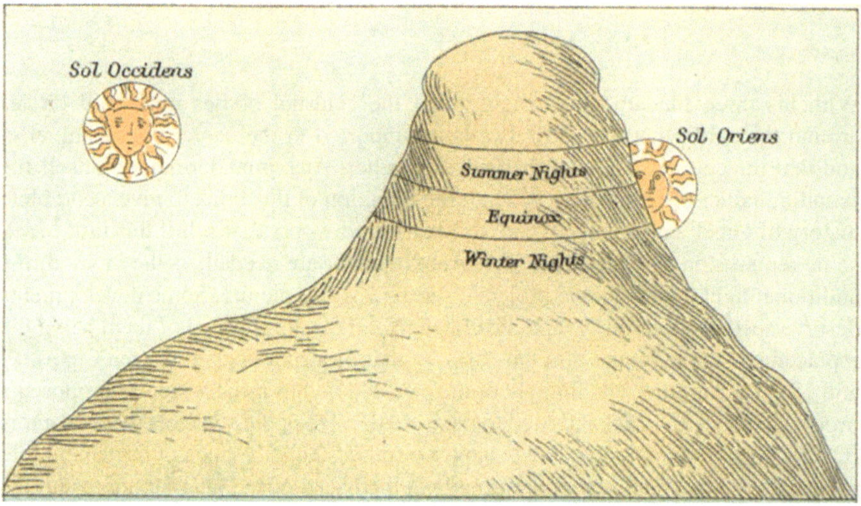

Fig. 13.1 The sun disappears at night behind the northern mountain, according to Cosmas Indicopleustes (Murray 1895: Plate Vb)

Xenophanes, too, did not believe that the sun goes under the earth. He thought, just like Heraclitus, that the sun was born every day anew at the eastern horizon (DK 21A33). Maybe he wanted to solve the problem why the sun does not get extinct in the course of the years. A fire has to be fed, and in the case of the sun it is not clear from where its fuel should come. If there is a new sun every day, this problem does not exist, although others arise, for instance where the new sun comes from, and how the sun manages to rise every day exactly at the right place. Apparently, these problems were less serious to Xenophanes than that of the sun going underneath the earth. As the whole story is from a very late source, I am inclined to take it with a pinch of salt. Heraclitus' version, on the other hand, is handed down by Aristotle (*Meteorologica* 355a13 = DK 22B6) and is hinted at by Plato (*Republic* 498a5–8, not in DK). My guess is that Heraclitus did not have the intention to express anything astronomical at all, but rather something like our encouraging saying "tomorrow comes another day." In the same sense, Democritus says that the rising sun by its light brings about new thoughts and actions in men (DK 68B158).

Perhaps this is the right place to mention the curious theory, ascribed to Empedocles, that the sky is divided into two hemispheres which turn around the earth, one "fiery" for the day and one "airy" for the night (DK 31A30, 31A51, and 31A56). The sun is, according to this text, is not made of fire but a reflection, caused by the earth, of the fire of an invisible sun in the other hemisphere. Furley calls these texts "a mystery I cannot penetrate" (1987: 93), whereas Dicks complains that "very little sense can be made of this" (1970: 55). At least, it shows – if these reports are trustworthy at all – how difficult it was to master the new cosmological paradigm and to free it from all kinds of strange (mis)conceptions.

Anaximander considered the celestial bodies as rings or wheels, turning around the earth. This conception was put aside rather soon. Usually, the celestial bodies were conceived of as consisting of fire, as is already the case with Anaximenes [DK 13A7(4)]. However, Parmenides seems to have revived somehow Anaximander's wheels, for he speaks of the cosmos as a system of concentric bands (literally "wreaths" or "garlands": στεφάναι). Unfortunately, the texts in question – a literal quotation from his poem and a free rendering by Aëtius – are barely intelligible (DK 28B12 and DK 28A37).[2] Anaxagoras, on the other hand, said that the celestial bodies were made of fiery stone. This idea that the Athenians looked upon as blasphemous, and that was suppressed immediately by Aristotle, is treated in Chap. 14. These new conceptions urged the need to formulate other answers to the question what makes the celestial bodies orbit in circles around the earth without falling upon the earth. A favorite explanation was to hold the celestial vortex (δίνη or περιχώρησις) responsible for the circular movement of the celestial bodies, on the analogy of tornados or whirlpools. This was, for instance, Empedocles' and Anaxagoras' conception. Simplicius describes it thus: "Some people say that a physical mechanism keeps the sky from falling – namely the action of a vortex which holds it up – since the downward pull on the heavens is less than the force exerted by the vortex. Empedocles and Anaxagoras say this" (*In Aristotelis De caelo commentaria* 374.32, not in DK).

Elsewhere, it is said that according to Anaxagoras the intense whirling of the surrounding fiery aether snatched up stones from the earth, set them on fire, and turned them into celestial bodies. There they are kept in their orbits by the force of the rotation that prevents them from falling on the earth (DK 59A71 and DK 59A12). However, this mechanism is not completely safe, as was proven by the huge stone that fell from heaven at Aegospotamoi (see next chapter). I think Gershenson and Greenberg are too critical when they write that "the tradition concerning Anaxagoras' use of rotational motion is one of the most confused and contradictory of all traditions about his doctrines." Their main problem seems to be "that the same rotational motion that brings heavy bodies to the center can keep heavy bodies up in the heavens" (1964: 349). However, once Anaximander's conception of the celestial

[2] Morrison has tried to give an interpretation of these texts, which is severely criticized by Dicks (Morrison 1955: 59–68; Dicks 1970: 226 n. 59).

bodies as wheels rejected, and not accepting the theory of the celestial bodies as attached to celestial spheres, it is hard to think of another mechanism that keeps them in a circular movement around the earth than on the analogy of the vortex that is seen in whirlpools and tornados. Moreover, Gershenson and Greenberg do not take into account that in Anaxagoras' conception the celestial bodies, although they are fiery stones, are smaller and lighter than the earth, as they are stones snatched up from the earth by the vortex. In whirls, the heaviest objects stay in the center, whereas the lighter objects are thrown to the periphery. And sometimes, when for some reason the whirl's power decreases for a moment, heavier objects fall back to the center, as must have been the case with the stone of Aegospotamoi.

As discussed more extensively in Chap. 18, a more metaphysical solution was defended by Aristotle, who maintained that the celestial bodies move in circles because this is the only perfect kind of movement. This also means that the bodies orbiting around the earth cannot be made of one of the so-called elements, for their movement is rectilinear (e.g., fire moves upward to the periphery and earth downward to the center). According to Aristotle, then, the celestial bodies do not consist of one of the four classic elements but of a fifth element of higher nature, the aether (*On the Heavens* 269a2 ff.). Additionally – one would almost say: to be sure – Aristotle thought it wise to attach the celestial bodies to spheres (or better sphere shells) to prevent them from falling upon the earth (*On the Heavens* 286b10 ff., 287a7 ff., 289b30 ff., and especially 293a5–12). Therewith, in a way, he copied Anaximander's solution, as the natural movement of a sphere is circular, just like that of a ring or wheel.

Anaximander's free-floating earth unavoidably raised the frightened question why the earth does not fall down, carrying along its inhabitants in its unbridled fall. In his *On the Heavens*, Aristotle renders an account of the highlights of the discussion that was released by Anaximander's ideas (*On the Heavens* 294a19 ff.). According to Aristotle, as we saw in Chap. 8, Anaximander reassured his cocitizens by arguing that the earth has no reason to fall because it is in the center of the universe, "everywhere at the same distance from the periphery." Apparently, however, this argument did not suffice to take away the fear of falling. Later Greek thinkers came with alternative explanations. Xenophanes thought, according to Aristotle, that the roots of the earth go down infinitely so that we live at its upper surface like Simeon the Stylite on his pillar (*On the Heavens* 294a21 = DK 21A47). Anaximenes, Anaxagoras, and Democritus thought that the earth, which they conceived of as flat, rests on the air "like a lid" (*On the Heavens* 294b13 = DK 13A20). According to Empedocles, the celestial vortex that we can observe in the orbits of the stars prevents the earth from falling (*On the Heavens* 295a13 = DK 31A67). Plato seems to have appreciated the power of Anaximander's argument, if he was not himself its *auctor intellectuelis* (*Phaedo* 108e; see also Chap. 8). It was Aristotle who delivered a new and brilliant argumentation why the earth does not fall that remained valid for many centuries. This is discussed more extensively in Chap. 18.

Gradually, all kinds of lacunae in Anaximander's new world picture were filled in, and additions proposed. Anaximander, for instance, taught the curious order of the celestial bodies: stars, moon, and sun. As was argued in Chap. 8, it is not as easy as it seems to give the right order because we cannot *see* whether a celestial body is relatively far away or nearby. It is told that Metrodorus of Chios and Crates held the same opinion as Anaximander, whereas Leucippus is said to have defended the order: moon, stars, sun [DK 12A18 and 67A1(33)]. Who was the first to place the celestial bodies in the right order is no longer known.

The determination of the distances of the various celestial bodies to the earth also remained difficult for the Presocratics – and long after that. Chapter 16 will explain that Anaxagoras, saying that the sun is about as big as the Peloponnesus, probably must have had an idea of the distance of the sun to the earth (always presupposing that the earth is flat). Empedocles estimated the distance from the moon to the sun twice as big as that from the earth to the moon (DK 31A61). Anaximander had expressed the distances between the celestial bodies in numbers. He chose multiples of 3, probably with the intention to indicate that the wheels of the celestial bodies were respectively far, farther, and farthest from the earth, as we saw in Chap. 8. The Pythagoreans, and probably already Pythagoras himself, followed Anaximander's example, but they took other numbers and another starting point, also correcting the order of the celestial bodies. Being convinced that everything in the world is ruled by numbers, Pythagoras chose the three basic musical harmonies, the octave, the fifth, and the fourth. As Guthrie says: "With Anaximander's scheme before him, he transformed its equality of distances into a dynamic mathematical relationship. In this scheme only three orbits are in question, those of the moon, sun, and stars" (1962: 300). In a late source, other numbers are mentioned that show a remarkable resemblance with Anaximander's numbers, without, however, a possible meaningful interpretation, as is the case with Anaximander. There the diameter of the central fire is taken to be 1, the distance to the counter-earth 3, to the earth 9, to the moon 27, to Mercury 81, to Venus 243, etc., always powers of 3 (Plutarch, *On the Generation of the Soul in the Timaeus* 1028b, not in DK). Further, in this chapter, the curious Pythagorean cosmological system, including the counter-earth and the central fire, is discussed in somewhat more detail.

Aristarchus tried to measure the relative distances of the sun and the moon to the earth. He argued as follows: At first and last quarter of the moon the three bodies, earth, moon, and sun make a rectangular triangle with its right angle at the moon. From the shape of the triangle EMS (in Fig. 13.2), he derived that the distance from the earth to the sun must be about 19 times greater than that from the earth to the moon (*On the Sizes of the Sun and Moon*, propositions 7 and 9, in Heath 1913: 377 and 383).[3] That this outcome, although the method was right in principle, was only a fraction of the real distance of the sun is due to the facts that it is difficult to determine exactly the time of the first or last quarter of the moon and that Aristarchus' measurement of the angle EMS was far beyond its real size (87° instead of 89°52′).

[3] Aristarchus' argumentation is here simplified. For more information see, e.g., Ferguson (1999: 18–21; see also Pannekoek 1961: 118–120, and North 2008: 90–91).

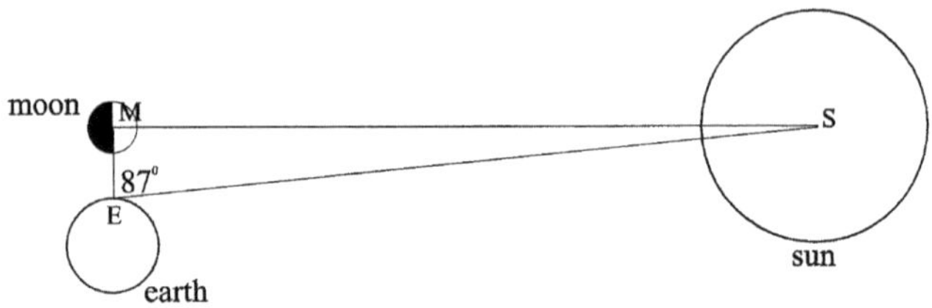

Fig. 13.2 Aristarchus' method for measuring the relative distances to the moon and the sun

Estimating the sizes of the celestial bodies remained a big problem. According to Heraclitus the sun was a man's foot big (DK 22B3). Almost as strange, at first sight, is Anaxagoras' comparison of the size of the sun with the Peloponnesus [DK 59A1(8), 59A42(8), and 59A72]. Chapter 16 will show that this estimation is not so bad as it seems, given his supposition that the earth is flat. As we have already seen, Aristarchus, knowing of the sphericity of the earth, had figured out that the distance from the earth to the sun was about 19 times as big as that from the earth to the moon. Accordingly, from the fact that the sun and the moon have the same angular diameter he concluded that the real diameter of the sun must be also about 19 times that of the moon. From his study of lunar eclipses, Aristarchus derived that the size of the earth's shadow is twice that of the moon. From these data, he calculated that the diameter of the sun has to the diameter of the earth a ratio between 19/3 and 43/6, which was of course also wrong, although the method used was right (*On the Sizes of the Sun and Moon*, hypothesis 5 and proposition 15, in Heath 1913: 353 and 403).

That the moon has no light of itself but derives its light from the sun seems to have been a discovery of Anaxagoras, as is already reported by Plato (*Cratylus* 409a9–b1 = DK 59A77, cf. 59B18). This discovery also means that Anaximander's explanation of the phases of the moon and lunar eclipses as the opening and closing of the apertures in its wheel could not be right. Accordingly, Anaxagoras probably was the first to give the right interpretation of the eclipses, not only of the moon but also of the sun, as is attested by Valerius Maximus (*Facta et dicta memorabilia* 8.11 ext.1, not in DK), who says that Pericles had learned this from his teacher Anaxagoras, and by Hippolytus [DK 59A42(9)]. More on this subject in Chap. 14.

Anaximander probably did not recognize the planets as a separate group of celestial bodies, which is clear from the fact that he did not bestow them with separate celestial wheels. Probably, he regarded them as a wandering stars (which is the literal translation of the Greek word πλανήτης), moving between the other stars. It is not clear who discovered that they are another kind of bodies than the stars and were more akin to the sun and the moon, but after that discovery in the Greek conception of a geocentric universe, the word "planets" included not only Mercury, Venus, Mars, Jupiter, and Saturn, but also the sun and the moon. This is also the case in much later renditions of the geocentric universe, as for instance those of Cellarius in the seventeenth century A.D. (see e.g., Cellarius 1660, plate 3). Their order remained a problem, and especially the position of the inner planets Mercury and Venus that always remain in the vicinity of the sun was problematic.

Plato held the following order: moon, sun, Venus, Mercury, Mars, Jupiter, and Saturn (*Epinomis* 987b-c; *Timaeus* 38d-39b).

All these "planets" move through specific constellations that make up a belt called the Zodiac. The sun's orbit within that belt is called the ecliptic. In my reconstruction of Anaximander's world picture in Chap. 10, I argued that according to Anaximander all celestial bodies move from east to west, with different velocities and that Pliny's report that Anaximander knew the obliquity of the ecliptic has to be considered incorrect. How little the relation between the ecliptic and the monthly path of the moon was really understood appears from the alternative explanation of lunar eclipses, ascribed to Anaximenes, Anaxagoras, Diogenes of Apollonia, and the Pythagoreans. They are said to maintain that below the celestial bodies other, invisible bodies revolve that sometimes cut off the light of the moon. These invisible bodies were postulated to account for the greater frequency of lunar than solar eclipses, as Aristotle reports. They were also presented as an explanation for meteorites like the stone of Aegospotamoi [DK 13A7(5), 13A14, 59A42(6) and 59A42(9), 64A12, and Aristotle, *On the Heavens* 293b21–25, not in DK, but see footnote at 59A42(6)].

Several sources ascribe the discovery of the ecliptic (or the Zodiac) to Oenopides, who lived about one century after Anaximander and was a younger contemporary of Anaxagoras (DK 41A7). This makes Oenopides a serious candidate for the discovery of the sphericity of the earth as well, as the ecliptic must be thought of as inclined to the celestial equator, which is the projection of the equator of a spherical earth on the celestial sphere. We may assume that in his time a vehement discussion about the shape of the earth was going on because we know that Anaxagoras put forward an argument for its being flat, which we will discuss in Chap. 15.

At closer inspection, the movements of the planets appeared to be rather capricious, as they sometimes seem to stand still among the stars, then to return with an elegant curve, and to take up again their original course from west to east. This is illustrated in Fig. 13.3 in a recent example.

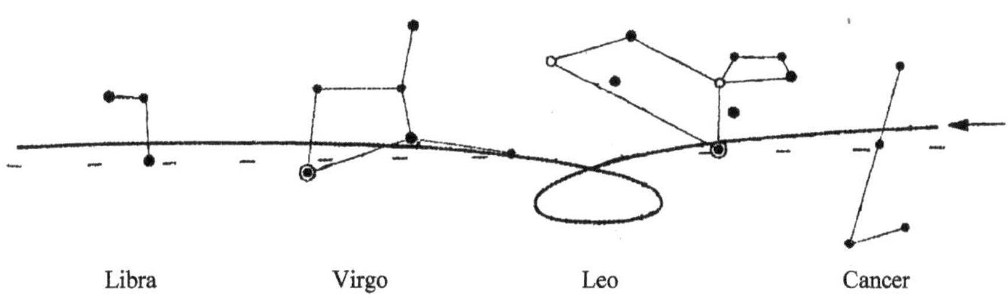

Fig. 13.3 The movement of the planet Venus through Cancer, Leo, Virgo, and Libra from June until December 1935 (the *dotted line* is the ecliptic) (adapted from Kobus & Raimond 1936: 100)

This led finally to the ingenious system of Eudoxus, who, perhaps on Plato's demand (see Simplicius, *In Aristotelis De caelo commentaria* 488.18 ff.), tried to explain the complex movements of the planets (and those of the sun and the moon) by introducing a number of spheres for every celestial body, all of which having their own movement. For the stars, only one sphere was needed, turning from east to west. Sun, moon, and planets all have not only such a sphere, but also another one, turning from west to east around the earth, for each of these celestial bodies with a different speed. The equator of these last spheres is the ecliptic, which results in them turning aslant to the first sphere. To deal with the irregular movements of sun and moon, Eudoxus needed a third sphere for both of them, and for the movements of the other planets three extra spheres each. This resulted in a sum total of 27 spheres (one for the stars, three for the sun, three for the moon, and four for each of the five planets).[4] These moving and countermoving spheres were supposed to result in the observed movement of the planets against the background of the sphere of the stars. Although this system was rather meager as a description of the observed movements of the planets, yet it has been influential, as it was adopted in a modified version by Aristotle.

Aristotle and Achilles Tatius mention another displacement at the heavens. Aristotle tells that according to some so-called Pythagoreans the Milky Way is the circle in which the sun once moved (see DK 41A10; see also 58B37c).[5] Achilles Tatius ascribes this idea to Oenopides, adding some details: it was in disgust of the banquet offered to Thyestes (who inadvertently consumed his own children) that the sun turned away from the Milky Way and followed its course in the opposite direction along the Zodiac. Obviously, this concerns another tradition than that of the tilting of the heavens that was discussed in Chap. 5 above. For in this last theory, the sun is supposed originally to have turned around the horizon, as explained in that chapter, whereas in the story of the Milky Way the sun's path originally must have been about through the zenith, where we now see the Milky Way. The story of the Milky Way, then, is not meant to explain the tilting of the heavens, but to explain the contrary movement of the sun (from west to east) along the Zodiac.

Apparently, this story of the sun changing and reversing its course is the background to the reports that ascribe to Oenopides the discovery of (the obliquity of) the ecliptic. The intention seems to be that the sun, horrified by the banquet, not only changed its path but also, as a kind of throwing up in nausea, reversed its course. Mark that the velocity of the sun in its original orbit from east to west along the Milky Way must have been another than that of the stars, for otherwise there would not be the "path" of the Milky Way. Mark, too, that the sun's new path along the Zodiac is not visible, whereas its former path is still visible as the Milky Way. For Anaxagoras' explanation of the phenomenon of the Milky Way, see Chap. 14.

The Pythagoreans developed an idiosyncratic and rather curious cosmology that is generally associated with the name of Philolaus. According to them, in the center of the universe an enormous fire is situated, also called the Hearth of the universe

[4] A detailed description of Eudoxus' system can be found in Dicks (1970: 177–188), as well as in Neugebauer (1975, vol. 2: 675–685).

[5] For an extensive treatment of the early conceptions of the phenomenon of the Milky Way, see Jaki (1972: 1–12).

(‘Εστία τοῦ παντός), or the Watchtower of Zeus (Διὸς φυλακή). As Aristotle reports, their reasons for not placing the earth at the center of the universe were not observational or astronomical, but purely metaphysical or a priori (ἐκ τῶν λόγων): “For they think that the most honorable body ought to occupy the most honorable place, and that fire is more honorable than earth and a limit more honorable than what lies between limits, whereas the center and the outer boundary are limits. Arguing from these premises, they think it is not the earth which is in the center of the sphere (of the heavens), but rather the fire” (*On the Heavens* 293a30–b2 = DK 58B37).

This central fire is not the sun, as one is tempted to think at first sight. The earth turns around the central fire, but just as the moon has always its same side turned toward the earth, so the earth always turns its same side to the central fire. We, however, live on the side that is turned away from the central fire, and that is why we never see it. Between the earth and the central fire the counter-earth (ἀντίχθων) is situated that for the same reason is invisible for us. Moon, sun, and planets also orbit around the central fire.

Maybe the Pythagoreans invented the counter-earth to complete the number 10 for the spheres of the celestial bodies. At least, this is what Aristotle surmises (Aristotle, *Metaphysics* 986a4–12, not in DK). Kingsley has argued that this cannot be right, because there are eleven celestial bodies in the Pythagorean system (1995: 174). I think he is mistaken, for only the moving celestial spheres are involved in the celestial harmony (the stars, Saturn, Jupiter, Mars, Venus, Mercurius, the sun, the moon, the earth, and the counter-earth), whereas the central fire is supposed to stand still in the center of the universe and thus does not produce a tone. At another place, Aristotle suggests that the counter-earth had to account for some eclipses of the moon, just like the invisible celestial bodies mentioned above (*On the Heavens* 293b20–24 = DK 58B36). However, it is hard to see how the counter-earth that is supposed to orbit between the earth and the central fire could ever cause an eclipse of the moon, as Dicks rightly remarks (1970: 67).

The implication of the earth being removed from the center is that the movements of the planets, including the earth, had to be described as circling around the central fire from west to east, whereas the sphere of stars stands still, as Huffman rightly explains: “All of the bodies moving around the central fire have one circular motion from west to east” (1993: 253). This is the way Alcmaeon of Croton, a Pythagorean, is said to have described the movements of the planets: they travel contrary to the movement of the fixed stars, from west to east, and not from east to west, somewhat slower than the stars, as in the geocentric system of Anaximander and the other Presocratics (DK 24A4). This report is usually, and perhaps already by its author Aëtius, misunderstood as a description of the contrary movement of the planets along the zodiac as is appropriate in the geocentric system (see e.g., Dicks 1970: 74–75; Burkert 1972: 52 and 282; Dreyer 1953: 39). It makes more sense, however, to understand it as a description of the movement of the planets, including the earth, in the Pythagorean system. Not only the direction of the movement has to be described opposite to that in the geocentric system but also their velocities: “The sun, moon, and planets then each have one circular motion from west to east which is much slower (in angular velocity at least) than that of the earth" (Huffman 1993: 253). The stars, as already said, must consequently be thought in this system as standing still.

As a whole, the movements of the celestial bodies thus described are the exact opposite of the way they can be described in the geocentric system, where all the planets move around the earth from east to west with increasing velocities, and finally the sphere of the stars in 24 h. As elucidated in Chap. 10, this was in principle the way Anaximander and other Presocratics explained the movements of the celestial bodies. Accordingly I do not understand why Huffman praises the Pythagorean system as "postulating just one motion for each body, with all the motions being in the same direction" (1993: 254). The same can be said of the early geocentric systems, with the movement in the opposite direction. Huffman's conclusion that "Philolaus (...) presents us with a much more coherent model of the cosmos than any other fifth-century thinker" (1993: 259) is, according to me, completely unfounded. This is the more so as he has to admit that the Pythagorean system "would make most sense if the fixed stars had no movement," whereas the textual evidence leads to the conclusion that the sphere of the stars in the Pythagorean system must be considered as having a very slow motion as well (1993: 256 and 257). Both systems fail to account for the fact that the planets' movement (from east to west in the geocentric system, from west to east in the Pythagorean system) does not progress uniformly. In other words, they ignore the fact that there are stationary points and points where the planets show retrograde movements (from west to east in the geocentric system, from east to west in the Pythagorean system). "This phenomenon is not explained by any Presocratic theory, as Huffman rightly remarks" (1993: 254; cf. Fig. 13.3 above).

According to the Pythagoreans, the movements of the celestial bodies resulted in the "harmony of the spheres," of which we are not aware, because we hear it during our whole lives. It was told, however, that Pythagoras was the only human being that was able to hear this celestial music. This story is typical for the almost godlike status that Pythagoras obtained in Antiquity. As a matter of fact, however, the theory of the cosmic harmonies "was indeed a scientific blind alley" for cosmology, as Guthrie dryly remarks (1962: 302). Burkert calls it "mythology in scientific clothing, rather than an effort, in accord with scientific method, to 'save the phenomena'" (1972: 342). And more recently, Kahn writes: "the system of Philolaos, taken as a whole seems less like scientific astronomy than like symbolical speculation, an imaginative expression of the view that the order of the universe is a function of musical harmony and meaningful numbers" (2001: 26–27). And even Huffman has to admit that the introduction of the central fire, which caused the earth to revolve around it, "had nothing to do with astronomical phenomena, but arose of a priori notions of order and fitness." Somewhat further he writes: "most of the rest of his (i.e., Philolaus') astronomical system can be seen as trying to square the obvious phenomena with this initial postulate," and "a similar emphasis on a priori notions is found in the reasons for introducing the counter-earth" (1993: 244, 245, and 246). As already said in the Introduction, as regards its meaning for cosmology, I would rather call it fantasy than creative imagination, and certainly not "a triumph of thought over mere appearance" (Huffman 1993: 259).

Generally speaking, the Pythagorean system was mainly based on a priori reasoning and was hard to bring into agreement with the observation of the celestial phenomena. According to Kingsley the reason why the Pythagoreans invented the central fire and the counter-earth was not to make the number of the celestial bodies ten, but their identification of the central fire with Tartarus and the counter-earth with Hades (1995: 185–187). However, also in Kingsley's interpretation the arguments for the existence of counter-earth and central fire are purely mythical. Dreyer remarks that the Pythagorean system "does not appear to have won any adherents outside the philosophical school in which it originated" (1953: 49). Actually, it was a sterile a priori way of arguing, based on preconceived ideas like that of the analogy of the musical harmonies with the distances of the celestial bodies, the honorable place of the center and the fringes of a spherical cosmos, and the divinity of the number ten. Dreyer is, however, too optimistic as regards its lack of influence. Not much later, Plato, in a context that shows clearly Pythagorean influences, banned all observation from astronomy: "we shall pursue astronomy, as we do geometry, by means of problems, and ignore the visible heavens" (*Republic* 530b5–7). In Heath's words: "Plato conceives the subject-matter of astronomy to be a mathematical heaven of which the visible heaven is a blurred and imperfect expression in time and space; and the science is a kind of kinematics, a study in which the visible movements of the heavenly bodies are only useful as illustrations" (1913: 138).

In the context of this book, the Pythagorean ideas are interesting for two reasons. In the first place, within the group of the Pythagoreans probably for the first time the idea rose that the celestial bodies have spherical shapes. I return to this in Chap. 17. In the second place, the Pythagoreans were the first to dispose of the idea that the earth is the center of the universe, albeit that the argument they put forward for it was of a metaphysical kind, as we have seen. Aristarchus agreed to this thought, but he made the sun the center of the universe and let the earth and the other celestial bodies turn around it. Maybe for him yet another argument played a role: he may have argued that it is less strange to let a smaller body orbit around a bigger one than the other way round. However this may be, it was a revolutionary idea that made the earth no longer hold the center of the universe but lets it turn around the sun. How Aristarchus (and the Pythagoreans as well) coped with the problem why we do not fall off the earth, when it is not in the center, or why we are not hurled off the earth, when it circles around the sun (or around the central fire), we do not know. We have to wait until Copernicus before someone dared to suggest the idea of a heliocentric universe again. Aristarchus also taught the rotation of the earth. This idea was developed earlier by Heraclides Ponticus, but in a geocentric world picture. His ideas deserve treatment in a separate chapter (Chap. 19), as they have direct impact on the question whether the universe is finite or infinite. Strictly speaking it is true that already in the Pythagorean system the earth turns around its own axis because it has always the same side turned to the central fire, as Huffmann notes (1993: 250), but I doubt whether they would have recognized that as an axial rotation.

The notion of the celestial bodies as fiery masses of stone, the proof that the earth is a sphere, the discovery that the planets are a separate category of celestial bodies, the conception of the ecliptic, the clarifying of the correct order of the celestial bodies, the right explanation of the phases of the moon and of the lunar and solar eclipses, and the idea of the earth turning around its own axis were all great achievements. But they were only possible as additions to and modifications or improvements of the world picture introduced by Anaximander. Without his pioneering work, they are unthinkable.

Chapter 14
With Fear for His Own Life: Anaxagoras as a Cosmologist

The dates of Anaxagoras' life are a matter of dispute. Usually they are given as 500–428 B.C. but the dates 533–462 B.C. have also been defended. According to Diogenes Laertius, Anaxagoras was an apprentice of Anaximenes (literally: he heard Anaximenes), whose death can be dated 524 B.C., and only the latter dates make this at least possible (DK 59A1(6), see Unger 1884: 511–550; Cleve 1973: 2–3). Anaxagoras was born in Clazomenae in Asia Minor, and he bore the nickname "Brains" (Greek: νοῦς) because of his quick mind and great knowledge of the natural phenomena (DK 59A15). The philosophy of Anaxagoras has been subject of many studies, but in the context of this book, I confine myself to his opinions on astronomy and cosmology that have received relatively little attention.[1] Anaxagoras himself did not consider these subjects as the least part of his work. The story says that when someone wondered why he exerted himself spending whole nights outdoors, he looked up to the stars and answered: "to study the cosmos" (Philo, *De incorruptibilitate mundi* 2, not in DK). On his grave was written: "Here lies Anaxagoras. His image of the order of the universe came closest to the truth" (DK 59A1(15)).

Anaxagoras is said to have been a pupil of Anaximenes, but as regards his cosmology, he stands more in the tradition of Anaximander.[2] He takes over Anaximander's world picture and develops it further on the basis of his own new insights. We do not know whether Anaximander made known his new world picture outside his home city Miletus. His visit to Sparta probably had to do with his ability in using the gnomon, and in Apollonia at the Euxine, he executed a colonial mission on behalf of the government of Miletus. Anaxagoras, on the other hand, traveled intentionally from Ionia to Athens to proclaim the new doctrine (DK 59A7). Themistocles, Pericles, and Euripides sat at his feet, Democritus admired him and wanted to become his pupil but was refused, and even Socrates thoroughly read his

---

[1] For a survey of Anaxagoras' teachings, see Gershenson and Greenberg (1964), Cappelletti (1984), and less reliable Cleve (1973).

[2] Cf. Cappelletti: "La astronomía y la meteorología de Anaxágoras continúan directamente la doctrina de los filósofos de Mileto" (1984: 271). Aravantinou speaks about "Anaxagoras, the representative of the old Milesian tradition in Athens" (1993: 99).

D.L. Couprie, *Heaven and Earth in Ancient Greek Cosmology*,
Astrophysics and Space Science Library 1, DOI 10.1007/978-1-4419-8116-5_14,

book (DK 59A1(10), 59A1(1)12, 59A1(14), 59A15, 59A17, 59A21, 59A32, 59A35, 59A73, 59B21a, and 30A3).

Just like Anaximander, Anaxagoras thought that the earth is flat and shaped like a drum and resides motionless in the center of the cosmos (DK 59A42(3), 59A87, and 59A88), although he offered another explanation for its position. Following his teacher Anaximenes, he said that the earth does not fall because it is supported by the air like a lid (DK 59A20b, 59A88, and 59A89). He added three remarks that are apparently meant as clarifications: that the earth floats on air is owing to its big size, the thickness of the air, and because there exists no vacuum (DK 59A20b). Maybe Anaxagoras argued that when a leaf and a bird, both of which are heavier than air, can be lifted up or fly in the air, the same can hold for the earth as well. In a similar way, Aristotle assumes that an argument like this ("flat bodies apparently drift on the air") is behind the conception of the earth being supported by air (*On the Heavens* 294b13 = DK 13A20). Or maybe we have to imagine that the natural inclination of the air to rise up and that of the earth to fall down hold each other in equipoise. Yet in this situation the earth is not completely safe, for earthquakes are caused, according to Anaxagoras, because of air from above the earth tumbling into the air underneath the earth, thus causing the earth to move (DK 59A42(12)). Elsewhere, he explains how the air from above the earth gets underneath it: the air is dragged along with the rain that penetrates through the porous earth. Once arrived under the earth, the air gives in again to its natural inclination to rise, and when it arrives in the subterranean cavities it makes the earth shake (DK 59A89).

We may call it remarkable that Anaxagoras clung to the conception of a flat earth, although he was acquainted with the revolutionary idea that the earth is a sphere. At any rate, he felt compelled to bring forward at least one proof for his conception of a flat earth, which means that there were others who defended the opposite idea. In the next chapter, we discuss more thoroughly the argument that has recently been shown to belong to Anaxagoras. This contradicts, by the way, Fehling's apodictic words: "Ferner pflegte man in der Generation des Anaxagoras noch nicht zu polemisieren; man schrieb rein dogmatisch" (1994: 142). Who were these upholders of the conception of a spherical earth is not exactly known, but it is possible to make a reasonable guess, as we see in Chap. 17.

Just like Anaximander, Anaxagoras taught, as is explicitly handed down, that the celestial bodies make full circles and thus go also underneath the earth (DK 59A42 (8)). From this, he derived a remarkable explanation of the phenomenon of the Milky Way: when the sun is under the earth, some stars lie in the shadow of the earth and thanks to this they are better visible. This band of stars we call the Milky Way. Some of the stars outside the Milky Way are invisible because they are outshined by the light of the sun shining from behind the earth (59A1(9), 59A42 (10), 59A80, and 68A91). Strange as this theory may be, its merit is at least that it considered – perhaps for the first time in history – the Milky Way as an accumulation of stars.

Anaxagoras shares with Anaximander the opinion that the celestial bodies lie behind each other, but he has come to the conclusion that the right order must be as follows: moon, sun, and stars (DK 59A42(7)). The main point, however, in which he differed from his great predecessor, is that he no longer conceived of the celestial bodies as wheels with apertures through which the internal fire shines, but as fiery masses of stone (DK 59A1(8), 59A2, 59A3, 59A11, 59A12, 59A19, 59A20a, 59A35, 59A42(6), and 59A72). Apparently, Anaxagoras had reached the conclusion that Anaximander's conception of the celestial bodies as wheels was untenable. The reasons that led him to this conviction are handed down to us. Anaxagoras realized that the moon receives its light from the sun (DK 59A77, 59B18). Although this discovery has been attributed to others as well, most probably Anaxagoras has to be credited with it (see Wöhrle 1995: 244–247).

To be precise, Anaxagoras seems to have adopted a kind of position in between, for he also maintains that the moon has some light of itself, like glowing cinder. This is handed down in a late and thus not wholly reliable source: we may observe such an afterglow at a total lunar eclipse, when the moon is still visible as a pale orange disk (Olympiodorus, *In Aristotelis Meteora commentaria*, 67.33, not in DK). Perhaps we have to understand that the moon does not only receive its light from the sun, but also warmth that causes an afterglow, just like glowing ashes. In a very early source, a comparable phenomenon is described in the same way and also attributed to Anaxagoras, namely, the faint visibility of the lunar disk between the horns of the moon sickle shortly after new moon, the so-called "ashen glow." There this afterglow is said to be due to "old light" of the sun of the previous month (Plato, *Cratylus* 409a6 = DK 59A76). Remarkably, although since Leonardo Da Vinci this phenomenon is known to be due to the reflected light of the earth ("earthshine"), it is still called not only "the Da Vinci glow" but also "the Old Moon in the new moon's arms." It is nice to see how such an ancient theory as that of Anaxagoras survives in a modern saying. Cosmologically speaking, the discovery that the moon receives its light from the sun implies that the moon cannot be a light-emitting aperture in a celestial wheel, as Anaximander thought (Fig. 14.1).

Moreover, Anaxagoras realized that lunar eclipses occur when the shadow of the earth falls upon the surface of the moon and that solar eclipses take place when the moon slides before the sun [DK 59A42(9) and 59A77, see also 59A42(6)]. Eclipses could not, therefore, be due to the closing of the aperture in the wheel of the moon or of the sun, as Anaximander thought. The sources also offer an alternative explanation for the origin of a lunar eclipse, viz., that they are caused by invisible celestial bodies that exist between the earth and the moon. According to Gershenson and Greenberg, however, a confusion of Anaxagoras with Anaximenes is at stake here (1964: 351).

Anaxagoras probably did not conceive only of the earth but also of the other celestial bodies as flat disks or cylinders, as he is explicitly said to have taken the moon as another earth, with hills and ravines, and even inhabitants (DK 59A1(8)). When the moon is so much alike the earth, it must be disk- or drum-shaped as well. Although he must have studied the moon to reach his conclusion that it receives its light from the sun, Anaxagoras apparently did not realize that the shapes of the phases of the moon never can occur on a flat disk, but only on a sphere that is shined upon by the sun. This is strange because tradition has it that he not only explained

Fig. 14.1 The old moon in the new moon's arms (photograph by Steve Jurvetson)

the eclipses of the moon, but also its phases (DK 59A42(9) and 59A42(10)). Not much later, Aristotle uses this argument to prove that the moon must be a sphere (*On the Heavens* 291b18 ff.).

Anaxagoras' name remains connected with the enormous meteorite that in the year 467 B.C. in broad daylight fell from heaven at Aegospotamoi in Thracia, which caused great arousal in the civilized world of that time. At the same time, a comet was seen at night. Pliny tells that in his time, that is the first century of our era, this stone was still there and that it was as huge as a cartload (cf. Fig. 14.2). According to Plutarch, in the second century A.D., the stone was still present and was handled with awe by the local people (DK 59A11 and 59A12). It is told that Anaxagoras had predicted the fall of the stone from heaven (DK 59A1(10), 59A6, 59A10, 59A11, and 59A12). Bicknell has speculated that his prediction was based on the observation of a big sunspot. Anaxagoras could have observed it by studying the reflection of the sun on the surface of oil in a bowl, as described in Chap. 2. He could have interpreted the sunspot as a piece of stone loosening from the surface that fell upon the earth at Aegospotamoi (see Bicknell 1968: 87–90). I think, however, that it is more simple to connect this story with Anaxagoras' rather strange idea that a comet is the result of a conjunction of planets (DK 59A81). One may assume that at a collision of planets, a piece of stone was broken off and fell upon the earth. But it is also possible that the story of Anaxagoras' prediction found its origin in his unheard-of theory that the celestial bodies are masses of fiery stone, and that the prediction is ascribed to him in retrospect. However this may be, Anaxagoras must have welcomed the falling stone at Aegospotamoi as a confirmation of his theory that the celestial bodies are stones.

Fig. 14.2 The author with a meteorite about as big as that of Aegospotamoi (the Mundrabilla meteorite in the South Australian Museum, Adelaide) (photo by Heleen Pott)

Anaxagoras' theory that the celestial bodies are fiery stones of course immediately raised the question why they do not fall down on the earth. The stone of Aegospotamoi made this fear palpable. When it appeared to be possible that a big stone fell from heavens, who or what could guarantee that not one of these days the sun or the moon themselves, or even the entire heaven, would fall upon the earth? Anaxagoras' answer to this question was that the celestial bodies were kept in place by the cosmic whirl that prevented them from falling. According to Simplicius, Anaxagoras meant that the heaven does not fall because the force of the cosmic whirl is stronger than the downward force (*In Aristotelis De caelo commentaria* 374.32, not in DK). Should the celestial vortex ever stop for one reason or another, then they unquestionably would fall down (DK 59A1(12) and 59A42(6)). There had to be, then, something else that was responsible for the fall of the stone at Aegospotamoi, such as the conjunction or even collision of two planets mentioned above. The cosmic whirl was also considered the cause of the origin of the celestial bodies, when, long ago, lumps of stone were torn off the earth and whirled around, which caused them to glow as a result of friction with the aether (αἰϑήρ) (DK 59A71, 59A42, and 59A12(6)).

Perhaps one would expect that the earth, being at the center of the vortex, would itself turn around as well, but this is not the case. Our texts do not give a decisive answer, but apparently the earth is too heavy and ponderous for the vortex to have any hold on it, apart from tearing off the stones that have become the celestial bodies. Commentators disagree, however, about the question whether Anaxagoras used the image of the vortex (δίνη) for the cosmic revolution. Ferguson denies it, whereas Furley assumes that he did (Ferguson 1999: 105–106; Furley 1989: 96–97).

Anaxagoras may have seen in the stone of Aegospotamoi a proof of his belief that the celestial bodies are masses of stone, but this did not mean that he had convinced the Athenians thereof. Socrates, for instance, vehemently opposed during his trial the accusation that he should have taught that the sun is a stone and the moon a kind of earth: "Do I not believe, just like other people, that the sun and the moon are gods? (. . .) The books of Anaxagoras of Clazomenae are full of such utterances (viz. that the sun is a stone and the moon an earth)" (Plato, *Apology of Socrates* 26d = DK 59A35). According to Xenophon, Socrates himself argued as follows to show that Anaxagoras was wrong: "We are able to look into the fire, but not into the sun. In the second place we get browned by the sun but not by the fire. In the third place the sun grows the plants, whereas the fire burns them. And finally, a stone put into the fire gives no light, but is even consumed by the fire, whereas the sun is an everlasting light. Therefore the sun is not a fiery stone" (DK 59A73).

At the end of *the Laws*, the work of his old age, Plato fulminates against those who think "that the whole moving heaven is crowded with stones, earth, and other soulless bodies" (*Laws* 967c). And even centuries later, discussing the idea that the stone of Aegospotamoi has its origin from the sun, Pliny said: "Anaxagoras knows nothing of the natural phenomena, and everything is confused when we should have to believe that the sun is a mass of stone or that there should ever have been a stone on the sun" (DK 59A11). The Athenians agreed with Socrates that the conception of the celestial bodies as fiery masses of stone was blasphemous. So Anaxagoras was brought to justice, among others by Thucydides. If he had not flown to his birthplace Clazomenae, with the help of Pericles, he would have had to pay for it with his life (DK 59A1(12), 59A1(13), 59A1(14), 59A3, 59A17, and 59A19).

It is remarkable not only that predecessors of Anaxagoras were allowed to state publicly and freely their ideas in Miletus and other places in Asia Minor and were even honored as famous citizens, but also that he who had come to Athens to make known the new world picture was sentenced to death on the accusation of blasphemy. However, may be we must say that the Athenians saw sharper than the Ionians what an enormous revolution this new world picture really meant. When all natural phenomena can be explained by natural causes, there is no longer place for the gods. And how could people still feel safe in a universe that was disposed of the gods and in which only fiery stones orbited around the earth? Actually, Anaxagoras triggered a discussion that goes along with the scientific way of explaining things until this day. His trial reminds us of Giordano Bruno's or Galileo Galileï's, many centuries later. Today's controversy about Intelligent Design versus Darwinistic evolution is a recent version of the same discussion, although nowadays it is not science but religious belief that is in the defense.

Chapter 15
The Sun at the Horizon, Anaxagoras' Proof of the Flatness of the Earth

Most Presocratics believed that the earth is flat, shaped like a drum, as we see in Chap. 4. In *On the Heavens* 293b34 ff. Aristotle argues with those who maintain that the earth is flat, until he finally, in 298b20, considers the case for a spherical earth as settled. Reading those pages, one can still feel how vehement and also how complicated the discussion must have been. As nobody was able to go up into outer space and see what the earth really looks like, one had to take refuge to arguments, varying from common sense to the evidence of the senses, alleged physical laws, and even metaphysics. As a matter of fact, the defenders of a flat earth had common sense on their side: what would prevent us, and especially our antipodes, from falling off a spherical earth? A particular difficulty was that sometimes both parties shared their presuppositions. One party argued, for instance, that the earth must be flat because it is at rest, motionless at the center of the cosmos, all the other celestial bodies circling around it (*On the Heavens* 294a10 and 294b13 ff.). The idea was, obviously, that a spherical earth could easily roll away, as Simplicius remarks (*In Aristotelis De caelo commentaria*, 520.13–14). This argument was a tough one, as the defenders of a spherical earth generally shared the supposition that the earth is immovable at the center of the cosmos. In this chapter, I discuss more thoroughly another argument, in which Anaxagoras in defense of the flatness of the earth appeals to a fact from sense experience. I intend to show that the argument is more sophisticated than it seems at first sight. In trying to understand what is meant, I refer to Simplicius' commentary that is sometimes, although unfortunately not always, elucidating. Thereupon, I discuss Aristotle's counterarguments, one of which proves to be unsatisfactory, and the other correct.

In Aristotle's rendition, the argument that, as Dmitri Panchenko has shown, must have been used by Anaxagoras, sounds like this: "Some think it (sc. the earth) spherical, others flat and shaped like a drum. These latter adduce as evidence the fact that the sun at its setting and rising shows a straight instead of a curved line where it is cut off from view by the horizon, whereas were the earth spherical, the line of section would necessarily be curved" (*On the Heavens* 294a1–4).[1]

[1]Panchenko compares the argument mentioned by Aristotle in *On the Heavens* 293b24 with another one by Martianus Capella that has remained unnoticed thus far, and in which the same argument (albeit not very well understood by this author) is ascribed to Anaxagoras (1997: 175–178).

D.L. Couprie, *Heaven and Earth in Ancient Greek Cosmology*, Astrophysics and Space Science Library 1, DOI 10.1007/978-1-4419-8116-5_15,

Aristotle sets himself the rather modest task to show that "this phenomenon (...) gives (...) no cogent ground for disbelieving in the spherical shape of the earth's mass" (*On the Heavens* 294a8). Nevertheless, as we will see, Aristotle has some difficulties in refuting the argument. At first sight, the argument looks rather peculiar, and it is even not quite clear what could be meant. In an attempt to understand it, I will start to explore the notion of "horizon" (Fig. 15.1).

Fig. 15.1 Sunset at San Diego (author's photograph)

The word "horizon" may mean different things. The *true horizon* is the apparent line that separates earth from sky. Usually, however, we do not see the true horizon, but the *visible horizon* (or the *skyline*), which is the intersection of earth and sky as it is obstructed from free view by trees, buildings, mountains, and so forth. The word "horizon" may also mean the *astronomical horizon*, which is the horizontal plane through the eyes of the observer. And finally, in perspective drawing, the horizon is the straight line toward which parallel lines converge. Although all these meanings of "horizon" are used, when in the next pages I speak of the horizon without further indication, I mean the *true horizon* (Fig. 15.2).

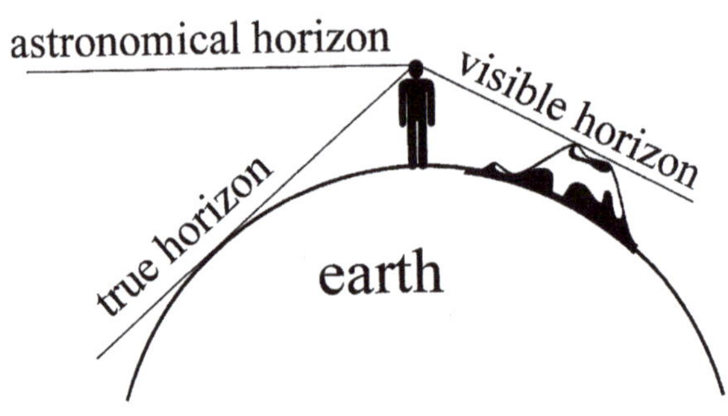

Fig. 15.2 The true horizon, the visible horizon, and the astronomical horizon

The best way to think about the horizon is to imagine being at full sea, when the weather is bright and I have an unhindered view all around, so that I can see the true horizon. When I turn around my axis for the full 360°, I will recognize that the horizon makes a full circle around me. The horizon is a visual phenomenon resulting from my eye being above the earth. The horizon determines the relatively small part of the world that is visible to me. When my eyes are at 1.7 m above the surface, the distance to the horizon is about 4.66 km, according to the formula: $d = \sqrt{h(2r + h)}$, where d is the distance to the horizon, h the height above sea level, and r the earth's radius.[2] The circle of the horizon is shown in Fig. 15.3. If my eye were on the ground, there would be no horizon, or in other words, in the case of zero height, the distance to the horizon would be zero as well.

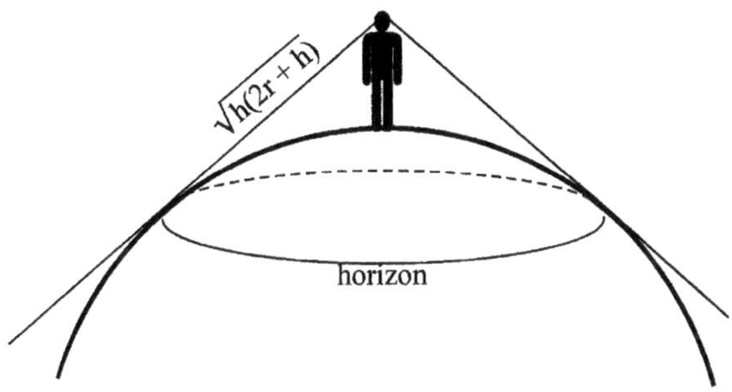

Fig. 15.3 The circle of the horizon

Standing at sea level, I am the top of a cone, the length of its side being 4.66 km, and its base being the circular plane of the horizon. As the surface of the earth is curved, the plane of the horizon cuts the earth. At my height of 1.7 m, this is also 1.7 m under my feet.[3] This means that the height of the cone with my eye at top and the plane of the horizon as its base amounts to 3.4 m.

---

[2]See http://newton.ex.ac.uk/research/qsystems/people/sque/physics/horizon/, where also a calculator for the distance to the horizon at different heights can be found. For the derivation of the formula, see next note.

[3]The derivation of the formula for the distance d to the horizon is found by Pythagoras' theorem. In the right-angled Δ CHO in Fig. 15.4, where C the center of the earth, r is the earth's radius, O the observer's eye, $OP = h$ the distance from the observers' eye to the earth, and H the horizon. Then: $d = \sqrt{(r+h)^2 - r^2} \Rightarrow d = \sqrt{r^2 + 2rh + h^2 - r^2} \Rightarrow d = \sqrt{2hr + h^2} \Rightarrow d = \sqrt{h(2r+h)}$. When we insert the radius of the earth (6,378 km) for r, and the height of the observer (0.0017 km) for h, then $d \approx 4.66$ km.

By comparing the two similar right-angled $\Delta\Delta$ CHO and CQH in Fig. 15.5, we may find x (the distance from the observer's feet to the plane of the horizon HH'). In addition to Fig. 15.4, P is the point where the observer's feet stand on the ground, h the height of the observer's eye above the ground, x the distance from the observer's feet to the plane of the horizon, HH' is the plane of the horizon, and Q is the point where the line OC cuts that plane. Note that $x + QC = r$.

Then, we get the following equation:

$$(r+h) : r = r : (r-x) \Rightarrow r^2 = r^2 - rx + hr - hx \Rightarrow hr - rx - hx = 0 \Rightarrow hr$$
$$= rx + hx \Rightarrow hr = x(h+r) \Rightarrow$$
$$x = \frac{hr}{h+r}$$

When we now insert the values 6,378 km for r and 0.0017 km for h, we get $x = 0.0017$ km.

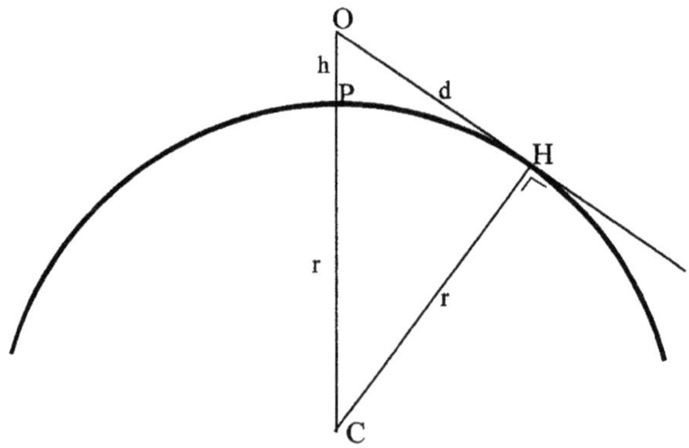

Fig. 15.4 Calculation of the distance *d* from the observer's eye to the horizon

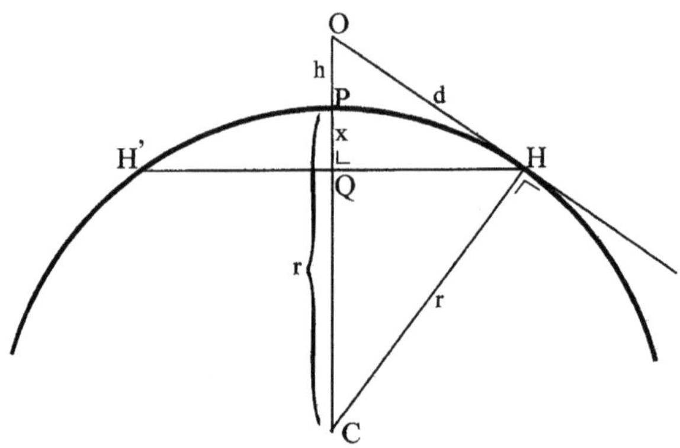

Fig. 15.5 Calculation of the distance *x* from the observer's feet to the horizontal plane

When I stand still, I see only a part of the horizon, which I perceive as a straight line separating earth and sky. Now, let us make a little thought experiment. Suppose that I am 100 m tall (or at a height of 100 m above sea level). Then, the distance from my eye to the horizon amounts to 36 km. And when I imagine being 10 km tall (or being in an airplane at that height), the distance to the horizon is 357 km. Growing taller and taller, at a certain moment, I will see the horizon clearly as curved, and finally I will be able to see the whole circle of the horizon in one view. Being tall enough, or far enough away, the horizon will practically coincide with the circumference of the earth. When I, at a certain height, see the horizon as curved, I will also observe that the rising sun shows a curved line where it is cut off from my view by the horizon. Now, I may conclude that this line has always been curved, although it was not discernable for me as I was too small (or too near the earth's surface). This is actually the same thought experiment as that which Simplicius had in mind, reaching the same result: "Perhaps one should say that if we were outside the earth and saw the sun partially obstructed by the earth, the sections would always appear to us to be curved" (*In Aristotelis De caelo commentaria*, 519.33–520.2, transl. Mueller 2005: 60).

All this holds for the horizon on a spherical earth. However, Anaxagoras' argument starts from a horizon on a flat earth. On a flat, drum-shaped earth, we can do the same thought experiment as we made above for a spherical earth. When Anaxagoras would imagine himself growing to 100 m and finally to many kilometers above his flat earth, at a certain moment, he would observe the curvature of his horizon, and finally he would see the full circular surface of his drum-shaped earth, the horizon coinciding with the full circle of the earth's rim. And he would also see this surface cut the rising or setting sun with a curved line. In other words, what he would see would not differ much from what I saw growing on a spherical earth. He would have been forced to conclude, then, that on a flat earth the sun is cut off at the horizon by a curved line as well, although we see it as straight. In other words, it seems that Anaxagoras' presupposition that the rising or setting sun on a flat earth is cut off by a straight line is easily shown to be false, being an optical illusion, just like in the case of a spherical earth.

We may wonder, however, why Anaxagoras took refuge to a visual phenomenon of which he knew that it was an optical illusion (the Greek text has φαίνεται), that turns out to be the same whether the earth is flat or spherical. This would have been a poor argument for someone who was called "Brains" by his contemporaries. Moreover, it would hardly be understandable why Aristotle paid so much attention to so childish an argument. And finally, Vitruvius explicitly mentions that Anaxagoras wrote about perspective drawing, where he had to do with another concept of the horizon, as mentioned above (*De architectone* VII, prol. 11). The implication of this is that Anaxagoras cannot have meant something trivial with his argument that the sun is cut off by the horizon with a straight line. It is again Simplicius, who explains what Anaxagoras probably meant: "*if the arc of a circle is placed in the same plane as the eye*, it will appear to be a straight line, as has been proved in the *Optics*" (*In Aristotelis De caelo commentaria*, 519.22–23, transl. Mueller 2005: 60, my italics). Mueller (2005: 115, n. 283) notes that he is hinting at Euclid's *Optics*. There we read in theorem κβ′ (22): "If the periphery of a circle is placed in the same plane as that of the eye, then the line appears to be straight" (ed. Heiberg 1895: 32, my translation).

Simplicius identifies the "plane of the circle" with the plane of the horizon: "the horizon is the plane extended through the surface of the earth and our eye (. . .) but a circle which is in the same plane as our eye is seen as a straight line" (*In Aristotelis De caelo commentaria*, 520.6–9; Mueller 2005, 60). Strictly speaking, what Simplicius here refers to is the *astronomical* horizon (as explained above), and not the true horizon, which is the result of my being somewhat above the plane of the horizon. The obvious idea is that *practically speaking* I am in the plane of the horizon. Euclid lived about one and a half century after Anaxagoras. However, Anaxagoras, as said, wrote on perspective drawing, where all parallel lines converge in one point on the straight line of the horizon. This presupposes the idea of an infinite horizon, as parallel lines cut each other in the infinite. So he could very well have pondered about the circumference of an infinite circle (the astronomical horizon) being a straight line as well. If Simplicius is right, Anaxagoras' argument was more subtle than we supposed it to be. In that case, he would readily agree what we would see from far above the earth, but he would stress that the argument is not about what we see when we are far above the earth, but about what we see while standing on the earth.

We may use the same thought experiment in the opposite direction to show what is meant. When we are far above Anaxagoras' drum-shaped earth, we see the small circle of its circumference. When we come nearer, this circle grows bigger and bigger, until we cannot see the complete horizon in one view. Finally, when we are in the plane of the horizon that is the surface of the flat earth, the circle of the horizon is infinite. And an infinite circle equals a straight line. Mark that, whereas my eye on the ground of a spherical earth would result in a zero horizon, as we saw, on a flat earth it would result in an infinite circle that equals a straight line. In other words, Anaxagoras was applying a sophisticated mathematical principle to the phenomenon of the horizon.

The gist of his argument can be formulated as follows: "Although strictly speaking we are not in the same plane as the earth's flat surface, but some 1.7 m above it, practically speaking we are in that plane, as our height is negligible in regard to the size of the earth's surface. Consequently, for the horizon on a flat earth holds the same as for the circumference of a circle that is placed in the same plane as the eye. Thus on a flat earth the horizon is seen as a straight line. On a spherical earth, on the contrary, the eye is not in the same plane as the horizon, which is confirmed by the fact that if, on a spherical earth, my eye were on the ground, there would be no horizon at all."

Let us now return to Aristotle's text. Aristotle has some pains in refuting the argument that the horizon cuts the rising or setting sun in a straight line, which according to Anaxagoras proves that the earth is flat. Having analyzed what this argument probably really was about, we may understand better what Aristotle's problems were. What he tried to do is to show that on a spherical earth the setting sun must be perceived as cut off with a straight line by the horizon as well. In other words, he must make clear that, after all, an optical illusion, and not the mathematics of an infinite circle, is at stake. If he were successful, he would be able to conclude that "this phenomenon (...) gives (...) no cogent ground for disbelieving in the spherical shape of the earth's mass" (*On the Heavens* 294a8). Aristotle has two counterarguments that are somewhat mixed up in the text.

The first is that those who say that the horizon cuts the rising sun with a straight line "fail to take into consideration (...) the distance of the sun from the earth" (*On the Heavens* 294a5). Two lines further Aristotle adds some words that Guthrie in a rather puzzling translation renders as: "(they fail to take into consideration) the appearance of straightness which it naturally presents when seen on the surface of an apparently small circle a great distance away." The Greek text has: ὥς ἐν τοῖῒς φαινοῖμένοις μικϱοῖῒς κύκλοις εὖθεῖα φαίνεται πόϱϱωθην (*On the Heavens* 294a7). Simplicius is not helpful here, for the way he renders these words distorts their meaning: "or that the circles in apparently small bodies appear to be straight lines from a great distance" (ϰαὶ ὅτι οἶΐ ἐν τοῖς μικϱοῖς φαινομένοις σώμασι κύκλοι ἀπο πλείονοῖς διαστήματῖς εὐθεῖαι φαίνεται). And the way he explains them is beside the point: "For spherical surfaces are judged to be plane from far away, as in the case of the sun and the moon" (*In Aristotelis De caelo commentaria*, 519.27–29; Mueller 2005: 60).

A more literal translation of Aristotle's text would sound: "such as in the case of circles that appear to be small: a straight line appears when they are seen from far." I think this translation provides a better understanding of what Aristotle meant: the "little circles" that are "seen from far" are those of the rising and setting sun; the "straight line" is that part of the horizon that cuts these little circles. Thus translated, these added words relate back to Aristotle's first counter-argument that is about the sun being far away. Obviously, he means that the apparent diameter of the sun at the horizon is too small to see the cutting line as curved. For a very small part of a curved line will be seen as straight. In other words, the optical illusion of a straight line cutting the setting sun is due to the relative smallness of the sun disk in regard to the horizon. If the sun were closer to the earth, while keeping its absolute size, it would look bigger. We can imagine the apparent diameter of the sun being so big as to cover, for instance, one third of the visible horizon. Then, we would see, Aristotle would argue, that the line that cuts the rising or setting sun is curved. However, this argument is not convincing, as this line would still be seen as straight, even if it would cut a huge rising or setting sun. For, as we saw in the thought experiment, we do not perceive the curvature of the horizon, unless we are far above the surface of the earth. On our height, the horizon looks like a straight line, and so it does where it cuts the sun, however big the sun might be.

Aristotle's second argument, separated from the first by the word καὶ, is that those who say that the horizon cuts the rising sun with a straight line "fail to take into consideration (...) the size of the earth's circumference" (*On the Heavens* 294a6). This argument is basically sound. Instead of imagining myself to grow, as I have done above, I may also imagine the earth on which I stand to shrink. Then, too, at a certain moment, I will see the horizon as curved, or even as a full circle. This is, apparently, what Aristotle has in mind: if the earth were small enough in relation to my length, I would be able to see the horizon as curved. In other words, the optical illusion of the horizon cutting the sun with a straight line is due to the fact that in reality I am so small that I am almost in the plane of the horizon of a big spherical earth.

We might reformulate Aristotle's argument as follows, also elaborating it somewhat: "To understand the phenomenon of a horizon, the line that separates earth and sky, it is essential to recognize that this phenomenon is due to the fact that our eye is at some distance above the earth's surface. Therefore, the issue of the horizon is not the mathematics of an infinite circle with our eye in the plane of that circle. Our eye has to be always at some height above the ground for there to be a horizon at all. This is the same for a flat and for a spherical earth. If our eyes were on the ground, there would be no horizon. On a spherical earth, this can be expressed by saying that the horizon would be zero, and on a flat earth by saying that the horizon would be an infinite straight line. Practically speaking, the height of our eye above the plane of the horizon is almost the same on a flat and on a spherical earth. On a flat earth, where the plane of the horizon coincides with the surface of the earth, the height is 1.7 m at sea level, whereas on a spherical earth this is 3.4 m.[4] These distances are so tiny in regard to the circle of the horizon that the visual result in both cases is the same: the horizon is perceived as a straight line." The result is that *the issue of the shape of the earth cannot be settled by the argument of the sun at the horizon*. The straightness of the horizontal line cutting the sun "gives them no cogent ground for disbelieving in the spherical shape of the earth's mass" (*On the Heavens* 294a8).

Finally, we can speculate on what the phenomenon of the horizon would be on a flat earth like that of Anaxagoras. We are acquainted with the horizon as the limit of our view of the earth because of the curvature of the earth's surface. Standing on a flat earth, however, we would be able to look much further than the lousy 4.66 km that limits our visual field on a spherical earth. If Anaxagoras' earth were completely flat, without mountains, trees, houses, and so forth, and without any atmosphere that could reduce the range of our view, the horizon would coincide with the rim of the earth. Anaxagoras' earth, however, is not completely flat, but has mountains, etc. Moreover, atmospheric and weather conditions would prevent us from seeing until the end of the flat earth. In other words, on a flat earth only the visible horizon could be perceived, at different distances, depending on seeing conditions. The true horizon of a flat earth would always remain imperceptible. At full sea on a flat earth, and under favorable weather conditions, we must be able to see as far as 50 km or even more. However, because of the atmospheric conditions, on a flat earth the horizon would always appear to be vague and never as sharply cut as it sometimes is under favorable conditions at 4.66 km distance on our spherical earth (see Fig. 15.1). For the same reason, on a flat earth the rising or setting sun would never be perceived as cut at the horizon by a straight line, but always by a blurred and vague line. Perhaps if Anaxagoras had realized how small the distance to the horizon really is, he would have recognized that the simple fact of the existence of the horizon, and especially the sometimes sharp horizon, proves that the earth is not flat, but curved, convex, or even spherical.

[4]As is well known, Aristotle in *On the Heavens* 298b16 estimated the circumference of the earth too big (400,000 stadia, which equals 63,000 km,). Using the calculations of note 110 above, the horizon on an earth of this size would be at about 5.8 km distance, and the plane of the horizon would cut the earth 1.7 m under my feet as well.

Chapter 16
The Sun Is as Big as the Peloponnesus

Plutarch, Hippolytus, and Diogenes Laertius report that Anaxagoras compared the size of the sun with the Peloponnesus (see Fig. 16.1). The aim of this chapter is to show that Anaxagoras was not crazy when he said this but that it was a fair estimate, from his point of view, which was that of a flat earth. More precisely, I show that with the instruments (gnomon, clepsydra, sighting tube) and with the geometrical knowledge (the properties of similar triangles, simple equations, Pythagoras' theorem) available, he must have been able to use the procedures and perform the calculations needed to reach approximately his result.

It is important to state in advance that Anaxagoras' point of view was that of a flat earth, as is already clear from Aristotle (*On the Heavens* 294b14). As we are not used to think in terms of a flat earth, the fact that Anaxagoras believed the earth to be flat yields a special problem when we try to understand what he meant with his comparison of the sun with the Peloponnesus. Both the geography and the astronomy that are based on the presupposition that the earth is flat lead to conclusions that differ essentially from those of a spherical earth.

Let us first take a look at the sources. Plutarch says that, according to Anaxagoras, "the sun is much bigger than (πολλαπλάσιον) the Peloponnesus"; Hippolytus, that "the sun surpasses (ὑπερέχειν) the Peloponnesus in size"; and Diogenes Laertius: "the sun is bigger than (μείζω) the Peloponnesus" (DK 59A72, 59A42(8), and 59A1(8)). Gershenson and Greenberg classify these reports as "late traditions whose validity is uncertain" (1964: 352). However, the comparison between the sun and the Peloponnesus is so unusual and surprising that it is hardly believable to be inserted by a doxographer. We must not forget that Plutarch and the other doxographers lived at a time when it was known that the sun is very far away and very big. They lived a considerable time after Aristarchus, who for the first time tried to measure the distance between sun and earth, and concluded that "the distance of the sun from the earth is greater than 18 times, but less than 20 times, the distance of the moon from the earth," and that "the diameter of the sun is greater than 18 times, but less than 20 times, the diameter of the moon" (see Chap. 13). Therefore, the doxographers had no reason to make up such a strange comparison as that between the sun and the Peloponnesus. Moreover, according to Diels these reports go back to Theophrastus, who probably still had access to Anaxagoras' writings. We may conclude that the comparison of the sun with the Peloponnesus, in one way

D.L. Couprie, *Heaven and Earth in Ancient Greek Cosmology*,
Astrophysics and Space Science Library 1, DOI 10.1007/978-1-4419-8116-5_16,

Fig. 16.1 The sun as big as the Peloponnesus

or another, was made by Anaxagoras himself (Diels 1879: 138). This was also Fehling's conclusion: "Anaxagoras' Angabe sieht nicht wie aus der Luft gegriffen aus, wie ein bloβes Synonym zu 'ungeheuer groβ', sondern eher wie eine Schätzung auf Grund vernünftiger Annahmen" (1985b: 209).

When we look somewhat closer at these texts, another feature might strike us. There is some oddity in saying that the sun is bigger than the Peloponnesus, without a further addition. If one would say something that makes sense, it would be something like: "The sun is bigger than the Peloponnesus, *but smaller than Greece,*" or: "The sun is *ten times* bigger than the Peloponnesus," or even "the sun is *a little bit* bigger than the Peloponnesus." It is noteworthy that this last is, explicitly or implicitly, how most authors whose commentary I will treat, Dreyer, West, O'Brien and Fehling, read it.[1] They do not explain, however, why we should read "*somewhat* bigger than," whereas the texts have "(much) bigger than." Yet I tend to go even further and doubt whether the qualifications like "bigger than" and "much bigger than" that the doxographers added to the word "Peloponnesus," reflect what Anaxagoras really said. These additions rather seem to express the uneasiness the doxographers must have felt when they read that Anaxagoras compared the size of the sun with the Peloponnesus. They sound like an attempt to make the strange comparison more acceptable. For these reasons, I think that

[1] E.g., Dreyer: "the sun (. . .) greater than the Peloponnesus, *and therefore not at a great distance from the earth*" (my italics), which implies that the sun must be rather small, see Fig. 9.9 in Chap. 9 (1953: 31). Fehling: "vielfach so groβ nach Aëtius (to whom Plutarch's text goes back, according to Diels, D.C.), *ihm war das Richtige nicht groβ genug*" (1985b: 209, my italics). O'Brien: "For if the sun is smaller than the earth" (1968: 124).

Anaxagoras, comparing the size of the sun with that of the Peloponnesus, originally must have said something like: "The sun is about the size of the Peloponnesus." This is also the way in which Gobry reads the text, without, however, giving any reason (2000: 171).

Another report of Plutarch that is not in DK states that, according to Anaxagoras, "the moon is as big as the Peloponnesus" (*On the Face in the Moon* 19. 9 (932a), not in DK). Two features of this text catch the eye. The first is that it is the moon, and not the sun, that is compared here with the Peloponnesus. The second is that the indication "(much) bigger than" is missing, or, to put it positively, the moon is said to be exactly the size of the Peloponnesus. At first sight now, everything seems to be clear: Anaxagoras had measured the absolute size of the moon (as big as the Peloponnesus), and inferred from that to the relative size of the sun (bigger than the moon, that is, bigger than the Peloponnesus).[2] We may wonder, however, whether there is any way in which Anaxagoras could have measured the absolute size of the moon, unless we assume, like West, that he argued erroneously. West suggests that Anaxagoras had gathered the reports of the solar eclipse of 19 May 557 B.C. with its path of totality about 80 km crossing the Peloponnesus from west to east, and concluded wrongly "that the moon's shadow must be the same size as the moon" (1971: 233 n. 1). This is highly improbable, not only because that eclipse occurred more than 50 years before Anaxagoras was even born, as West himself admits, but also because it presupposes that he had not the slightest idea of the laws of perspective. Vitruvius, however, mentions explicitly that Anaxagoras wrote about perspective (Vitruvius, *De architectone* VII, prol. 11). On the other hand, Anaxagoras' knowledge of perspective apparently was not sufficient enough to make him recognize that if the sun is smaller than the earth, "then the moon would be eclipsed night after night" (O'Brien 1968: 124).

My guess is that in Plutarch's last quoted text we have a trace of Anaxagoras' original words. When we read "sun" instead of "moon," then according to Anaxagoras, the sun is as big as the Peloponnesus. In other words, I agree with Fehling, who holds that Plutarch simply transferred the report on the size of the sun to the moon (1985b: 209 n. 38). However this may be, the least we could say is that Anaxagoras, thinking about the size of the sun, somehow chose the Peloponnesus as his point of reference.

Fehling takes for granted that the sun, according to Anaxagoras, is somewhat bigger than the Peloponnesus and states, quite arbitrarily, that its diameter is about 250 km. Accordingly, he calculates the diameter of its orbit around the earth as lying between 15,000 and 60,000 km, depending on the estimation of the angular diameter of the sun, which, as he says, varied from 0.5 to 2° (1985b: 209–210). Fehling's text reads as a commentary on Dreyer's lapidary remark: "therefore (the sun is) at not a very great distance from the earth" (1953: 31). However, what reasons Anaxagoras could have had for placing the sun relatively near the

[2] This is how Görgemanns (1970: 135 (24)) and West (1971: 233 n. 1) read it. See also Panchenko (2002a: 333 n. 24).

earth, they do not tell (in spite of the word "therefore" used by Dreyer), let alone that they explain why the sun should be the size of the Peloponnesus.

Sider uses an argumentation analogous to that of West. He argues that Anaxagoras could have estimated the minimal size of the sun with the help of the solar eclipse of 30 April 463 B.C. The width of the path of this eclipse (that is, the moon's shadow on the surface of the earth) that passed through Greece in an east–west direction, was 219 km. Anaxagoras "may have reached a similar figure by asking as many people as he could to ascertain who saw a full and who a partial eclipse" (1973: 129). Knowing the laws of perspective, Sider says, he may have concluded that the sun was bigger than 219 km (133 miles), which is, bigger than the Peloponnesus. This sounds like an elegant argument. However, except for a text in Empedocles, which shows that he has no idea of perspective, there is no evidence that the Greeks bothered about the width of the moon's shadow during a solar eclipse (DK 31B42). The path of the eclipse, as can be seen from recent calculations (see Fig. 16.2), ran from Thessaly until into Macedonia, and not through the Peloponnesus, as Sider hoped (1973: 129 n. 12). Accordingly, it was rather difficult

Fig. 16.2 The path of the solar eclipse of 30 April 463 B.C.

for Anaxagoras to gather the information needed. Therefore, I agree with Fehling, who calls this attempt "der kaum realistische Versuch von D. Sider" (1985b: 209 n. 40).[3]

To make a second step, we have to look somewhat later in time and in another part of the world, where astronomers, starting from the same presupposition of a flat earth, wrestled with the same questions of the distance and size of the sun. In the third chapter of the hybrid philosophical book *Huai nan tzu*, about 120 B.C., it is told how Chinese astronomers set up a gnomon (AB in Fig. 16.3), the length of which was 10 *chi* (Chinese feet). On the day of the summer solstice, at noon, they observed that their gnomon cast a shadow (BX) of 2 *chi*. They supposed that at the same moment a second gnomon (CD), put at a distance of 1,000 *li* (Chinese miles) due south of the first one, cast a shadow (DY) of 1.9 *chi* (1 *li* = 415.8 m; 1 *chi* = 1/1,500 *li* = 27.72 cm).[4] They concluded that when for every thousand *li* southward the shadow shortened by 1 *cun* (Chinese thumb; 1 *chi* = 10 *cun*, so 1 *cun* = 2.772 cm), there must be a point T, at a distance of 20,000 *li* to the south of the first gnomon, where a gnomon would cast no shadow at all. At that point, the sun must be right in the zenith. As the proportions of the triangle XAB are the same as those of the triangle XST, and AB:BX = 10:2 = 5:1, they could measure the length of ST, being 5 × 20,000 *li* = 100,000 *li*, which is 41,580 km (see Needham 1959: 225. The drawing resembles that of Thurston (1994: 91), who uses somewhat different figures). In this relatively easy way, they managed to calculate the distance of the sun from the earth.

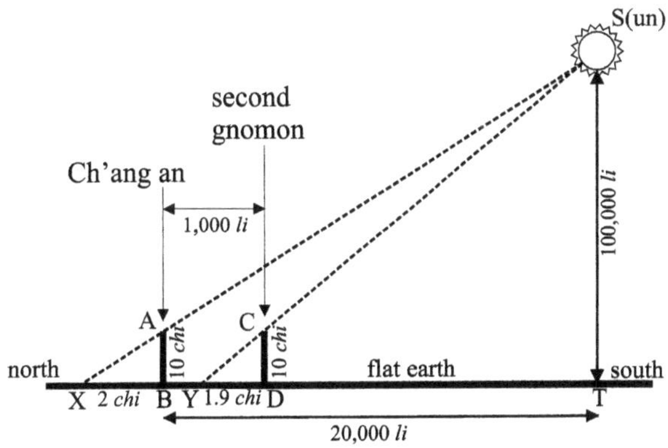

Fig. 16.3 How Chinese astronomers measured the distance of the sun (not to scale)

As in the *Huai nan tzu* the relation between the length of the gnomon and that of its shadow at the summer solstice was 5:1, we may calculate with the help of a little trigonometry that the angle at $X = 78.7°$ and the angle at $A = 11.3°$. This is the case when we measure at the summer solstice at 34.8°N (11.3° plus the inclination of the ecliptic, which is 23.5°). We may conclude that the astronomers

[3] For Fig. 16.2, I consulted NASA's *Five Millennium Catalog of Solar Eclipses* on the internet. More specifically http://eclipse.gsfc.nasa.gov/SEsearch/SEsearchmap.php?Ecl=04620430.

[4] Usually, 1 *li* is said to equal about 500 m, but Dubs has calculated the *li* of the Han dynasty to be 415.8 m (1955: 160 n. 7).

of the *Huai nan tzu* had their observation posts at 34.8°N. Perhaps they were somewhat north of *Ch'ang an* (at 34.3°N, the present-day city of Sian).

Surprisingly, the Chinese astronomers did not *measure*, with how many *cun* the shadow of a gnomon diminishes for every 1,000 *li* southward. They simply took that to be 1 *cun* per 1,000 *li*. The text suggests that this was a kind of revelation. However, when we try to calculate the real difference between the two shadows, completely other figures result. It appears that 1,000 *li* (415.8 km) to the south of 34.8°N is approximately at the 31st parallel of latitude. On that latitude, the angle at C of the *Huai nan tzu* gnomon at the time of the summer solstice is 7.5°, and accordingly the angle at $Y = 82.5°$. The length of the shadow (DY), then, is about 1.3 *chi*. The discrepancy with the supposition of the Chinese astronomers that the shadow shortens for 1 *cun* per 1,000 *li* is so significant that their number cannot be the result of observation. The same conclusion also follows from the strange consequence of these Chinese measurements that at the summer solstice, one has to go 20,000 *li* (8,316 km) to the south, according to the *Huai nan tzu*, to find a place where the sun is in the zenith (or even 60,000 *li* (24,948 km), at an equinox, according to the *Zhou bi*).[5] In reality, however, the Tropic of Cancer runs through the south of China at a distance of about 1,200 km from *Ch'ang an*, where the astronomers were supposed to live. One might try to solve this problem by introducing another value for the *li*, as Tzuong-Tsieng Moh has done in an article on the internet (see Bibliography). His value of 77 m for the *li*, however, is a deduction from his calculations, and is not based on any Chinese source, as far as I know. Moreover, this value leads to an unacceptable value for the *chi* as well, as he himself indicates. Recently, Panchenko has made the more promising suggestion that the method used by these astronomers "was established somewhere outside China and that, in the process of the transmission, the Chinese *li* was substituted for a foreign measure" (2002b: 252). Panchenko argues that "somewhere outside China" must have been Greece.

A final remark on these ancient Chinese calculations. Exactly speaking, the Chinese did not measure *the* distance of the sun from the earth, but only one of its many possible distances. On a flat earth, the distance of the sun differs according to time and place. Not only is (strangely enough) the sun in winter closer to the flat earth than in summer but it is also closer in the morning and in the evening than at noon. This is easily seen when we imagine a flat earth with a diameter that is almost as big as the diameter of the sun's orbit around it. The Chinese astronomers seemed to have been aware of this, as the *Huai nan tzu* that is supposed to have measured at the summer solstice gives another distance (100,000 *li*) than the *Zhou bi* (80,000 *li*) that probably measured about the time of an equinox.[6] And still another text says that at the winter solstice the sun is 20,000 *li* above the land (see Cullen 1996: 189). The reason for these various figures is that on a flat earth the distance to the sun not

---

[5] The final redaction of the *Zhou bi*, a collection of ancient Chinese texts on astronomy and mathematics, was probably in the first century B.C., but contains older material.

[6] In the *Zhou bi*, the gnomon has a length of 8 *chi*, its shadow 6 *chi*, which is also said to shorten 1 *cun* for every 1,000 *li* southward. In this text, BT = 60,000 *li*, and ST (the distance of the sun to the earth) = 80,000 *li* (see Cullen 1996: 78 and 178).

necessarily coincides with the distance from the observer to the sun, as is the case on a spherical earth. On a flat earth, this is only the case when the observer has the sun in the zenith (in Fig. 16.3, when the observer is at T).

Although we will need the procedure of the Chinese astronomers in determining the way Anaxagoras could have measured the size of the sun, their false results do not impinge on the argument of this chapter. At the end of this chapter, we see what will result when we calculate with the real instead of a supposed shortening of the shadow of a gnomon that is placed at a certain distance to the south of another one. Also, the fact that the Chinese imagined the earth as square does not matter here.

In principle, Anaxagoras too could have used the procedure of the Chinese astronomers. We do not know whether he did so, but it was certainly not beyond his possibilities. The use of the gnomon was known since Anaximander introduced the instrument into the Greek world. Moreover, the method, used by the Chinese to measure the distance of the sun was essentially the same as Thales is told by Plutarch to have used for measuring the height of a pyramid, as we saw in Chap. 2, Fig. 2.17. This picture, I concluded, almost invites you to draw a line perpendicular from the sun to the earth to measure its distance.

The curious thing about the procedure of the Chinese astronomers is that it is analogous to the famous experiment, by which Eratosthenes measured the circumference of the earth. The only difference is that Eratosthenes, knowing that the earth is spherical, managed to measure the circumference of the earth, whereas the Chinese astronomers, supposing that the earth is flat, measured the distance of the sun. As is well known, Eratosthenes noticed that at the moment when the sun at Syene did not cast any shadow, a gnomon at Alexandria (which he assumed to be on the same meridian as Syene) cast a small shadow, by which he could measure that the sun was 7° off zenith. He knew that the distance of the two cities was 5,000 stadia. So he concluded that the circumference of the earth must be $360/7 =$ approximately $50 \times 5{,}000 = 250{,}000$ stadia. Figure 16.4 illustrates this.

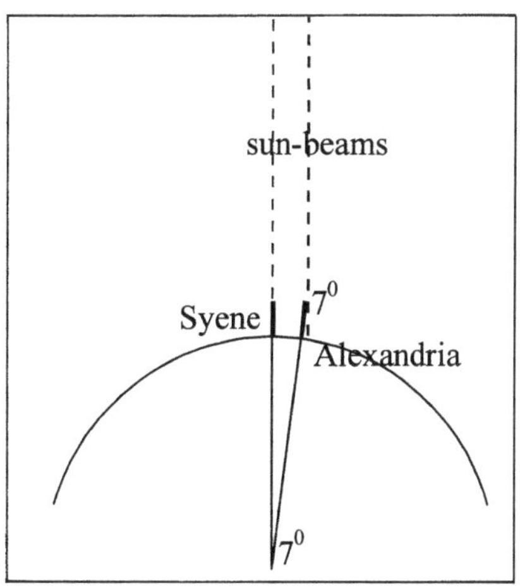

Fig. 16.4 Eratosthenes' method of measuring the circumference of the earth

As already stated, Anaxagoras must have been able to use the procedure of the Chinese astronomers. I would even suggest the possibility that Eratosthenes used the setting of a procedure like that of the Chinese astronomers, set up by a now forgotten Greek astronomer, perhaps Anaxagoras, while replacing the supposition of a flat earth by the assumption of a spherical earth. This would mean that we may construe a possible line from Thales' measurement of the height of a pyramid, via Anaxagoras' measurement of the size of the sun, to Eratosthenes' measurement of the circumference of the earth.

Before we enter into specific calculations, it is appropriate to underline that they are based on methods that are very far from exact, although the figures used may give the impression that they are rather precise. At every step, however, the calculations below will have to be looked upon as rough approximations. The ancient Greek arithmetic operations were complex and laborious, especially where fractions were involved.[7] In Anaxagoras' time, people will probably have rounded off broken numbers, not to make calculations too complicated. Moreover, "in the absence of all but the most basic trigonometry (...), the measurement of angles was not the most obvious of ploys" (Lewis 2001: 41). Accordingly, I present calculations (the "Chinese" method) in which no angles will have to be measured, except those of the apparent diameter of the sun, which can be measured indirectly with the help of a water clock. And, finally, the instruments used hardly made any exact measurement possible. The gnomon, e.g., was not so easy to put exactly perpendicular, and as the sun has a certain width, "accurate measurements of the shadow-lengths were difficult to obtain in practice" (Dicks 1954: 77). The calculations given below, therefore, indicate an order of magnitude, no more and no less, but this will suffice fully for the aim of this chapter, which is to show that on a flat earth it makes sense to say that the sun has about the size of the Peloponnesus. At the end of this chapter, I return to the subject of inexactitude of the measurements.

As already said, we do not know whether Anaxagoras executed a procedure with gnomons like that of the Chinese astronomers. But let us see what happens when we suppose he did it, and additionally, let us suppose that he used a real calculation of the shortening of the shadow of the gnomon, instead of a fictional one. The Greeks considered Delphi as the navel of their circular flat earth. Let us suppose that Anaxagoras erected one gnomon (AB) of 200 cm length at Delphi (38.5°N, 22.5°E). And let us suppose that he erected a second gnomon (CD) of equal length in the heart of the Peloponnesus, at Sparta (37.1°N, 22.5°E), about 156 km due south of Delphi (see Figs. 16.2 and 16.5). We might even imagine that he made use of the gnomon that is said to be erected there by Anaximander (DK 12A1(1)). He could have observed that at the time of the summer solstice at noon, the shadow BX of the first gnomon was about 53.6 cm long, and the shadow DY of the second one

[7] See http://www-history.mcs.st-andrews.ac.uk/HistTopics/Greek_numbers.html. Cf. also Boyer: "It was in the use of fractions that the (Greek notation) systems were weak" (1968: 11).

48.4 cm.[8] Subsequently, he could have extrapolated that for every 156 km the shadow shortens by 5.2 cm, and he could have concluded that at about 1,608 km to the south of the first gnomon the point (T) must be, where the sun stands right in the zenith.[9] This number, he could have noticed, fitted reasonably into the information he could have gathered from travelers to the south of Egypt.[10] Using the properties of the similar triangles XBA and XTS, the distance from the earth to the sun (TS in Fig. 16.5) follows from the equation $53.6{:}200 = 1{,}608{:}x$ and is 6,000 km. Of course, Anaxagoras did not calculate in (kilo)meters, but in units like feet and stadia or parasangs (the distance that could be traversed on foot in an hour), but that does not matter here.

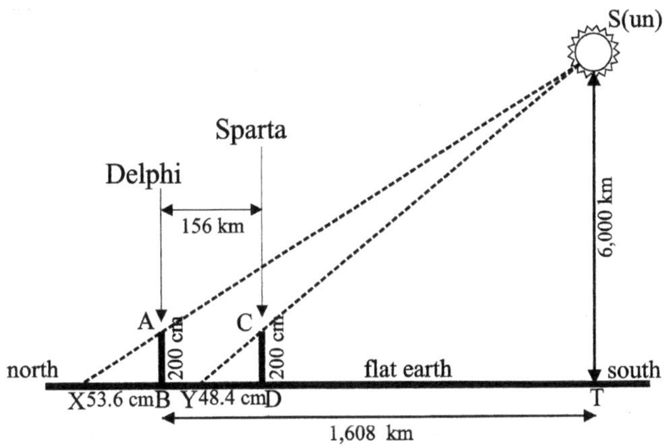

Fig. 16.5 How Anaxagoras could have measured the distance of the sun (not to scale)

The next step he could have taken is that he calculated the distance of the sun to Delphi, which is the hypotenuse XS of the triangle XTS in Fig. 16.5. According

[8] Anaxagoras could have observed this. We, however, can also calculate it with the help of trigonometry, knowing that the angle at the top of the gnomon at the summer solstice at Delphi is $38.5 - 23.5 = 15°$, and hence the shadow at the bottom of the gnomon 75°. The length of the shadow is then $200{:}\tan 75 = 53.6$ cm, and at Sparta $200{:}\tan 76.4 = 48.4$ cm. If one might think that Sparta is not far enough to the south to discern any significant difference between the shadows of the gnomons at Delphi and Sparta, because of the range of uncertainty that is inherent to measuring the shadow of a gnomon, one could imagine the second measuring point at Cape Tainaron in the farthermost south of the Peloponnesus, at 36.4°N and 22.5°E, where the shadow of the gnomon is 45.8 cm (a learned man like Anaxagoras would not have been afraid as the Greeks thought there to be one of the entrances to Hades).

[9] As the circumference of the earth, measured over the poles, is about 40,000 km, the difference between two successive grades of latitude = about 111.13 km. The distance between Delphi and Sparta is $1.4 \times 111.13 =$ about 156 km.

[10] The real distance between Delphi and the tropic of Cancer is 1,670 km.

to Pythagoras' theorem, the hypotenuse is $\sqrt{1,608^2 + 6,000^2} = 6,212\,\text{km}$. This corresponds well with a supposed diameter of Anaxagoras' flat earth of 5,000 km and allows even the moon to circle in between the sun and the earth.[11] If the big figures involved in squaring and extracting the square root might have yielded a problem to find out the length of the hypotenuse XS, it is not necessary to make use of Pythagoras' theorem. The same result could be reached with the help of similar triangles. Since in Fig. 16.5 AB and XB are known, XA can be measured with a measuring rope as about 207 cm. The length of XS, then, is the result of the equation XA:AB = XS:ST, thus 207:200 = XS:6,000, and this makes XS = 6,210 km.

To understand the last step, which is the determination of the size of the sun, we must remember that Delphi was the center of the flat earth and could also be considered as the center of the sun's orbit around the earth. The radius of this orbit is, as we have seen, 6,212 km, and thus the complete orbit of the sun around the earth $2\pi \times 6{,}212$ km = 39,031 km. The tradition has it that already Thales discovered that the angular (or apparent) diameter of the sun is 1/720 its orbit. As we have seen in Chap. 2, this attribution is too optimistic and certainly false, but it makes good sense to ascribe this discovery to Thales' successor Anaximander, since he was the first to describe the sun's orbit as a full circle around the earth. This means that Anaxagoras could have been acquainted with it. He could have calculated the sun's angular diameter with the help of one of the oldest time-measuring instruments, a water clock or *clepsydra*, as explained in Chap. 2 as well. As the angular diameter of the sun is about 0.5°, the real diameter of the sun must be 1/720 of 39,031 = about 54 km.[12]

The last step in the procedure can be carried out in another way as well. To show this, let us return, for one last time, to the Chinese astronomers. The ultimate intention of the section of the *Zhou bi* quoted above was to measure the diameter of the sun. As they used fictional numbers for the distances involved, as we have seen, this measurement as well went wrong. Their result was that the diameter of the sun = 1,250 *li* (520 km) (see Cullen 1996: 78). However, their method was not only original but also right in principle (always given the presupposition of a flat earth). They took a hollow bamboo tube (essentially a sighting tube as described in Chap. 2) of 8 *chi* with an internal diameter of 1 *cun* (=0.1 *chi*) and found that the sun exactly fitted into the bore. Then they "worked things out in proportion," as the text says. "Working things out in proportion" must mean that they calculated, again, with two

[11] Cf. Fehling: "Nun betrug die gröβte Entfernung innerhalb der damals bekannten Erde (von den Säulen des Herakles bis Babylon) ca. 5000 km" (1985b: 210).

[12] In an analogous way, one could imagine Anaxagoras to have estimated the distance and size of the moon. The moon at its highest point due south is about 5° higher than the sun. Accordingly, the point on earth where the moon at its highest stands in the zenith is at 28.5°N. If we take S in Fig. 16.6 to be the moon at its highest point due south, then the angle XST = 10°. Let us suppose that Anaxagoras estimated the distance XT from Delphi (at 38.5°N) to the point on earth where the moon stands in the zenith to be 1,000 km. Then XS = 5,817 km, and the size of the moon = 51 km. However, as far as I know, there exist no reports of the use of the gnomon for measuring the shadow of the moon, although this may be done, especially at full moon.

similar triangles: OPQ and OYZ in Fig. 16.6. The length of the tube is the perpendicular from O on PQ (the diameter of the tube), the extension of which is also the perpendicular on YZ (the diameter of the sun). Now, when x is the diameter of the sun, the following equation holds: $80:1 = 100{,}000:x$. As we have seen, the figure of 100,000 *li* (41,580 km) is much too big, according to the false supposition that the Chinese astronomers made, as explained above. When we convert, however, the Chinese *chi* and *cun* into centimeters, and take 6,212 km as the distance from the eye of the observer to the sun, as found above, instead of the wrong number of 41,580 km, the resulting diameter of the sun is $221.76:2.772 = 6{,}212:x$, so $x =$ about 78 km. Given the inaccuracies of the methods and instruments used, this number may be taken as lying in the same range as the 54 km we found earlier. As we have seen in Chap. 2, the sighting tube was known to Aristotle, but it must have been a much older instrument so that it could have been used by Anaxagoras as well.

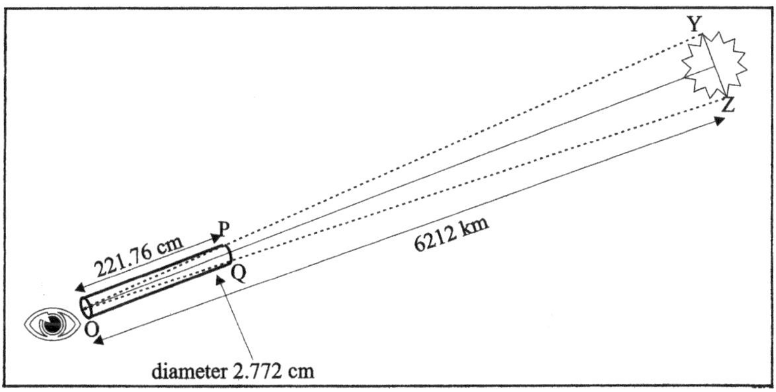

Fig. 16.6 A Chinese method for measuring the diameter of the sun (not to scale)

Anaxagoras compared the size of the sun with the Peloponnesus. The smallest east–west width of the Peloponnesus, measured through Sparta, is about 100 km. The size of the sun (that is, its diameter), when the earth is supposed to be flat, can be calculated as about 54 km, or about 78 km, depending on the method used, as we have seen. We will have to realize, however, that all the calculations involved were necessarily rather rough and inaccurate in those ancient days so that Anaxagoras' measurements easily might have resulted in a (somewhat) bigger figure than we have found with either method. To give just one example: the ancient estimations of the angular diameter of the sun varied from 0.5 to 2°.[13] If we take this last figure that is, surprisingly, used by Aristarchus, the size of the sun would become about 216 km. This figure equals approximately the greatest width of the Peloponnesus. The calculations in this chapter, then, give no more than an indication of

[13] For this and other variations in the methods and results of calculations concerning the size of the sun, see Heath (1913: 311–113).

the range of magnitude that nevertheless appears to be compatible with that of the Peloponnesus.

Supposing that Anaxagoras had his reasons to compare the size of the sun with the Peloponnesus, I have tried to bring forward circumstantial evidence to show that he had at his disposal the means to mathematically support his statement. However, whether he made an experiment like that of the Chinese astronomers or not, whether he measured the angular diameter of the sun with the help of a clepsydra or with any other method or not, or whether he simply made a reasonable guess, we may conclude that Anaxagoras was quite right, from his point of view, when he compared the size of the sun with the Peloponnesus.

However, Anaxagoras was fighting a lost battle. Plato, and definitely Aristotle, argued that the earth is spherical, and this became the prevailing opinion. As a consequence of the sphericity of the earth, the sun was, as it were, catapulted into the heavens and became much bigger than the defenders of a flat earth could have ever imagined. Aristotle seems to have understood this very well. When he was trying to counter Anaxagoras' argument in defense of a flat earth that we discussed in the preceding chapter, he remarked that the defenders of a flat earth "fail to take into consideration the distance of the sun from the earth" (*On the Heavens* 294a5). As an objection against Anaxagoras' argument, this remark is not convincing, as was shown in the preceding chapter. Aristotle's words, however, witness the recognition that the shape of the earth and the distance of the sun are two subjects that are linked together. At the very end of Chap. 14 of the second book of his *On the Heavens* (which is the chapter in which he proves the sphericity of the earth), he states, in obvious opposition to Anaxagoras' view that "it (sc. the earth) is not large in comparison with the size of the other celestial bodies" (*On the Heavens* 298a20). And elsewhere he says, more precisely, that "astronomical researches now have made it clear that the earth is far smaller even than some of the celestial bodies" (*Meteorologica* 339b7-9). We will discuss Aristotle's arguments for the sphericity of the earth more extensively in Chap. 18.[14]

[14] The article by D.W. Graham and E. Hintz, "Anaxagoras and the Solar Eclipse of 478 B.C.", *Apeiron* 40 (2007) 319–344, came to my attention after I had finished the manuscript of this book. I intend to devote a separate article to it, but essentially Sider was already right when he wrote: "The eclipse of 478 B.C. was annular (i.e. there would be no umbra)" (1973: 129, n. 10).

Chapter 17
The Dodecahedron, or the Shape of the Earth According to Plato

The discovery of the spherical shape of the earth is the finishing touch of the new world picture that was introduced by Anaximander. Strictly speaking, it is not right to use the word "discovery," since the idea of a spherical earth, as we will see in this and the next chapter, was the product of metaphysical speculations, rather than a discovery based on verifiable observations. Such observations, of which we now would say that they should have led to the conception of a spherical earth, were met with disbelief. Herodotus, for instance, tells that in the far north, there are people who sleep 6 months of the year and that sailors who navigated around Africa had the sun at their right when sailing to the west (*Histories* IV 25 and IV 42). Herodotus rejects these stories as typical bluff of sailor men. Panchenko has made plausible that Anaxagoras, being convinced that the earth is flat, argued with those who held that it is spherical (1997:175–178, see also above, Chap. 15). How hard it must have been to evaluate empirical evidence is proven by the fact that Anaxagoras was credited with the true explanation of eclipses but nevertheless apparently did not conclude from the round shadow cast upon the moon that the earth must be spherical (DK 59A42). When and by whom the earth for the first time was considered as a sphere has got lost in the mist of times. In Plato's *Phaedo*, which contains the earliest known exposition of the sphericity of the earth, it is not treated as a new knowledge, but rather as an accepted fact, as Cherniss has pointed out (*Phaedo* 108e3–5, cf. 97d. Cherniss 1964: 395; see also Tarán 1965: 296–297). According to some authors, the Pythagoreans were the first to teach the sphericity of the earth; others think it was Parmenides, to whom it is ascribed by Diogenes Laertius. For a discussion of Parmenides' claim, see Chap. 4. In a still interesting study, Frank holds that it was Archytas, who was the first to teach the sphericity of the earth (1923: 186–187). The followers of Pythagoras will have adduced mathematical-speculative arguments concerning the sphere as the most beautiful of the solid figures [DK 58C3(35)]. For Parmenides, metaphysical considerations will have played a role, on the analogy of the "Sphere of Being" (so, e.g., Burkert 1972: 305). Probably, they referred to empirical observations as additional arguments as well. In Chap. 13, I already mentioned my personal guess, Oenopides, who was also the one who is said to have discovered the ecliptic. The notion of a spherical earth entails that of an earth-equator, of which the celestial

D.L. Couprie, *Heaven and Earth in Ancient Greek Cosmology*, Astrophysics and Space Science Library 1, DOI 10.1007/978-1-4419-8116-5_17,

equator is the representation on the celestial sphere. Given the celestial equator, the ecliptic can be conceived as being inclined to it.[1]

However this may be, one would expect that Plato, if he were consistent, would have maintained that the shape of the earth is a cube. In the book of his old age, the *Timaeus*, he explains that all that exists consists of four elements, earth, water, air, and fire. These elements are tiny parts, invisible to us, and have the shapes of regular polyhedra. Fire has the shape of a tetrahedron, air that of an octahedron, and water that of an icosahedron. Earth has the shape of a hexahedron or cube (*Timaeus* 54e–55b). The reason why Plato chose the cube as the shape of the element earth is clearly that it is the most stable and immovable of all, as Aristotle explains (*On the Heavens* 307a8). So we might expect that the earth as a whole, the very word already expresses it, consisting mainly of the element earth and being immovable at the center of the cosmos, has the shape of a cube as well (Fig. 17.1).

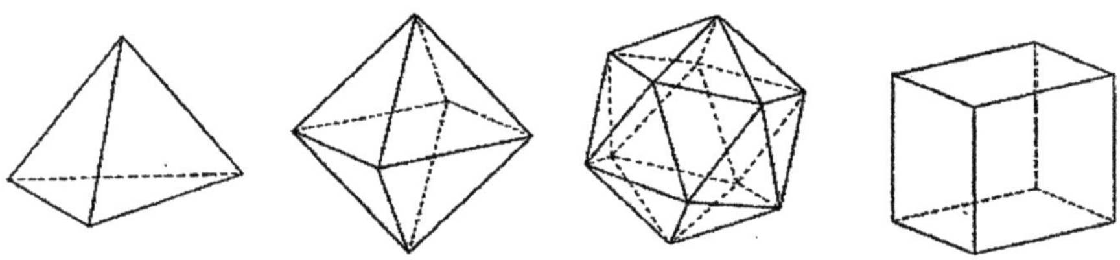

Fig. 17.1 Tetrahedron (fire), octahedron (air), icosahedron (water), hexahedron or cube (earth)

Yet it was Plato who, in the *Phaedo* which is heavily indebted to Pythagorean thinking, by the mouth of Socrates and as the first record in history, proclaims the sphericity of the earth. When it comes to a description of the earth, Plato uses a rather surprising image. The earth is, he says, like a multicolored ball that children play with, made of 12 pieces of leather (*Phaedo* 110b). Although Plato does not mention the shape of these leather pieces, scholars agree that he is hinting at a dodecahedron, which is a polyhedron made of 12 regular pentagons (Fig. 17.2).

There is, however, something strange. On the one hand, Plato lets Socrates explicitly say that he is convinced that the earth is round, meaning spherical, but on the other hand he compares the earth with a dodecahedron, which of all polyhedra comes closest to the sphere, but which of course is not a sphere. This is perhaps why some authors have argued that Plato in the *Phaedo* does not defend

[1] Burkert's suggestion that Hippocrates was acquainted with the sphericity of the earth because he "had projected the celestial circles onto the earth" is dubious (1972: 305). All Aristotle does in *Meteorologica* 343a is describing Hippocrates' opinion that the tail does not belong to a comet. In this context, he also uses the words "between the tropics" denoting a region of the heavens. However, as we see in Chap. 10, the τροπαὶ ἡλίου καὶ σελήνης can be described perfectly without using the concept of the ecliptic. Moreover, the word ὑπολείπεσθαι, used to describe the "falling behind" of the comet is the same as that used in early cosmologies to describe the apparent retrograde movement of the planets without the notion of the ecliptic. See also Lee's edition of the *Meteorologica* (1962: 40–41, note *f*).

Fig. 17.2 The fifth regular polyhedron, the dodecahedron

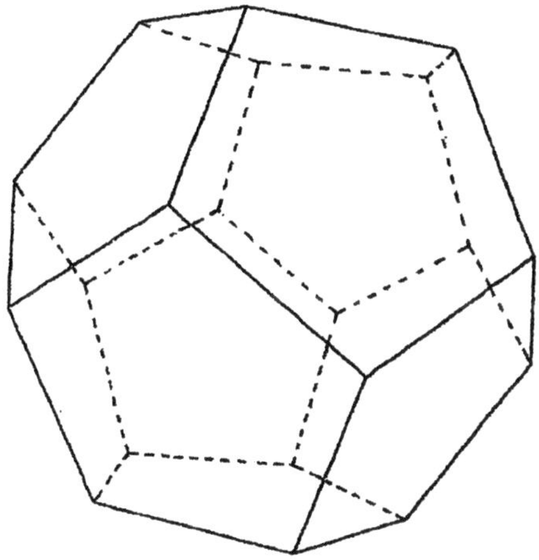

the sphericity of the earth at all.[2] Plato uses the words στρογγύλος and περιφερής, both meaning "round," to indicate the shape of the earth. If he really meant "spherical," why did Plato not use the word σφαιροειδής that was at his disposal, so they ask. This is not the place to go deeper into this matter, but the arguments pro seem to me to be more acceptable. I briefly mention them here: (1) In the section that we are concerned with Plato uses the word στρογγύλη ("round") as opposed to πλατεῖα ("flat"), so here it must mean "spherical." (2) Plato's statement that the earth, seen from above, looks like a *ball* made out of 12 pieces of leather, makes clear that he is not concerned with the defense of the Ionian conception of a disk-shaped earth. (3) Socrates' argument that the earth does not need support as a result of the homogeneous nature (ὁμοιότης) of the heavens all around it and its own equilibrium (ἰσορροπία) presupposes that he conceives of the earth as spherical.[3] (4) In a sufficiently clear context (and it is argued in 1–3 that such is here the case), the normal Greek word for spherical is στρογγύλος, just like we say in English that the earth is round, meaning that it is spherical. (5) And finally, the whole dramatic and serious context – Socrates' last words before he empties the poisoned cup – implies that Socrates would not have chosen to defend something so trivial as the conception, already found in Homer, of a flat and circular earth.

Somewhat earlier in the dialogue Socrates tells how disappointed he was when he thought that he could find in Anaxagoras' writings the answer to the questions whether the earth is flat or spherical, why it is better (ἄμεινον) for it to have this

[2] Rosenmeyer (1956), Morrison (1959), and (Fehling 1985b) are the most important defenders of the view that Plato in the *Phaedo* does not teach the sphericity of the earth.

[3] Ebert thinks that the argument also holds for a cylinder or a cone: "Was Sokrates hier sagt, trifft nicht nur auf kugelförmige Gegenstände zu. Auch ein Zylinder oder ein Kegel würde dieser Beschreibung entsprechen" (2004: 435). It is not clear to me what he might mean.

shape rather than the other, and why it is better for the earth to be at the center of the cosmos rather than elsewhere (*Phaedo* 97e). This last question, which is equivalent to the question why the earth does not fall, has become urgent since Anaximander had proclaimed that the earth floats unsupported in the center of the cosmos. Socrates explicitly answers this question: "I am convinced that, if the earth is spherical and dwells in the center of the heavens it needs neither the air nor any such force to keep it from falling, but the all-sided symmetry of the heavens and the equilibrium of the earth itself are sufficient hold it in its place. For whatever is in equilibrium and is set at the center of a homogeneous medium has no reason to incline in one direction rather than another, but being neutral it will remain immobile" (*Phaedo* 108e3–109a8). As we have seen in Chap. 8, a similar argument is ascribed already to Anaximander as well. It is almost tragic that the argument, put into the mouth of Socrates just before he is to empty the poisoned cup, is fallacious, for the simple reason that the earth is not at the center of the universe.

It is not right to say, as Ebert does, that the only reason why Plato puts forward the spherical shape of the earth is to explain why the earth does not fall (2004: 435). For Plato, two different things are important: why it is better for the earth to be spherical and why it is better for the earth to be at the center of the cosmos, in other words, how the earth manages not to fall. Notwithstanding the fact that he just had rebuked Anaxagoras for not telling why it should be better for the earth to have the shape he advocates, Socrates explicitly refrains from giving arguments for the spherical shape of the earth. This is, he says, because he is not able to do that, and even if he were it would take more time than he has left before his imminent death (*Phaedo* 97d–e and 108d).

This attitude seems justified when one sees the trouble even Aristotle still will have in fighting some empirical arguments for the flatness of the earth, as will be discussed in the next chapter. Yet we may tell with some certainty what the general sense of Socrates' answer would have looked like. The kind of proof for the sphericity of the earth Plato would like to deliver would not be based, as we perhaps might expect, on empirical observations, but would have been of quite another, metaphysical kind. The proof of the sphericity of the earth should make us understand why it is better for the earth to have precisely this shape. I think the image of the dodecahedron indicates the direction in which we have to look for an answer. When we take this image seriously, it means that Socrates on his deathbed should have given an exposition on regular polyhedra such as we can read now in the *Timaeus*. From that account should have appeared that four of the five polyhedra were used for the elements so that the fifth, the dodecahedron, remains available for the shape of the earth. And finally, he should have had to explain how the dodecahedron relates to the spherical shape of the earth. An exposé like this would, in Plato's terms, provide an answer to the question why that shape is the best for the earth. Moreover, perhaps this discourse would have contained some of Plato's "secret teachings," as will be argued hereafter.

I surmise, however, that something else is at stake as well. Perhaps the reason why Socrates is said not to be able to answer the question he had posed himself, why it is better for the earth to be spherical (c.q. has the shape of a dodecahedron), is that

Plato himself could not bring the argument to an end. In the *Timaeus*, where he takes all the time to describe the four regular polyhedra of fire, air, water, and earth, and to explain the world as a whole, in order to fulfill the promise he had made in the *Phaedo*,[4] the shape of the earth is not even mentioned. In the *Timaeus*, it is the universe that is said to have the shape of a dodecahedron: "Only one construction remained, the fifth. And god used it for the whole, making a pattern of animal figures thereon" (*Timaeus* 55c). It seems that Plato borrowed this last idea from the Pythagoreans, as Theophrastus relates (DK 44A15). With the pattern of figures, apparently the constellations of stars are meant. But why it should be better for the whole (τὸ πᾶν) to have the shape of the fifth body, which is the dodecahedron, Plato does not tell. Moreover, earlier in the *Timaeus* he had assigned to the whole the shape of the sphere, and there he had added an argumentation in the indicated sense: "The suitable shape (σχῆμα) for it is that which all other shapes encloses in itself. Therefore he made it spherical (σφαιροειδής) – equidistant everywhere from the center to the periphery – the most perfect and uniform of them all. For he judged that uniformity is immeasurably more beautiful than its opposite" (*Timaeus* 33b).

It looks as if Plato here fails to harmonize two lines of argumentation. On the one hand, a meaningful role has to be found for the dodecahedron, and he does not know otherwise than to assign it to the shape of the whole, but on the other hand he had already given the whole the shape of the sphere, and argued for it. It is an analogous ambiguity between sphere and dodecahedron as we met already in the *Phaedo* in relation to the shape of the earth. Perhaps Plato later found a way out by taking five elements instead of four by adding the aether (*Epinomis* 981c. See also Cornford 1937: 221). The fifth regular polyhedron (the dodecahedron) could then be linked to the fifth element (the aether), and the ambiguity of assigning both the spherical shape and that of the dodecahedron to the earth and to the whole could be solved by assigning to them the spherical shape, as the shape of the dodecahedron was already given to the fifth element. However, this has to remain a speculation.

Let us return to the shape of the earth in the *Phaedo*. Several authors who have tried to visualize Plato's earth completely ignore the image of the dodecahedron.[5] Some imagine Plato's earth as a kind of potato that is only "roughly speaking spherical, a corrugated sphere," as Pecker expresses it (2001: 57, Fig. 2.1; see also Baensch 1903: 190, Fig. 1, here reproduced as Fig. 17.3). Why they choose this irregular shape will be clear in due time, but already now we may say that it is hard to believe that Plato, being fond of regular mathematical bodies, would have meant such a big potato when he described the earth as a sphere or a dodecahedron. In the picture below also, the mythical subterranean channels

[4] Cf. Cornford: "Socrates in the *Phaedo* says that this distinction (between the true reason and aimless action of elements, D.C.) ought to be applied to the explanation of the world as a whole, but that he himself had been unable to attempt that task. It is the task which, many years afterwards, Plato set himself to accomplish in the *Timaeus*" (1937: 175).

[5] For visualizations of Plato's earth, see Baensch (1903: 190), Fehling (1985b: 197), Frank (1923: 24 and 25), Friedländer (1914: 99 and 104), and Pecker (2001: 57).

inside the earth are rendered that Plato describes somewhat further on, but that does not concern us here (*Phaedo* 112d ff.).

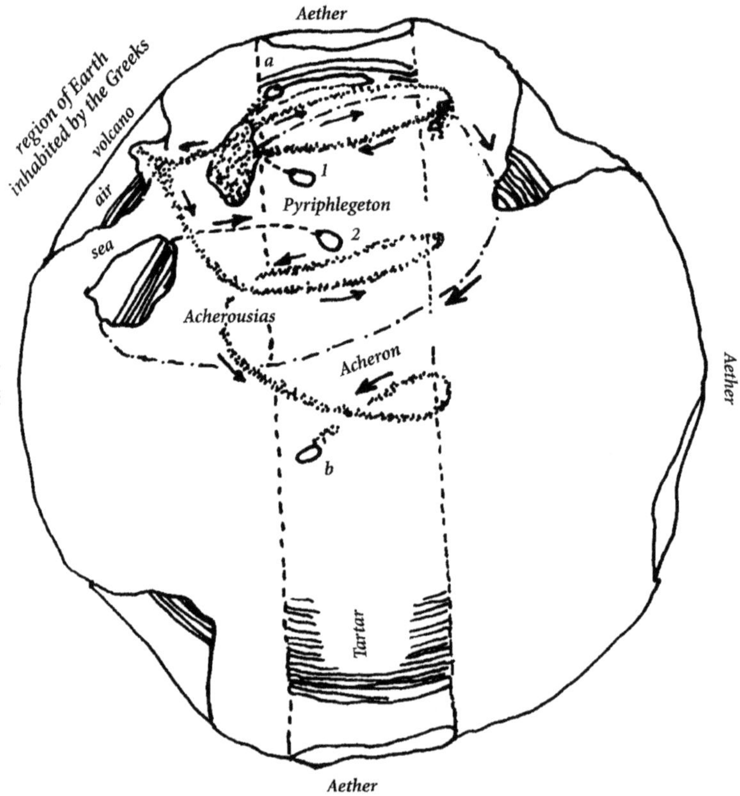

Fig. 17.3 Plato's earth according to Pecker (2001: 57, Fig 2.1, after Baensch 1903: 190, Fig. 1)

Another example is Friedländer, who draws a kind of moon landscape with craters, also ignoring the image of the dodecahedron (1914: 99, Abb. 1, here reproduced as Fig. 17.4).

A leather dodecahedron, stuffed like the balls that Socrates mentions, makes a rather perfect sphere. This is how Ebert understands the comparison, when he says that the pentagons that make up the faces of this dodecahedron must not be thought of as flat, but as curved (Ebert 2004: 438; see also e.g. Tredennick 1955: 168 n. 54; Dicks 1970: 232 n. 122). This is obviously also how most commentaries tacitly understand it, without paying further attention to it, as it seems to solve elegantly the ambiguity between sphere and dodecahedron. Figure 17.5 may elucidate this interpretation. In the following sections, I explain why I think this interpretation is not right as well.

As already said, I think that Plato's comparison of the earth with a dodecahedron is not just incidental. From the *Timaeus* it is clear that for Plato the regular polyhedra

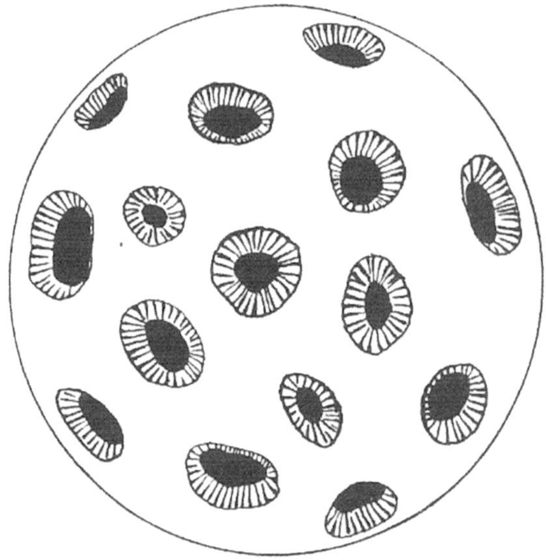

Fig. 17.4 Plato's earth according to Friedländer (1914: 99, Abb. 1)

Fig. 17.5 A dodecahedron blown up onto a sphere (author's photograph)

were very special mathematical objects. When he uses the dodecahedron for his description of the shape of the earth, he must have done so intentionally. Moreover, he must have realized that although a dodecahedron is the regular polyhedron that comes closest to a sphere, of course it *is not* a sphere. The dodecahedron is, so to speak, a sphere with dents or hollows. Just before introducing the comparison of a ball with 12 faces Plato describes the earth as follows: "(the earth) is very large, and between the Phasis and the Pillars of Hercules we inhabit only a small part of it around the sea, like ants or frogs around a pool, while many others live elsewhere, in many similar regions" (*Phaedo* 109a–b). This description is referring to the classical conception of the basin of the Mediterranean Sea, around which the lands of the οἰκουμένη are grouped. However, according to Plato, this οἰκουμένη is not the whole earth, but only a small part of it. As far as I know, nobody has drawn the simple consequence that the many hollows he is speaking of must be the 12 pentagonal faces of the dodecahedron.[6] With his use of the image of a dodecahedron, Plato multiplies, as it were, the classical image of the flat earth. And in doing this he combines, or reconciles, the ancient conception of the flat earth and the new one of a spherical earth.

Maybe one reason why this has remained unnoticed is that Plato speaks of "many hollows," whereas a dodecahedron has only 12 faces. One must remember, however, that when Plato in *Phaedo* 109b describes the hollows or cavities of the earth he has not yet introduced the shape of the earth as a dodecahedron, which he does in *Phaedo* 110b. This is why he still speaks rather vaguely of "many" hollows. Another reason why nobody has identified Plato's hollows in the earth with the dents that make the difference between a dodecahedron and a sphere might be that the sentence just quoted is followed by the words: "of all sorts of shape and size." And elsewhere: "and in the earth, in the cavities all over its surface, are many regions, some deeper and wider than that in which we live, others deeper but with a narrower opening than ours, while others again are shallower than this and broader" (*Phaedo* 109b and 111c–d). This seems hardly applicable to a dodecahedron that consists of identical regular pentagons. Undoubtedly, these texts are the reason why some authors depict Plato's earth as a kind of potato, as we have already seen. I think, however, that we have to read these words keeping in mind the already quoted description of our part of the earth that is one of these hollows. Of course, the other regions are not all shaped like the lands around the Mediterranean Sea, but they are, so to speak, "filled up" in different ways by lands and seas. In that sense, they will show "different shapes and sizes," some being "deeper and wider," others "shallower and broader." Roughly speaking, however, these different *oikoumenae* (the faces of the dodecahedron) can be called flat, just like the disk-shaped earth of the Presocratics is called flat.

Now, we can also understand why Plato depicts the earth as a multicolored dodecahedron. Seen from above, some hollows, lying north of our regions, will

---

[6] A dodecahedron with circular openings in the pentagons, dated ca. 200 B.C., which is preserved in the Rheinisches Landesmuseum in Bonn, unintentionally illustrates Plato's image very nicely (see Emmer 1993: 216).

look bright white because they are covered with ice and snow. Others, being south of us, will look purple and gold, as they are burnt by the sun or desert-like. Our own moderate zone around the Mediterranean Sea will look green and blue.[7] Or, in Socrates' words: "One part is purple (...), and one is golden, another one is white (...), and in a similar way the earth is made up of other colors. (...) For those hollows of the earth that are filled with water and air present a manifestation of glittering color among the various other colors, so that the whole reveals one continuous multicolored image" (*Phaedo* 110b–c).

However, this interpretation of the earth as a dodecahedron with 12 "hollows" is not the whole story. In a well-known text, Plato makes a distinction between the apparent earth (ἥδε ἡ γῆ) and the real earth (ἡ ὡς ἀληθῶς γῆ): "For there are many holes everywhere around the earth (...), into which the water, the mist, and the air have run together. But the earth itself is pure and is situated in the pure heaven where the celestial bodies are, and which those who study these things usually call the aether. Its sediment is that which always flows together into the hollows of the earth. We do not realize that we are living in the earth's hollows, and think that we live at its surface, just like someone who dwells in the middle of the bottom of the sea [would believe to live upon the surface of the sea and], seeing the sun and the other celestial bodies through the water, would think that the sea is the sky (...). The same is the case with us. We dwell in a kind of hollow in the earth, thinking that we live on its surface, and we call the air the heaven, as if it were the heaven where the celestial bodies move (...). If someone should reach the top of the air or got wings and fly up, he would emerge and see. Just like fishes, emerging from the sea, would see this world here, so he would see that world there. And then (...) he would recognize that that is the real heaven and the real light and the real earth. For this earth of ours (...) is injured and corroded (...)" (*Phaedo* 109b4–110a3).

Let us first look at the strange clause between square brackets where Plato compares us – the dwellers of the hollow that is formed by the basin of the Mediterranean Sea – with someone who lives in the middle of the bottom of the sea and believes to live upon the surface of the sea. As far as I know, nobody has noticed that something is wrong with this text. First of all, how do we have to imagine someone who lives under water to think that he lives at the surface of the water? Second, the comparison does not fit. Plato compares us, who live at the bottom of the air sea, with a man who lives at the bottom of the water sea. We believe that the air is the sky, just as the sea-dweller believes that the sea is the sky. He does not notice that the sky, in which the celestial bodies move, is above the surface of the sea. In the same way, we have no idea that the real sky is above the surface of what we call the sky, but which is only the air.

---

[7] Plato probably did not yet realize that on a spherical earth there are two moderate zones, on either side of the tropical zone around the equator, just like there are two cold regions at the both poles. This means that he still started from the Ionian scheme of a flat earth, on which it becomes colder as one goes to the north, and warmer as one goes to the south (see Chap. 6). The division of the earth globe in climate zones is for first time described in Aristotle's *Meteorologica* 361a ff.

So far so good. We do not believe, however, that we live at the surface of our air sea, as the parallel with the sea-dweller would require, but we assume that we live at the surface of the earth (οἴεσθαι ἄνω ἐπὶ τῆς γῆς οἰκεῖν). Something has gone wrong here. Rather than thinking that Plato had a blackout, I would suggest to omit the clause οἴοιτό τε ἐπὶ τῆς θαλάττης οἰκεῖν καὶ ("would believe to live upon the surface of the sea and") at 109c5–6, which looks to me as the insertion of a copyist, who unsuccessfully tried to improve the text. What remains makes perfect sense and contains the comparison Plato must have thought of. The passage in question, then, will read like this: "We do not realize that we are living in the earth's hollows and think that we live at its surface, just like someone who dwells in the middle of the bottom of the sea seeing the sun and the other celestial bodies through the water, would think that the sea is the sky (etc.)" (*Phaedo* 109c3–7).

The whole text (*Phaedo* 109b4–e8) looks like an anticipation of the famous allegory of the cave in the *Republic*. The apparent earth is what we would call the empirical earth, and what Plato calls the real earth we would rather call a mythical earth. The consequence of Plato's simile is that what we are used to call the heaven must have an upper surface as well that separates the air from the aether, just like the surface of the sea separates it from the air. The air, or what we call the heaven, is as it were the transparent roof of the hollow in which we live, and above this roof the real heaven, the aether, is situated. We are, however, not aware of this, just like the imaginary dweller on the bottom of the sea does not know that there is air above the sea.

I think that the apparent earth has to be identified with the dodecahedron, as described above, the stereometric body with hollows, one of which is the Mediterranean basin with the lands around it. This apparent earth is far from perfect, just like a dodecahedron is not a perfect sphere. If I am right that the hollows in the earth, of which Plato speaks, are visualized in the 12 faces of the dodecahedron, then the easiest way to imagine the boundary between the air and the aether, and thus the real earth itself, is the circumscribed sphere of that dodecahedron. When we draw the circumscribed sphere of the real earth on the dodecahedron-shaped earth, then every hollow or cavity (every pentagon of the dodecahedron) has its own curved coverture of air, as is easiest seen at the pentagon on top of Fig. 17.6.

The construction of a dodecahedron in a sphere was a hot topic in Plato's days. It is told that the Pythagorean Hippasus was thrown into the sea and drowned because he had given away the secret of how it has to be done (DK 18A4).[8] Perhaps this aspect of "secret teaching" is the reason why Plato is somewhat reluctant to exemplify completely the coexistence of the two images of the sphere and the dodecahedron. Nevertheless, we can try to imagine what a real spherical earth circumscribed on an apparent dodecahedron-shaped earth looks like. What we call the air is, as it were, the sea of what Plato calls the real earth. Or better, one of its

[8] Later, Euclid at the end of his *Elements* (Book XIII, Proposition 17) explains how to construct a dodecahedron and how to inscribe it in a sphere.

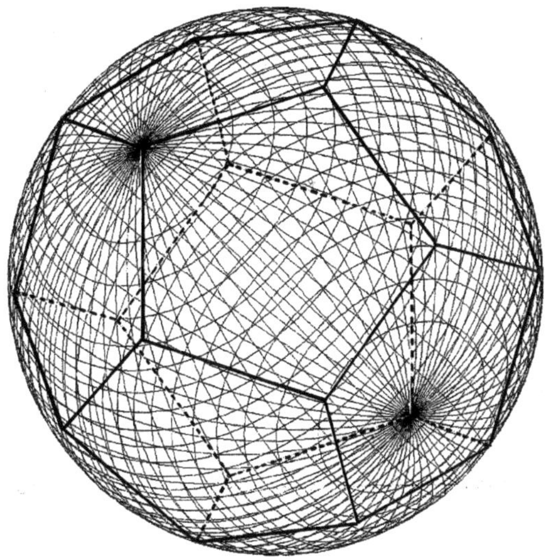

Fig. 17.6 Dodecahedron with circumscribed sphere (drawing by Ton Lecluse)

12 "air seas." The habitable regions of the real earth will not be right above our heads, for there we can look through the air and through the aether and see the stars, just like the imaginary dweller on the bottom of the sea sees the celestial bodies through the water. The habitable regions of the real earth will rather be at the fringes of the cavities that make up the apparent earth, where mountains reach above the air into the aether. As Heath says: "these hollows are separated by the ridges between them, and it is only the tops of these ridges that are on the real surface of the spherical earth" (Heath 1913: 147).

Here and there, even high mountains will rise up like islands in the middle of the air sea. Accordingly Plato speaks of people that live there "on the coasts of the air," or on "islands encompassed by air" (*Phaedo* 111a6–8). This means that the contours of the pentagonal holes are still visible when one looks from above upon the earth. This is what Plato describes when he says that the earth when seen from above looks like a leather ball of 12 pieces. Seen from above, and being in the aether on the real earth, everything looks more bright and beautiful than we are used to in our cavities covered with air. This holds especially "for those very hollows of the earth that are full of water and air. They present an appearance of color as they glisten amid the variety of the other colors, so that the whole produces a continuous multi-colored view" (*Phaedo* 110c8–d3).

This interpretation of the "real earth" as the sphere circumscribed on the dodecahedron of the "apparent earth" differs fundamentally from Kingsley's interpretation, in which (if I understand him well) the "real earth" is situated elsewhere in the universe, just like the "aitherial earth," or "celestial earth," or "Olympian earth," or the Counter-Earth in Pythagorean speculation (1995: 91 and n. 12).

As far as Plato is concerned, we cannot speak accurately of the "discovery" of the sphericity of the earth. For him, rather teleological speculations are at stake, answering the question what will be the best shape for the earth, or metaphysical-mathematical theories concerning the regular polyhedra and the sphere. At first sight, these two lines of argumentation seem to conflict. On the one hand, his speculations lead to the conclusion that the spherical shape is the best for the earth because it includes all the others (meaning all regular polyhedra). On the other hand, he imagines the earth as a dodecahedron, the fifth regular body that has not yet found an application after the four others were used for the four elements. It may be argued, however, that this twofold shape of the earth, far from being an inconsistency in the *Phaedo*, points to the very heart of Plato's philosophical intentions. When the shape of the dodecahedron is assigned to the apparent earth and the spherical shape to the real earth, this distinction is essential. Plato is not concerned with cosmology, but with metaphysics. He uses the cosmological theory of the sphericity of the earth, of which he says that he has heard it from "someone," not for cosmological ends but to illustrate his own philosophy. Elsewhere, he expresses the same intention by saying that the guardians of his ideal state must not occupy themselves with empirical observations of the celestial bodies and accordingly have to do away with them, since real exactitude is impossible to obtain in the physical world. Astronomical observations only convey the imperfect images of the perfect uniform circular movements. These perfect movements are expressed by ideal numbers and studied in "real astronomy" (*Republic* 530a–b).

Gregory vindicates a rather sophisticated interpretation of this text, according to which a distinction is meant between the sensible and intelligible parts of astronomy (1996: 459–462). As far as I can see one may with reason call the intelligible part of astronomy, which "leaves the things in heaven alone," "real" astronomy in Plato's sense. An example of such numbers is given in the *Timaeus*, where a celestial musical scale is constructed that has nothing to do with music, nor with the science of harmonics, as Cornford contends (*Timaeus* 35b–36d; Cornford 1937: 68). Frank expresses it thus: "Die Tonleiter, die er im Timäus (. . .) durch (. . .) Zahlenspekulation erhält, ist (. . .) eine metaphysische Konstruktion und hat mit den Tonleitern der wirklichen Griechischen Musik kaum etwas gemein" (1923: 13). Here, we have the same difference between empirical music and its scales, which is studied using the ears, and the real, perfect music and its scale which is studied by the mind, as Plato explains elsewhere (*Republic* 530c8–531c7). In the same sense, in the *Phaedo* the empirical, cosmological earth is not what finally matters, as for Plato the empirical earth is only the apparent earth, an imperfect copy of the real, metaphysical earth. This is why the empirical earth is a dodecahedron, whereas the real earth is a sphere.

Chapter 18
Fear of Falling: Aristotle on the Shape of the Earth

The fear that the heavens will fall down is a universal theme in mythology. Strabo tells that the ancient Celts did not fear anything so much as the possibility that the heaven would collapse (*Geographica* liber VII 3.8). Another example is a myth from East Siberia, in which it is told that the gods had built the heaven from stone, but the human beings on the earth became afraid that the heaven would fall down. Therefore, the gods blew air under the celestial vault so that it was hidden from human view (Holmberg 1922–1923: 41). In the famous trial of Horus and Seth, the goddess Neith writes to the Ennead (the nine gods of the court that had to decide whether Horus or Seth was to succeed Osiris): "If you don't give the office to Horus, I will become very angry and cause the heaven to touch the ground" (Borghouts 1988: 99–100, my translation). The same theme also appears in the *Gilgamesh Epos*. Gilgamesh tells his mother a dream: "Mother, last night in my dream I walked nervously up and down between the men. Then the stars of the heaven came down on me, the firmament fell upon me! I wanted to lift it up, but it was too heavy for me, I tried to remove it, but I was not able to put it away" (*Gilgamesh Epos*, transl. De Liagre Böhl 1952: 27). The archaic fear of a catastrophic collapse of the celestial vault can also be heard in some verses of Hesiod, when he describes the battle between the thunderbolt-throwing Zeus and the Titans: "The view and sound of all this was as if the wide heaven had fallen upon the earth. Like the thundering of the collapsing heaven and the crashing earth was the shock of the gods charging onto each other" (*Theogony*, 705–709). And much later Lucretius predicts that one day "the complex structure of the cosmos will tumble down (...) with horrifying cracking the universe will collapse" (*De rerum natura* 5.91–109).

Actually, the fear for the collapse of the heaven is the fear that the situation before the separation of heaven and earth will enter again. In Chap. 1, we have already seen that in the Egyptian representation Nut and Geb (heaven and earth) originally were entangled until they were separated by Shu. Myths about the separation of heaven and earth are known from cultures all over the world, as Staudacher shows in his book *Die Trennung von Himmel und Erde* (1968: 3–42). On the question how heaven and earth remain separated, various answers were given. The Laps, for instance, thought that the heaven was fixed by the "north nail" (*bohinavvle*), the polar star, around which the heaven turns. When at the Day of Judgment this nail will loosen because Arcturus shoots on it with his bow

D.L. Couprie, *Heaven and Earth in Ancient Greek Cosmology*, Astrophysics and Space Science Library 1, DOI 10.1007/978-1-4419-8116-5_18,

(the Pleiades), the heaven will fall down and will shatter the earth (Holmberg 1922: 10). Usually, people imagined a solid support in the shape of a huge tree or pillar, at the end of which the polar star resides (Holmberg 1922: 12). This image is present in all kinds of varieties with peoples from Scandinavia to Japan and the Indians of North America, but also, for instance, in ancient Mesopotamia. Frequently, the heaven itself was imagined as a big tent (Holmberg 1922: 19–20. See also Eisler 1910: passim). In ancient Greece, Atlas is sometimes imagined as not bearing himself the heaven, but as upholding one or more pillars that support the heaven (e.g., Homer, *Odyssey* I, 52–54). In Chap. 1, we looked at the Babylonian version of the tree of the heaven, as well as my interpretation of the Biblical story of the rainbow as a support of the heavens.

Anaximander of Miletus broke with the ancient image of the celestial vault as a kind of cupola or roof, the edges of which touch the earth at the horizon. One of the consequences of the new world picture he introduced was that it made possible a completely new answer to the question why the heaven does not fall upon the earth. The new answer was that the heaven of the archaic world picture, which is to say the celestial vault, simply did not exist. Yet the problem of falling returned in Anaximander's world picture as well, and in two different ways.

In the first place, it returned as the question why the celestial bodies do not fall upon the earth, which was the equivalent of the crashing of the celestial vault in the archaic world picture. Anaximander's answer to that question was, as we see in Chap. 8, that the celestial bodies are rings or wheels turning around the earth. Later thinkers on cosmology, who did not share Anaximander's opinion on the nature of the celestial bodies, had to answer this question once more. In Chap. 13, we saw that the usual answer was that the celestial bodies were kept in place by the celestial vortex. Whether this was a definite safeguard became dubious when in the year 467 B.C. in Aegospotamoi a stone fell from heaven. Anaxagoras, who held the opinion that that stone took its origin from the sun, may have seen in this event a proof of his conviction that the celestial bodies are fiery masses of stone (DK 59A1 and 59A11).

In his *Meteorologica*, Aristotle does his utmost to make clear that the stone of Aegospotamoi could not have originated from the sun, but must have been transferred by the wind at the time that a comet was seen in the heavens (*Meteorologica* 344b32). What Aristotle meant can be clarified by a text of Plutarch. In his *Life of Lysander*, he deals with the stone of Aegospotamoi, saying that some think that the stone has been blown from a mountain top by a storm of wind, has flown through the air, and fallen down on the earth again (*Life of Lysander* 12.1). Against Anaxagoras, Aristotle believed that the celestial bodies did not consist of one of the four elements and thus could not consist of stone, and not even of fire, but must consist of a fifth element that he called the "aether." The reason for his conviction is treated hereafter. At first sight, here, a dispute between Anaxagoras and Aristotle on the nature of the celestial bodies is at stake. But behind this controversy, the question is hidden whether we have to be afraid that the celestial bodies may fall upon the earth. When the sun can crumble off, as in Anaxagoras' opinion, what will be our guarantee that one day the sun as a whole will not fall down on earth? To Aristotle, it mattered a lot to show that that fear was unfounded.

In the second place, as a consequence of Anaximander's idea of the free-floating earth, the question why the earth itself does not fall had become urgent. In *On the Heavens*, Aristotle describes the highlights of the discussion that Anaximander's ideas gave occasion to (*On the Heavens* 294a19 ff.). He mentions how Anaximander with an ingenious argumentation that is discussed in Chap. 8 tried to reassure his fellow citizens that the fear of falling was unfounded. The earth, Anaximander argued according to Aristotle, has no reason to fall because it dwells in the center of the universe "being everywhere at the same distance from the periphery." Whether this sophisticated argument really has to be ascribed to Anaximander or not, it was obviously not sufficient to take away the fear of falling. Perhaps Fehling (1985b: 224) is somewhat exaggerating when he writes that the real subject of Aristotle's *On the Heavens* is the question why the earth does not fall, but at least the chapters xiii and xiv of Book II have the intention to settle this question that worried his predecessors.

Aristotle's arguments regarding the spherical shape of the earth are intertwined with his arguments why they earth does not move, and more specifically why it does not fall. More precisely, in *On the Heavens* the question of the shape of the earth is part of a triad of interdependent problems that are introduced at the beginning of book II, chapter xiii (*On the Heavens* 293a15). The other two are as follows: Where does the earth dwell in the cosmos (and why is it there and nowhere else) and Is the earth at rest or does it move? These three questions display a kind of hierarchy. The most important problem is whether the earth moves or not, as Aristotle himself clearly indicates. When he deals with this question he twice underlines its importance: "Everyone must be impressed by the difficulty of this problem" (*On the Heavens* 294a12), and again: "This question rightly has become a subject of common interest" (*On the Heavens* 294a19). The two other questions, regarding the place of the earth and its shape, are secondary as against the fundamental question whether the earth moves or not. This raises the question, why this was such an important problem at all.

When Aristotle discusses questions that we today would consider as belonging to the domain of cosmology, he is treating in fact questions of metaphysical, or if you want, existential importance. The ultimate question Aristotle had to answer was whether mankind could still feel safe in the universe, given the destruction of the archaic world picture of the flat earth with the celestial vault. For this task, in the end, empirical arguments were inadequate because they could be countered with other empirical arguments, of which the stone that had fallen at Aegospotamoi was not the least. Empirical arguments were not able to take away the fear of falling that was inherent to the new world picture. To cope with this task, Aristotle had to take refuge to metaphysical arguments. For centuries to come, it was not Aristotle's empirical argumentation, but the metaphysical vindication of his world picture that made people feel safe within a universe in which the earth is thought of as a free-floating sphere. It is typical for Dicks' misunderstanding of the meaning of Aristotle's world picture when he writes: "Aristotle himself can hardly be blamed for the fact that some of the main features of his universe (...) lasted nearly 2000 years" (1970: 217).

As we will see hereafter, the question of the shape of the earth as well could not be finally settled by empirical arguments alone. A typical example of how the far-reaching dimensions of the questions Aristotle was had to answer can be misunderstood is given by a modern physicist's commentary. At the end of his discussion regarding the questions what is the place of the earth in the universe, whether it moves or not, and what the shape of the earth might be, Aristotle, as a kind of appendix, gives two empirical arguments for the sphericity of the earth (*On the Heavens* 297b23 ff.). The first one is that during an eclipse of the moon, the earth's shadow on the moon is always convex. The second is that when we travel north- or southward, we do not see the same stars. The famous scientist Stephen Hawking praises Aristotle for delivering "two good arguments for believing that the earth was a round sphere rather than a flat plate" (1988: 2–3. Another example is, e.g., Upgren 2002: 93). However, the case is not as easy as one is liable to think from a present-day perspective like Hawking's. Actually, Aristotle had to cope with the defenders of the conviction that the earth is flat, and these, too, were able to put forward empirical arguments for their standpoint.

When one reads the *On the Heavens* it is obvious that Aristotle feels uncomfortable with these empirical arguments and has pains to deal with them. In Chap. 15, we saw that Aristotle has difficulties in countering Anaxagoras' argument that the sun at sunrise and sunset is cut off at the horizon by a straight and not by a curved line (*On the Heavens* 294a2). Initially, Aristotle tries to counter this argument by saying that it does not take into account the distance of the sun to the earth (*On the Heavens* 294a5). This is not a very strong objection, as the sun's distance to the earth has little to do with the question whether the line of the horizon, where it cuts the sun, is straight or curved. Furley rightly remarks: "it is not clear what this has to do with it," and Heath says: "His answer is confused" (Furley 1987: 198; Heath 1913: 235). Aristotle's second reply is that the argument does not take into account the huge circumference of the earth, so that a small piece of a curved line looks straight, makes more sense (*On the Heavens* 294a6). This argument has been discussed extensively in Chap. 15. Elsewhere Aristotle argues that the proponents of a flat earth do not take into account that heavy bodies do not fall parallel to each other (*On the Heavens* 297b20), as would be the case on a flat earth, whereas on a spherical earth all falling movements are radial, pointing toward the center of the earth. This, however, he could not know from his own experience, and thus he presupposes what has to be proven. Or, in other words, he blames his opponents for not accepting his own theory of falling, which we will discuss below.

Another argument of those who oppose to the sphericity of the earth is that its immobility necessarily involves its flat shape (*On the Heavens* 294a10). The idea seems to be that a spherical earth could easily roll over, whereas a flat earth (a flat disk) has a natural stability. See Simplicius' remark, already quoted in Chap. 15 (*In Aristotelis De caelo commentaria*, 520.13–14). Aristotle's first objection to this is: "when the earth is not flat, its immobility cannot be due to its shape" (*On the Heavens* 294b23). Of course, this is not an answer at all, but a kind of begging the question. Aristotle also argues that not its flatness but its bigness is the cause of the immobility of the earth (*On the Heavens* 294b27). It is not quite clear what he means, and he seems to be dissatisfied himself with his answer, for he suddenly interrupts the discussion, saying that we need not argue over details (*On the Heavens* 294b32).

Perhaps the fact that Aristotle obviously had problems with this kind of objections is the reason why he postpones his own empirical arguments for the sphericity of the earth to the end of the discussion, knowing that these alone will not fully convince the reader. The more important reason, however, is, I think, that according to Aristotle the question of the shape of the earth cannot be decided on empirical grounds. His decisive proof for the sphericity of the earth is of a quite different kind that perhaps even can be called metaphysical, as Dreyer does: "In his general conception of the Kosmos Aristotle is guided by purely metaphysical arguments" (1953: 109). See also Dicks: "His arguments are largely *a* priori" (1970: 196). Kahn, too, says that "in Aristotle's demonstration of the earth's sphericity, general cosmological arguments take precedence over τὰ φαινόμενα κατὰ τὴν αἴσθησιν" (1985: 118). Aristotle himself expresses it thus: we must consider this question from a universal and more principal point of view: ἀλλὰ περὶ ὅλου τινὸς καὶ παντός, and ἐξ ἀρχῆς γὰρ διοριστέον πότερόν ἐστί (*On the Heavens* 294b33–34).

In a certain sense, Aristotle here stands in the tradition of Plato, who is disdainful of empirical arguments, and instead looks for an explanation that says why it is better for the earth to have a certain shape, as we saw in the preceding chapter. Nevertheless, Aristotle's arguments are definitely of a different kind than those of Plato and the Pythagoreans. Plato's arguments for the sphericity of the earth have, au fond, nothing to do with cosmology, but are the result of his considerations about the number of regular solids and his idea that the metaphysical earth is much more important than the empirical earth. Aristotle, on the contrary, tried to answer cosmological questions, and his arguments are generalizations of what in his time could count as empirical (or better: common sense) facts. Both the questions and the answers, however, had farther reaching, and even metaphysical consequences.

Let us try to reconstruct Aristotle's way of arguing in a few steps. In the first place, he maintains that the earth is situated in the center of the cosmos, in conformity with what had become part of the everyday experience since Anaximander taught that the celestial bodies turn around us. Actually, Aristotle's arguments why the earth holds the central position can be seen as a vindication of common sense. One part of that vindication consists in fighting those who held an opinion contrary to common sense. Consequently, Aristotle argues against the Pythagoreans, who placed not the earth, but the central fire in the center of the cosmos, as we saw in Chap. 13. His argument is a kind of early example of Occam's razor: the Pythagoreans imagine on aprioristic grounds the existence of two bodies, the central fire and the counter-earth, the existence of which it is impossible to demonstrate empirically and which are superfluous to explain the celestial phenomena (*On the Heavens* 293a17–293b17). These negative arguments to counter his opponents were, however, not sufficient to prove in a positive way that the earth dwells in the center of the universe and stays there immovably. In order to deliver such a positive argument, Aristotle had to approach the problem within a much broader context, which was an analysis of the meaning of "falling."

Aristotle's analysis again starts from common experience, by stating that the direction toward which heavy things move is "down" and the direction toward which light things rise is "up." What we see is that heavy things fall toward the earth and that light things rise up from the earth. In other words, falling is a movement that is principally in a straight line, only accidentally bent away under the influence of secondary forces. Rising up is the opposite movement, also in principle in a straight line. The earth seems to be the ultimate end of falling heavy things and the ultimate starting point from which light things rise up. Talking in terms of elements, earth is the heaviest element and thus tends to move toward the earth, whereas fire is the lightest element and tends to move up in the air. The two other elements, water and air, will tend to move toward somewhere in between. Here, Aristotle refers to the exposition in *On the Heavens* 268b17ff.[1]

A third phenomenon of common sense is that the movement of the celestial bodies is not in a straight line, but circular, as everyone can see who looks at the stars by night or at the sun by day. This circular movement, which we see in the orbits of celestial bodies, is more perfect than the rectilinear movements of the elements because it returns in itself. For Aristotle, the fact that the celestial bodies do not show a rectilinear movement but orbit at a great distance around the earth proves that they are not made of one of the four elements (earth, water, air, and fire), and certainly not of stone. The bearer of this circular movement has to be, then, a fifth element that Aristotle calls the aether (αἰθήρ) (*On the Heavens* 268b11–269b17 and 289a). Thus, he knocked down Anaxagoras' pioneering conception of the celestial bodies as fiery masses of stone. When the celestial bodies were stones, they would since long have fallen toward their natural place, which is to say upon the earth.

From the circular movement of the stars, it follows that the cosmos has to be finite and spherical. Aristotle argued that the stars cannot be at an infinite distance, for then they would circle around the earth with an infinite velocity, whereas nothing can move with an infinite speed (*On the Heavens* 272b28ff.). Aristotle's argument is somewhat more complicated, but according to Furley it boils down to this:

---

[1] As the natural movement of the four elements is always vertical, or better radial (ὀρθός), which is to say centripetal or centrifugal, one may imagine that Aristotle has difficulties with the phenomenon of the wind. This is clear from a text in the *Meteorologica* (360b23), where he explicitly mentions as a problem that the wind blows at an angle, or better horizontally (λοξός), around the earth. This means that, according to Aristotle, wind cannot be simply a moving stream of air, as some people think (*Meteorologica* 349a17 and 360a28), for air naturally moves in a vertical direction. Aristotle's solution for this problem is rather complicated. There are, he says, two kinds of exhalations that originate when the earth is warmed by the sun: one cold and moist (damp) and one heat and dry (smoke). What we call "air" consists of these two components (360a21). When a moist exhalation descends, it rains. The dry exhalation causes winds (360a13). The reason why they blow horizontally is that the air follows the movements of the heavens (361a25). Here, Aristotle does not mean the movement of the heavens from east to west, for then permanently an eastern wind would be blowing, but the movement of the sun during the year between the tropics. In summer, the sun stands more to the north and in winter more to the south. This is why most winds are north or south winds, says Aristotle (361a5-10). However, it does not become entirely clear why in the case of wind the air does not move in its "natural" way.

"If we imagine the radius of the sphere (sc. of the fixed stars) to increase, we must imagine the speed of the stars to increase proportionally. And if the radius increases to infinity, so does the speed" (1989, 1–2). The cosmos, then, is an immense but finite sphere with the earth as its center. Aristotle's world picture that would last for two millennia was that of a "closed world." Here, strangely enough, a Pythagorean principle plays a decisive role in Aristotle's cosmology: finite is good and infinite is bad, so it is better for the cosmos to be finite. In the end, this spatial finitude of the cosmos proved to be the weak spot of Aristotle's conception, as will be discussed in the next chapter.

The implication of the finitude or boundedness and sphericity of the cosmos is that the natural rectilinear movements of the elements have both a terminus a quo and a terminus ad quem: heavy bodies fall toward the center and light bodies rise in the direction of the periphery. The movement of heavy things is centripetal and that of light things is centrifugal. Aristotle also argues, against the atomists, that there can be only one cosmos. It would be absurd to have more than one center, for in that case one and the same movement would be "up" in relation to one center and "down" in relation to another (*On the Heavens* 276b14–22). According to Aristotle, there is only one way to define an absolute "up" and "down," namely, in one spherical universe, in which "up" means "away from the center" and "down" means "toward the center."

Now, we have come into the position to see the complete picture of Aristotle's answer to the problem of falling. Aristotle's argumentation is ingenious, as he does not consider falling as the problem, but as the solution. With one big stroke of genius, he solves the three above-mentioned big questions at a time. These questions were: what is shape of the earth, where does the earth dwell in the cosmos, and is the earth in rest or does it move? What we call "falling" is, according to Aristotle, nothing but the natural movement of that which is heavy toward its natural place. Precisely because the heavy element, earth, falls toward the center of the spherical-shaped cosmos, the earth stays there immovably (*On the Heavens* 295a24). This means that Aristotle also has an answer to the question why it is the earth and not another body that dwells in the center of the cosmos, a question that could not be answered by the argument he ascribes to Anaximander (*On the Heavens* 296b27). As we saw in Chap. 8, this might have been the reason why he ridiculed that argument. The conclusion must be that we do not have to be afraid that the earth will fall, since the earth is – apart from the celestial bodies – the only thing that cannot fall but toward which all heavy things fall.

The spherical shape of the earth is also the consequence of Aristotle's theory of falling, for the sphere is the stereometric body in which all parts of its circumference have the shortest possible distance to its center. The spherical shape of the earth is the product of the natural tendency of heavy things to fall as near as they can toward the center of the cosmos (*On the Heavens* 297a8–25). Aristotle's solution of the problem of falling means that the earth has come into a unique position compared with the celestial bodies. The celestial bodies are made of aether, whereas the earth is made of earth. This also provides an answer to the argument of the defenders of the flat earth, that we would immediately fall off the earth if it were spherical. We do not fall off the earth because we are mainly made of the element earth and thus tend to move in the direction of the center of the cosmos as well. Wherever we stand, on the northern or on the southern hemisphere, we will not fall off the earth. However, if we should stand somewhere on the moon, we would immediately fall down toward the earth. The other way round, we will never have to be afraid that the celestial bodies will fall down upon us, for they are not made of one of the four elements and thus not liable to their rectilinear movements. The celestial bodies are made of the fifth element, the aether, and they partake in the everlasting circular movement of the celestial spheres.

Characteristic for the elegance of Aristotle's solution is that the immovable earth appears to have the shape that at first sight is most predisposed to movement by rolling away. The unique position of the earth in the universe shows also in its shape, which is spherical, just like that of the cosmos as a whole, but for quite another reason. This spherical universe we see is the sphere of the fixed stars. According to Aristotle, the celestial bodies do not move themselves, but are attached to spheres that turn around the earth (*On the Heavens* 289b30–33 and 291a). The celestial bodies are spherical because they generate from the spheres onto which they are attached (*On the Heavens* 290a8 and 291b11). The celestial bodies are so to speak intrinsically spherical because they take part in the perfect shape of the celestial spheres. The earth has the shape of a sphere, just like the celestial bodies, but for a completely different reason. The spherical shape of the earth is a derivative or by-product of Aristotle's solution of the problem of falling. In a sense, Aristotle's solution is analogous to that of Plato: the earth has its shape willy-nilly. For Plato, the cube lay at hand as the shape of the earth, as argued in the preceding chapter. For Aristotle, the spherical shape is the privilege of those bodies that consist of the most perfect element, the aether, and that take part in the perfect shape of the celestial spheres, onto which they are attached, and the circular movement of which they share. The earth that does not move and consists of the least noble of all elements nevertheless has, as a consequence of Aristotle's theory of falling, the shape of a sphere as well.

In this chapter, I have left out Aristotle's discussion of another possible movement of the earth, viz., turning around its own axis. In the last chapter, we will deal with his contemporary who taught this, Heraclides Ponticus, and in that context Aristotle's arguments will be considered as well.

Chapter 19
Heraclides Ponticus and the Infinite Universe

In Chap. 8, it was explained how Anaximander, with one of his fundamental speculative insights, broke through the firmament of the archaic world picture by placing the celestial bodies at different distances from the earth. Yet the size of his cosmos is not very big, albeit much larger than in the archaic world picture, in which the celestial vault is at about 2,500 km distance from the earth's center (see Chap. 1). Anaxagoras' calculations resulted in a distance of about 6,000 km from the earth to the sun (see Chap. 16). Anaximander clearly did not perform calculations like those of Anaxagoras, for he estimated a much bigger distance to the sun, which he supposed to be the farthest celestial body. If we agree, in conformity with what was called in Chap. 9 the unorthodox variant of the standard interpretation, that the radius of Anaximander's sun wheel counts 28 earth diameters and if we take the diameter of the earth to be about 5,000 km (the greatest distance known at that time, between Babylon and the Pillars of Hercules), then the diameter of Anaximander's cosmos amounts to 280,000 km, and the sun is 140,000 km away. What is beyond the sun is not so obvious, although his concept of the *apeiron* makes some surmise that we have to imagine an infinite space out there (see Chap. 8).

Anaximander thought that of all celestial bodies, the stars were nearest to the earth. As we have seen in Chap. 13, astronomers at the time of Plato were already convinced that, on the contrary, the stars are farthest away from us. The observational evidence that the stars are at unchanging distances from each other led to the idea that they all must be at the same distance from us, as we can read in Aristotle (*On the Heavens* 289b1ff.). There, it is argued that if the stars were at different distances from the earth, they would move with different velocities, just like the planets. As we have seen in Chap. 18, Aristotle also argued that the cosmos must be finite, for if it were infinite the stars would circle around the earth with infinite speed. As everyone can see, this is not the case, and moreover it is impossible for anything to move with an infinite velocity (cf. *On the Heavens* 272b28ff.). Aristotle imagined the celestial bodies as attached to spheres (or better: sphere shells). The outermost sphere that completes the cosmos is that of the stars. Outside the sphere of the stars absolutely nothing exists, even not empty space, according to Aristotle (*On the Heavens* 275b9 and 279a6–13), or as Furley expresses it: "Outside the sphere, there was nothing at all, as Aristotle believed, or nothing of any interest" (1989: 1).

D.L. Couprie, *Heaven and Earth in Ancient Greek Cosmology*,
Astrophysics and Space Science Library 1, DOI 10.1007/978-1-4419-8116-5_19,

The inevitable consequence was that the archaic concept of the celestial vault was reintroduced, this time not as the hemispherical dome that arches over the flat earth and onto which all celestial bodies are glued, but as the all-embracing sphere of stars, in the center of which the earth dwells. On the screen of this heaven with its constellations that turn around the earth in 1 day, we can observe and describe the movements of sun, moon, and planets.

The introduction of a spherical cosmos with the outermost sphere of the stars also meant, however, the end of the notion of infinity that was at least implicitly present in Anaximander's conception of the universe. In other words, Anaximander's potentially open universe was closed again. We may rightly say that Aristotle replaced Anaximander's cosmic wheels of fire within condensed air by celestial spheres of aether. Therewith, however, he reintroduced in a new disguise the celestial vault of the archaic world picture that Anaximander had blown up. This new celestial vault is the ultimate sphere of the fixed stars that makes up the boundary of the spherical cosmos, outside of which nothing exists.

Moreover, the spherical and limited cosmos is a necessary ingredient of Aristotle's theory of falling, as explained in the preceding chapter. According to the Stoics, Aristotle's argument that outside our cosmos nothing exists was irrational. They said that outside the cosmos there is simply empty space. As a result of this, they had to confront the horrifying question what prevents the matter of the cosmos to spread itself through the entire infinite emptiness. Epicure seems to have drawn the consequence that our cosmos is nothing more than a temporary and transitory concentration of matter that in due time will be dispersed through the universe (see Hahm 1977: 103).

The question of the finitude or infinity of the cosmos is intertwined with that of the "infinite worlds." The atomists Leucippus and Democritus taught not only that the universe is infinite but also that within this infinite universe worlds infinite in number (ἄπειροι κόσμοι) exist [DK 67A1(31) and 67A21]. In this context, with "world" is meant a "cosmos," consisting of a central body, such as the earth, together with other celestial bodies orbiting around it. Our familiar cosmos of earth, moon, sun, planets, and stars is such a world, of which there are, according to the atomists, infinitely more. As several sources ascribe to Anaximander the doctrine of the "infinite worlds," an extensive discussion exists in the literature on Anaximander about the question, whether he too was an adherent of this doctrine (see DK 12A9, 12A10, 12A11, 12A14, 12A17, and Simplicius, *In Aristotelis De caelo commentaria*, 202.14-16, not in DK).

In a still authoritative article, Cornford concluded that the idea of simultaneously coexisting worlds is a typically part of atomistic lore that the doxographers erroneously read into Anaximander as well (Cornford 1934). Cornford keeps open the possibility that Anaximander could have meant worlds succeeding one another in time. As we saw in Chap. 9, some authors advanced the idea that possibly the equal distances between the celestial bodies in Anaximander's conception of the cosmos has been the source of the report on "unlimited worlds" at equal distances of one another. Some even see a connection with the text of Anaximander's fragment, in which things are said to generate from and go under into the *apeiron*.

Recently, McKirahan has broken a lance again for the coexistence of innumerable worlds in Anaximander, because of the relatively big number of sources that ascribes this doctrine to him, as well as because the notion of infinite worlds fits very well to the *apeiron* taken as a spatial infinity (2001: 49–65). In an earlier article that McKirahan seems not to be acquainted with, Finkelberg already attained similar conclusions (1994: 485–506). My conclusion is that the question whether Anaximander already taught the existence of innumerable worlds is still not convincingly answered.

Furley, in his book *The Greek Cosmologists*, contrasts two pictures of the world: the modern, open world picture ("the infinite universe"), in which the earth figures as a tiny planet of a little star in a universe of many galaxies, and the classical, closed world picture ("the closed world"), in which the earth is the center of the cosmos that is bounded by the sphere of the stars. Furley's thesis is that the modern, open world-picture, too, has its roots in Greek antiquity. His book is devoted to the earliest history of both systems, from Anaximander to the atomists, the supposed defenders of the open universe, and Aristotle, the ultimate exponent of the closed world picture. There exists, by the way, a strange change of positions between the two camps. Whereas Aristotle represents the world as finite in space but infinite in time (our world is eternal), the atomists look upon the worlds as infinite in space but finite in time (worlds come into being and perish again) (DK 68A40).

One may wonder, however, whether the difference between the two conceptions in Greek antiquity was really as big as it seems. In practice, the atomists too believed in a spherical cosmos, the boundary of which is determined by the fixed stars. As Furley says somewhere: "Both the Atomists, who believed in the infinite universe, and the Aristotelians, who did not, agreed that our world is itself a finite system, bounded by the sphere of the stars. The controversy was about what, if anything, lies beyond the starry sphere" (1987: 136). And elsewhere: "What they (viz. both the Aristotelians and the atomists, D.C.) saw in the night sky was not the beginning of the infinite universe: it was rather the boundary beyond which the infinite universe began. (...) The world is like a walled city with unknown country outside the walls. (...) no one in classical antiquity believed that the *world* is infinite. The controversy was not about the *existence* of a closed world, but about its status: is it all there is, or is there something else too?" (1989: 2).

In the discussion about the "infinite worlds" there is, on closer inspection, something quite strange. For Leucippus and Democritus, too, the stars make up the outermost sphere of the only world we are acquainted with. All the stars that we see belong to our own cosmos, the center of which is the earth. This means that the other worlds, infinite in number, lie beyond our horizon of perception. They are no more than a theoretical datum that principally never can be verified by possible observation. In other words, the discussion about the infinite worlds has a highly theoretical and speculative, not to say esoteric quality.

It seems to me that the fundamental difference between the cosmologies of Aristotle and the atomists lies somewhere else, viz., in how they deal with the problem of falling. As we have seen in Chap. 18, Aristotle's cosmology pivots on the problem of falling, for which he had found a beautiful solution. His arguments for the existence of only one world are also derived from his theory of falling.

His crucial argument is that all those alleged other worlds must be made up of elements, for otherwise they could not be called "worlds" at all. These elements necessarily move to their natural places, the earth (the heavy element) to the center and the fire (the light element) to the periphery, and the other two – water and air – in between. It is not possible, however, for more than one center to exist, for otherwise heavy things would fall downward in relation to one center and upward in relation to another. Mutatis mutandis, the same holds for the periphery. Therefore, no more than one world can exist (*On the Heavens* 276a23–b23).

The atomists, as far as we may gather from the scarce sources (unfortunately mainly Aristotle), were not able to solve the problem of falling in a satisfactory way. They probably held that movement is inherent to the atoms. They imagined the atoms to move originally at random in infinite space (DK 67A37 and 68A47). The sources differ regarding the question whether the atoms have weight, the one being heavier than the other. In 1981, O'Brien devoted an extensive study to the complex problem of the weight of the atoms. In a reaction to this, Furley concludes: "Democritean atoms have weight, meaning a tendency to fall downward" (1989: 91–102). What "downward" and "falling" of atoms in an infinite space may mean, however, remains vague, as well as the relation of their falling to their original haphazard movement.

The cosmic whirl (δίνη) also plays a certain role in the atomistic cosmology. Light objects are swung to the periphery, whereas heavy objects gather at the center, and by means of this process the earth comes into being [DK 67A1(31), 67 A1(33), and 67A24.]. From where, however, the cosmic whirl originates is not clear, nor how its relation to the falling movement of the atoms has to be understood, as two scholars confess: "A cosmos begins to take shape when a number of atoms at random, *by some unexplained mechanism*, happen to distinguish themselves from all others by joining together in a vortex (δίνη)" (Furley 1989: 79, my italics) and "Then *for some reason not stated*, a whirl or vortex was separated off from the whole" (Ferguson 1999: 101, my italics).

It seems, then, not completely unjustified when Aristotle blames the atomists for not telling what kind of movement atoms have nor what their natural movement is (*On the Heavens* 300b9–12, see also *Metaphysics* 1071b31). To explain why the earth does not fall, the atomists made use of another and also hardly convincing theory: it is not the cosmic whirl that prevents the earth from falling, but the air that supports the flat earth (DK 59A88 and 13A20). Aristotle, on the contrary, as we have seen in Chap. 18, was prepared to answer consistently all the questions that have to do with the problem of falling.

Maybe Furley is right in maintaining that in the development of Greek cosmological thinking, the controversy between the "open universe" of the atomists and Aristotle's "closed world" was important. But by focusing on this controversy, we tend to forget that an even more fundamental opposition is at the root of Greek cosmology, both Aristotelian and atomistic. This is the opposition between the archaic world picture of a flat earth arched by the celestial vault on which the celestial bodies are glued, as described in Chap. 1, and the new world picture, introduced by Anaximander, in which the celestial bodies orbit around the earth, whereas the earth itself floats unsupported in the center of the cosmos and the celestial bodies move at different distances from the earth. Both Aristotle's finite cosmos and the open universe of the atomists are tributary to the new world picture introduced by Anaximander.

Yet Aristotle, with his sphere of stars that encloses the cosmos, in a certain respect returned to an essential part of the archaic world picture. The weak point of this conception of the celestial sphere is precisely its finitude, the closing of the universe. This was laid bare by Archytas when he asked whether it would be possible at the limits of the heavens to extend his arm or staff outside (see Fig. 1.1). It would be absurd, he argued, when he couldn't, and this means that there cannot be a limit to the universe (DK 47A24). Being a Pythagorean, however, Archytas probably believed that although the universe is infinite, the cosmos is limited (because limit is positive and unlimited negative) so that outside the cosmos there is simply empty space. Perhaps this was the more common conception, against which Aristotle argued. At least, it is handed down that Melissus and Diogenes held that the All is infinite, whereas the cosmos is limited: τὸ μὲν πᾶν ἄπειρον, τὸν δὲ κόσμον πεπεράνθαι (DK 30A9 and 64A10).

Rather than to Archytas, or to the atomists, the credits of being an early advocate of the infinite universe belongs to Heraclides Ponticus (±388–310 B.C.), an almost exact contemporary of Aristotle. He was, as is told, a very remarkable man, effeminately dressed, and more than corpulent. In 338 B.C., at the nomination for the succession of Speusippus as head of the school of Plato, he was beat by Xenocrates with only a few votes. Disappointed, he went back to his native town Heraclea. Heraclides maintained, against common sense and, which is more, against the authority of Aristotle, that the earth does not rest motionless in the center of the cosmos, but rotates around its axis in 24 h from west to east (Wehrli 1969: 35, fr. 104 = DK 51A5). This is, by the way, the text quoted by Copernicus in the Dedication to Pope Paul II of his *De revolutionibus orbium coelestium*. Heraclides' "brilliant suggestion" (Kahn 2001: 68) of the axial rotation of the earth explains the apparent orbiting of the celestial bodies around the earth from east to west. In other words, in Heraclides' theory the stars stand still, whereas the earth turns around its axis.

It may be argued that Heraclides' idea was based on a famous and much-disputed text in Plato's *Timaeus*, quoted by Aristotle, in which the earth is said to "wind round," or to be "packed round" the celestial axis, depending on one's interpretation of the word ἰλλομένην (*Timaeus* 40b9–c1; *On the Heavens* 293b30–33, cf. 296a26).[1] I agree with those who maintain that "there can be little doubt that the weight of the evidence is against imputing to Plato the concept of a rotating earth" and also with those who hold that Aristotle's reference here is not to Plato but to Heraclides' (incorrect) interpretation of Plato's text (Dicks 1970: 134; Cherniss 1944: 551ff.). An indication might be that Aristotle explains the word ἴλλεσθαι (winding) as κινεῖσθαί περὶ τὸν πόλον μέσον (moving around the polar axis).

[1] For a thorough discussion of the different standpoints, see Dicks (1970: 132–137).

Aristotle's arguments to prove that the earth does not display an axial rotation follow from his general theory of movement and intend, as usual, to confirm the common sense interpretation of the celestial phenomena. This kind of circular movement, Aristotle says, cannot be the earth's natural and eternal movement, for that is linear and centrifocal (*On the Heavens* 296a27–34 and 296b8–23). Moreover, all other celestial bodies that show circular movements, except the stars, display contrary movements as well. The stars, however, show a regular circular movement without the irregularities that should be visible when the earth behaved like the planets (*On the Heavens* 295a35–296b7). Aristotle adds a piece of empirical evidence: a stone thrown right upward will fall down to its starting place, which proves that the earth does not turn around its own axis (*On the Heavens* 296b25). It is a pity that we do not know how Heraclides countered these arguments.

Aristotle's experiment could be generalized: why don't we notice the movement of the earth around its axis, or even why aren't we hurled off the earth? Heraclides also would have had to explain why the atmosphere with its clouds takes part in the rotation of the earth, and how the clouds manage to move sometimes contrary to the direction of that rotational movement (see Ferguson 1999: 25). Unfortunately, we do not have any information of Heraclides' answers to these common sense questions.

The theory of the axial rotation of the earth is ascribed to Hicetas and Ecphantus as well (DK 50A1, 51A1, 51A5). The first mentioned text (= Cicero, *Academia Priora* II. 39.123) is also mentioned by Copernicus in his Dedication to Pope Paul II. Tannery and others have put forward the hypothesis that Hicetas and Ecphantus must have been characters from a lost dialogue by Heraclides, but this theory has been challenged (See Tannery 1897: 127–137; Heath 1913: 187–188 and 251–252).[2] However this may be, according to Heraclides, the daily movement of the celestial bodies around the earth is nothing but a deceptive appearance caused by the rotation of the earth from west to east. One text that apparently has bearings on Heraclides' theory but is less straightforward has disturbed scholars. It says that "a certain Heraclides of Pontus has come forward and says that the apparent irregularity connected with the sun can also be explained if the earth is moved in a certain way (κινουμένης πως) and the sun stays still in a certain way (μένοντός πως)" (Simplicius, *In Aristotelis De caelo commentaria*. 292.15 = Wehrli 1969: 36–37, fr. 110). Gottschalk calls this "infuriatingly vague" and says that the clause "the earth is moved in a certain way" could mean almost anything, whereas the clause "the sun stays still in a certain way" seems to be virtually devoid of meaning because there is only one way of staying still. His conclusion is that "the whole sentence is nothing more than a rhetorical flourish" (1980: 63, 66, and 68).

I think, on the contrary, that it is possible to make sense of this text, making it a perfect illustration of Heraclides' intentions. That the earth moves "in a certain way" may mean that it does not orbit around the central fire, as the Pythagoreans thought, but that it moves around its own axis without changing its place, remaining in the center of the universe. When we take into account that in Heraclides' theory

[2] For discussion of Tannery's proposal see Gottschalk (1980: 44) and Burkert (1972: 341 n. 170).

the stars do not orbit around the earth but stand still, the expression that the sun stands still "in a certain way" may mean that the sun too does not orbit around the earth from east to west. In this sense, the sun may be said to stand still, just like the stars in Heraclides' conception. Unlike the stars, however, the sun has a movement of its own: from west to east along the ecliptic, and thus it moves "in a certain way" as well. The same holds, mutatis mutandis, for the moon and the planets, which have even more and irregular movements of their own.

As regards the subject of this chapter, it is important that Heraclides also taught that the universe is infinite and that every star is a world, consisting of an earth with its atmosphere in the infinite aether (Diels 1879: 328 and 343 = Wehrli 1969: 37, fr. 112, 113a, 113b, and 113c). Perhaps Heraclides took the radical consequence of an infinite universe from Archytas' question what would happen when he put his hand or a stick outside the ultimate sphere of the stars. The implication of Heraclides' words is that the stars are not all at the same distance from us but that they are *behind one another*. Otherwise than Melissus, Diogenes, and Archytas thought, the universe is not empty but crowded with stars.

In other words, Heraclides broke through the firmament of the fixed stars. This was the consequence of his assumption that the earth turns around its axis. As long as the stars were envisaged as moving around the motionless earth, one had to think of them as rotating in a sphere, all at the same distance from the earth. For otherwise, in an infinite universe, they would have to move with infinite speed, as Aristotle argued. But as soon as the stars are thought of as standing still, there is no need for a solid outermost sphere so that they can be at various distances from us, infinitely. Furley has seen this very well: "(Heraclides) had no need to posit a *sphere* of stars (. . .). (. . .) as soon as they are conceived to be stationary, there seems (. . .) to be no cogent reason why they may not *vary enormously in their distance from the earth*" (1989: 3, my italics). In Gottschalk's interpretation the argument goes, less correctly, the other way round: "If the universe is infinite, it is difficult to conceive of a daily rotation of the fixed stars, and the simplest way of explaining their apparent movement is to make the earth rotate" (1980: 83).

Heraclides' infinite universe crowded with stars reminds us of the "infinite worlds" of the atomists. There is, however, one essential difference. When Heraclides is talking about the other worlds, he means in the first place the stars we can see: each of these stars is a world in itself, comparable with our own earth. These visible worlds, each at a different distance from us, continue infinitely, even beyond the limits of our view. This conception is to be definitely distinguished from the atomists' infinite worlds beyond our horizon of experience, which means beyond the stars that together make up the spherical boundary of our world. Heraclides' worlds also extend into infinity, but they are a continuation of the worlds we can see as the stars. Otherwise stated: for the atomists, the stars we see are all at the same distance from us, just as they are for Aristotle, but for Heraclides Ponticus they are at different distances from us, continuing infinitely.

Accordingly, if we may consider the words ascribed to Ecphantus as those of Heraclides, then he has said that there is only one cosmos (DK 51A3), and not infinitely many, as the atomists said. Here a note on the terminology used in the doxography is needed. The word κόςμος in the doxographical accounts on Heraclides is used not only as an indication of the infinite universe (which is also indicated as "infinite aether"), of which it is explicitly stated that there is only one, but also as an indication of each of the stars ("worlds") with its own atmosphere. These two ways in which the word κόσμος is used seem to be incompatible. However, one may understand them very well as meant polemically against the atomists: the "infinite worlds" (ἄπειροι κόσμοι) of Leucippus and Democritus are, according to Heraclides, not outside our horizon of experience, but they make part of the one and only universe (κόσμος). The atomist's infinite universe was speculative, and their "infinite worlds" were beyond any possible experience, whereas Heraclides' infinite universe and his infinite worlds were simply the infinite continuation of the visible universe of stars.

As far as we know, Heraclides had only two followers. The first was Aristarchus, who has become known as a forerunner of Copernicus. Aristarchus taught not only, like Heraclides, that the earth rotates around its axis but also that the earth orbits around the sun. This is explicitly stated by Plutarch (*On the Face in the Moon* 922f–923a). Heraclides' system, on the contrary, was not heliocentric, but geocentric, as Eastwood has convincingly shown. The conclusion of his article is: "modern proposals for an ancient Heraclidean heliocentrism have come from post-Copernican expectations rather than from a dispassionate reading of the texts" (1992: 256).

The second follower of Heraclides was Seleucus (±150 B.C.), who is mentioned by Stobaeus together with Heraclides (*Eclogae physicae* I 21.5; see Diels 1879: 328 = Wehrli 1969: 37, fr. 112). He seems to have adduced the rotation of the earth for his explanation of the tides, which does not interest us here (see Diels 1879: 383). Probably, however, neither of them also adopted Heraclides' conception of an infinite universe with stars at different distances from us. Later on, Seneca somewhere mentions the rotation of the earth as a theoretical possibility, also without taking the consequence of an infinite universe (*Naturales quaestiones* 7, 2.3).

Apparently, Aristotle's arguments were so strong (nothing can move with an infinite velocity, and the solution of the problem of falling requires a center and thus a limited universe), that the finite celestial sphere remained the ruling paradigm for many centuries, even for the few people after Heraclides who taught the rotation of the earth. The infinite worlds in an infinite universe of the atomists were a theoretical construction, and they were not able to loose the problem of falling in an infinite universe. Heraclides Ponticus offered a solution for Aristotle's problem of moving with infinite speed, but he, too, was not able to solve the problem of falling in an infinite universe.

We had to wait centuries for Nicolaus Cusanus and Giordano Bruno to speculate again about an infinite universe, the center of which is everywhere and nowhere, and for Thomas Digges to draw, in 1576 a representation of an infinite universe, in which the stars are behind each other. Mark the words "infinitely up extendeth" at the top of the outermost ring in Fig. 19.1. The words "the most auncient doctrine of the Pythagoreans" in the text above the picture refer to Aristarchus, who was already considered a predecessor by Copernicus himself because he situated the sun in the center of the universe. Apparently, Digges thought it obvious that they should have concluded that the universe is infinite. Thus, some 2,000 years after

Anaximander, the promise of an infinite universe that was contained in his notion of the *apeiron* was finally fulfilled.

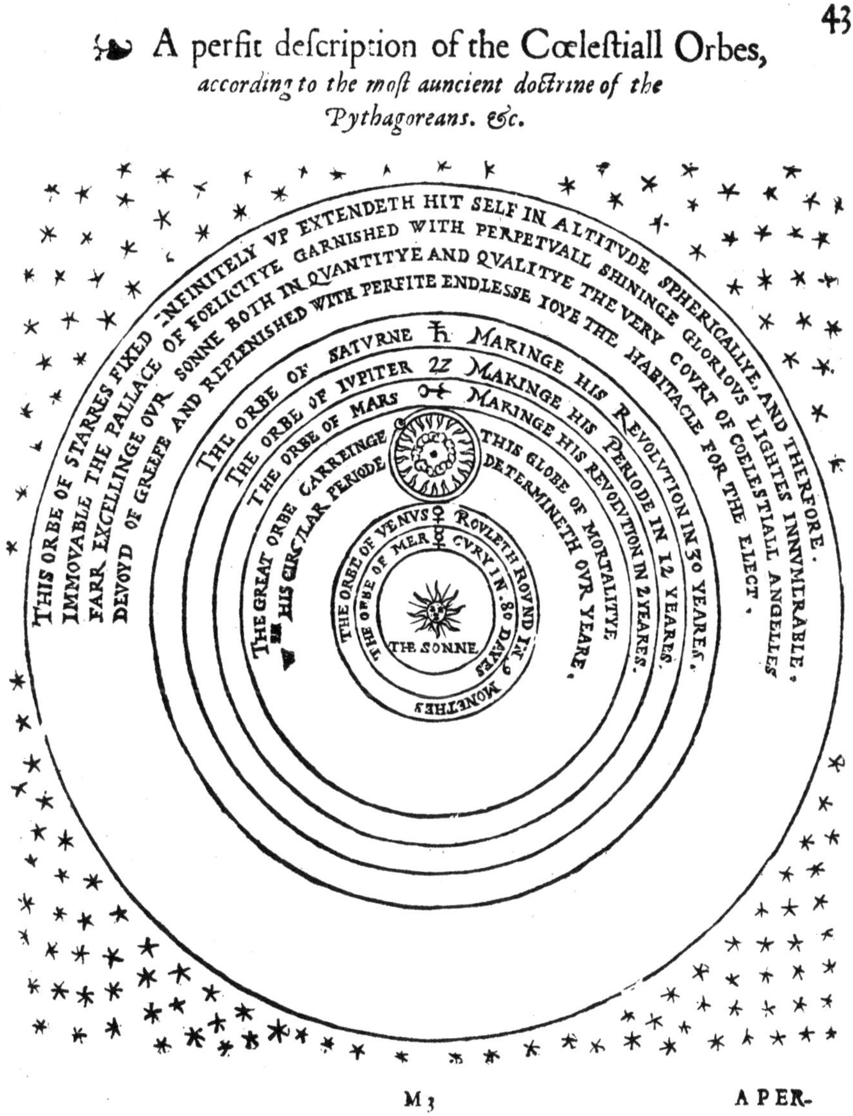

Fig. 19.1 The infinite universe according to Thomas Digges (Digges 1578: Folio 43. Photograph Collection of Owen Gingerich)

At Digges' time, no empirical proof existed of the idea that the stars are behind one another, His conception (and that of Cusanus and Bruno) was the result of creative imagination, just like, centuries ago, Anaximander's conviction that the celestial bodies are behind each other. It was only in the year 1838 that Friedrich Wilhelm

Bessel was able to measure the parallax of a star (61 Cygni), and there with its distance to the earth. When, some years later, Thomas Henderson and Friedrich Georg Wilhelm Struve had measured the parallax and distances of two other stars (α Centauri and Wega), it was finally proven that the stars are really behind one another or, in other words, that such a thing as a firmament or celestial vault does not exist.

However, the fear of falling that Aristotle so successfully had countered by proving that we do not have to be afraid that either the heaven will collapse or the that the celestial bodies will fall upon the earth, or even that the earth itself will fall, made a glorious comeback in the new world picture of an infinite universe and still haunts it until our days. Cawthorne (2004: 77–102) enumerates a list of possible cosmic disasters with which we are confronted and in which we recognize the fears of the ancients. An asteroid impact bigger than that which caused the extinction of the dinosaurs or a comet collision like Shoemaker–Levy 9 that hit Jupiter in July 1994 could herald the end of the world. We are reminded of the ancient fear caused by the meteorite of Aegospotamoi. If the slight tilt of the axis of the earth, which gives us the seasons and makes life on earth agreeable, for some reason like a cosmic collision would change notoriously, devastating climate changes would be the result. We are reminded of the mysterious tilt of the celestial axis in the archaic world picture. A supernova close enough to our solar system might destroy all life on earth. Or the gravity of an invading dwarf star could disrupt the solar system, with devastating effects. Our neighbor galaxy, the Andromeda nebula, is approaching us at 250,000 miles an hour and will eventually collide with our Milky Way. Inevitably, the sun will once attain the end of its lifespan and will absorb the planets. Ultimately, there is no escape, for the universe as a whole is doomed as well. When finally the gravitational pull overrules the expansion of the universe, the end will be the mirror image of the Big Bang that made it start, and in what is called the Big Crunch the universe will collapse into a single point. We are reminded of ancient predictions of the end of the universe.

Philosophers have expressed this "fear of falling" that inheres our modern world picture often in more existential wordings. Everyone knows the famous line of Blaise Pascal in which he expressed the horror of infinity: "Le silence éternel de ces espaces infinis m'effraie" (1962: 50, no. 91). Elsewhere, he calls the human being a feeble reed ("un roseau faible"), being crushed by the universe ("l'univers l'écraserait"). The only thing he finds comfort in is that man is a *thinking* reed (1962: 130, no. 264). Friedrich Nietzsche saw a relation between the fear of falling and the dead of God in a famous section of his *The Gay Science*, in which also the discussion between Anaxagoras and Aristotle on the sun at the horizon is placed in a completely new perspective: "Who gave us the sponge to wipe out the whole horizon? What were we doing when we loosened this earth from its sun? Where to does it move now? Where to do we move? Away from all suns? Do we not tumble incessantly? Backward, sideways, forward, in all directions? Is there still an above and a below? Do we not err around in infinite nothingness? Does not empty space breathe us in the face?" (1973b: 159, § 125, my translation).

The return of what I called "the fear of falling" is the price we have paid for abandoning the Aristotelian conception of a safe but finite universe.

Bibliography[1]

Aaboe, Asger. 1972. Remarks on the Theoretical Treatment of Eclipses in Antiquity. *Journal for the History of Astronomy* 3: 105–118.

Ammianus, Marcellinus. 1956. *History of Rome*. Rolfe, John C., ed. and transl., Bd. II. London: Heinemann, Cambridge, MA: Harvard University Press.

Andrae, W. 1933. *Die Ionische Säule. Bauform oder Symbol?* Berlin: Verlag für Kunstwissenschaft.

Apollonius, Rhodius. 1973. *Argonauticon Liber IV*. Enrico Livrea, ed. and transl., Firenze: La nuova italia.

Aravantinou, Anna. 1993. Anaxagoras' Many Worlds. *Diotima* 21: 97–99.

Aristarchus. 1913. On the Sizes and Distances of the Sun and Moon. In Thomas Heath, ed., *Aristarchus of Samos. The Ancient Copernicus*, 351–413. Oxford: Clarendon Press.

Aristophanes. 1973. *Ekklesiazusae*. Robert G. Ussher, ed., Oxford: Clarendon Press.

Aristotle. 1953–1957. *Problems, in Two Volumes*. Walter S. Hett, ed. and transl., London: Heinemann Ltd., Cambridge, MA: Harvard University Press.

Aristotle. 1957–1960. *The Physics, in Two Volumes*. Francis M. Cornford, ed. and transl., London: Heinemann Ltd., Cambridge, MA: Harvard University Press.

Aristotle. 1959–1962. *The Metaphysics, in Two Volumes*. Hugh Tredennick, ed. and transl., London: Heinemann Ltd., Cambridge, MA: Harvard University Press.

Aristotle. 1962. *Meteorologica*, Henry D.P. Lee, ed. and transl., Cambridge, MA: Harvard University Press, London: Heinemann Ltd.

Aristotle. 1979. *Generation of Animals*. Arthur L. Peck, ed. and transl., Cambridge, MA: Harvard University Press, London: Heinemann Ltd.

Aristotle. 1986. *On the Heavens*. William K.C. Guthrie, ed. and transl., Cambridge, MA: Harvard University Press, London: Heinemann Ltd.

Aveni, Anthony F. 1993. *Ancient Astronomers*. Washington, DC: Smithsonian Books.

Baensch, Otto. 1903. Die Schilderung der Unterwelt in Platons Phaidon. *Archiv für Geschichte der Philosophie* 16: 189–203.

Balls, Heinrich. 1949. *Antike Astronomie*. München: Ernst Heimeran Verlag.

Barnes, Jonathan. 1982. *The Presocratic Philosophers*, Vol. I: *Thales to Zeno*. London: Routledge & Kegan Paul.

Bauval, Robert G. 2001. A Brief Evaluation of Kate Spence's Article. *Nature* 408: 320–324, http://www.robertbauval.co.uk/articles/articles/spence.html, 16 Nov. 2000.

Bengtson, H., et al. 1963. *Grosser Historischer Weltatlas*, Vol. I: *Teil, Vorgeschichte und Altertum*. München: Bayerischer Schulbuch-Verlag.

Berger, Hugo. 1903 (1st impr. 1887). *Geschichte der wissenschaftliche Erdkunde der Griechen, Erste Abtheilung: Die Geographie der Ionier*. Leipzig: Verlag VonVeit & Comp.

[1]An extensive bibliography on Anaximander on my website http://www.dirkcouprie.nl.

D.L. Couprie, *Heaven and Earth in Ancient Greek Cosmology*,
Astrophysics and Space Science Library 1, DOI 10.1007/978-1-4419-8116-5,

Betegh, Gábor. 2002. Le problème des representations visuelles dans la cosmologie présocratique: pour une histoire de la modélisation. In André Laks and Claire Louguet, eds., *Qu'est-ce que la philosophie présocratique?/What is Presocratic Philosophy?*, 381–415. Villeneuve d'Ascq (Nord): Presses Universitaires du Septentrion.
Bickerman, Elias J. 1980. *Chronology of the Ancient World*. London: Thames and Hudson.
Bicknell, P.J. 1968. Did Anaxagoras Observe a Sunspot in 467 B.C.? *Isis* 59(1): 87–90, No. 196.
Blanck, Horst. 1997a. Anaximander in Taormina. *Mitteilungen des deutschen Archäologischen Instituts (Römische Abteilung)* 104: 507–511.
__________. 1997b. Un nuovo frammento del 'catalogo' della biblioteca de Tauromenion. *La Parola del Passato* 52: 241–253.
Bodnár, István. 1988. M., Anaximander's Rings. *The Classical Quarterly* 38: 49–51.
__________. 1992. Anaximander on the Stability of the Earth. *Phronesis* 37: 336–342.
Boll, Franz. 1950. *Kleine Schriften zur Sternkunde des Altertums*. Leipzig: Köhler und Amelang.
Boquet, F. 1925. *Histoire de l'astronomie*. Paris: Payot.
Borchardt, Ludwig. 1899. Ein altägyptisches astronomisches Instrument. *Zeitschrift für Ägyptische Sprache und Altertumskunde* 37: 10–17.
__________. 1920. Die altägyptische Zeitmessung. Band I. In E. Von Bassermann-Jordan, ed., *Die Geschichte der Zeitmessung und der Uhren*. Berlin und Leipzig: Walter de Gruyter.
Borghouts, Joris F. 1988. *sagen en verhalen*. Houten: Fibula.
Bousset, Wilhelm. 1960. *Die Himmelsreise der Seele*. Darmstadt: Wissenschaftliche Buchgesellschaft. (Reprint of 1901, *Archiv für Religionswissenschaft* 4: 136–169 and 229–273).
Boyer, Carl B. 1968. *A History of Mathematics*. New York: Wiley.
Bremer, Dieter. 1994. Milesischer Anfang. *Philosophical Review* 41: 204–211.
Britton, John P. 1989. An Early Function for Eclipse Magnitudes in Babylonian Astronomy. *Centaurus* 32: 1–52.
Bronowski, Jacob. 1973. *The Ascent of Man*. Bodton: Little, Brown and Company.
Bruins, Evert M. 1965a. The Egyptian Shadow Clock. *Janus* 52: 127–137.
__________. 1965b. Egyptian Astronomy. *Janus* 52: 161–180.
Brumbaugh, Robert S. 1981. *The Philosophers of Greece*. SUNY: Albany.
Bunbury, Edward H. 1979. *History of Ancient Geography*. London: John Murray. (Repr. of 1879, Amsterdam: Meridian Publishing Co.).
Burch, George B. 1949–1950. Anaximander, the First Metaphysician. *Review of Metaphysics* 3: 137–160.
Burkert, Walter. 1963. Iranisches bei Anaximander. *Rheinisches Museum für Philologie* 106: 97.
__________. 1972. *Lore and Science in Ancient Pythagoreanism*. Cambridge, MA: Harvard University Press.
__________. 1999. Diels' Vorsokratiker. Rückschau und Ausblick. In W.M. Calder III et al., eds., *Hermann Diels (1848–1922) et la science de l'antiquité*, 169–197. Genève: Fondation Hardt.
Burnet, John. 1930. *Early Greek Philosophy*. New York: World Publishing Company. (1st impr. 1892. London: A. & C. Black).
__________. 1956. *Plato's Phaedo*. Oxford: Clarendon Press (1st impr. 1911).
Calder, III William M. 1958. The Spherical Earth in Plato's Phaedo. *Phronesis* 3: 121–125.
Cappelletti, Angel J. 1984. *La filosofía de Anaxágoras*. Caracas: Sociedad Venezolana de Filosofia.
Cawthorne, Nigel. 2004. *Doomsday. 50 Visions of the End of the World*. London: Arcturus Publishing Limited.
Cellarius, Andreas. 2006 (orig. 1660). *Harmonia Macrocosmica*. Repr. in *The Finest of the Heavens*, with introd. by Robert H. van Gent, Hongkong: Taschen.
Cherniss, Harold F. 1944. *Aristotle's Criticism of Plato and the Academy*. Baltimore: Hopkins.
__________. 1964. *Aristotle's Criticism of Presocratic Philosophy*. New York: Octagon Books, Inc. (1st impr. 1935.Baltimore: Hopkins).
Clagett, Marshall. 1995. *Ancient Egyptian Science*, Vol. 2: *Calendars, Clocks, and Astronomy*. Philadelphia: American Philosophical Society.

Classen, Carl J. 1986. *Ansätze. Beiträge zum Verständnis der frühgriechischen Philosophie.* Würzburg/Amsterdam: Königshausen & Neumann/Rodopi.
Clayton, Peter A. 1994. *Chronicle of the Pharaohs.* London: Thames and Hudson.
Cleomedes. 1891. *De motu circulari corporum celestium.* Hermannus Ziegler, ed., Lipsiae: Teubner.
Clère, J.J. 1958. Fragments d'une Nouvelle Représentation Égyptienne du Monde. *Mitteilungen des deutschen archäologischen Instituts, Abteilung Kairo* 16: 30–46.
Cleve, Felix M. 1973. *The Philosophy of Anaxagoras.* The Hague: Martinus Nijhoff.
Cohen, Morris R. and I.E. Drabkin. 1948. *A Source Book in Greek Science.* New York: McGraw-Hill Book Company, Inc.
Conche, Marcel. 1991. *Anaximandre. Fragments et Témoignages.* Paris: PUF.
Copernicus, Nicolaus. 1643. *De revolutionibus orbium coelestium.* Norimberga: Petrius.
Cornford, Francis M. 1934. Innumerable Worlds in Presocratic Philosophy. *The Classical Quarterly* 28: 1–16.
__________. 1937. *Plato's Cosmology.* London: Routledge & Kegan Paul (4th impr. 1956).
Cotterell, Brian, and Johan Kamminga. 1990. *Mechanics of Pre-industrial Technology.* Cambridge: Cambridge University Press.
Couprie, Dirk L. 1989. *De verordening van de Tijd. Interpretatie en vertaling van het fragment van Anaximander, met een appendix over zijn astronomie.* Delft: Eburon.
__________. 1995. The Visualization of Anaximander's Astronomy. *Apeiron* 28: 159–181.
__________. 1996. Het universum volgens Kant. *Algemeen Nederlands Tijdschrift voor Wijsbegeerte* 88: 18–30.
__________. 1999 (1998). "Hätte die Welt ein Ziel, [...] so wäre es [...] mit allem Werden längst zu Ende". Ein Beitrag zur Geschichte einer Argumentation. *Nietzsche-Studien* 27: 107–118.
__________. 2001a. Anaximander's Discovery of Space. In Anthony Preus, ed. *Before Plato. Essays in Ancient Greek Philosophy VI*, 23–48. SUNY: Albany.
__________. 2001b. πρηστῆρος αὐλος revisited. *Apeiron* 34: 193–202.
__________. 2001c. Greek Influence on the Representation of the Heaven in Ptolemaic Egyptian Art. In *The Great Book of Aesthetics. Proceedings of the 15th International Congress of Aesthetics, Japan 2001* (Sect. 'Four Elements of Nature and Beauty', We/2–9) Tokyo (CD-Rom).
__________. 2003. The Discovery of Space: Anaximander's Astronomy. In Dirk L. Couprie, Robert Hahn, and Gerard Naddaf, eds. *Anaximander in Context*, 164–254. SUNY: Albany.
__________. 2004a. Anaximander und die Geschichte des Griechische Weltmodells. Eine Auseinandersetzung mit Detlev. *Prima Philosophia* 17: 127–143.
__________. 2004b. How Thales Was Able to "Predict" a Solar Eclipse Without the Help of Alleged Wisdom. *Early Science and Medicine* 9: 321–337.
__________. 2005a. Some Remarks on the Earth in Plato's Phaedo. *Hyperboreus* 11: 192–204.
__________. 2005b. Anaximander. In James Fieser and Bradley Dowden, eds., *The Internet Encyclopedia of Philosophy*, http://www.iep.utm.edu/anaximan.
__________. 2006. Anaxagoras und die Größe der Sonne. *Hyperboreus* 12: 55–76.
__________. 2008. Anaxagoras and Aristotle on the Sun at the Horizon. *Hyperboreus* 14: 39–49.
__________. 2009a. Problems with Anaximander's Numbers. *Apeiron* 42: 167–184.
__________. 2009b. The Tilting of the Heavens in Presocratic Cosmology. *Apeiron* 42: 259–274.
__________. 2009c. Anaximander's Legacy and the Stability of the Earth (review of Carlo Rovelli, *Anaximandre de Milet ou la naissance de la pensée scientifique.* Paris, Dunod, 2009). *Hyperboreus* 15: 1–10.
__________. 2009d. Anaxagoras and the Size of the Sun. In Elizabeth Close a.o., ed. *Greek Research in Australia. Proceedings of the Seventh International Conference of Greek Studies, Flinders University June 2007*, 21–30. Adelaide: Flinders University.
__________. 2010. Review of Robert Hahn, Archaeology and the Origins of Philosophy. *Aestimatio*, http://www.ircps.org/publications/aestimatio/aestimatio.htm (volume 7, nr. 25).

Couprie, Dirk L. and H.J. Pott. 2001. Imagining the Universe. *Apeiron* 35: 47–59.

__________. 2003. Valangst: hemel en aarde in de antieke kosmologie. *Tijdschrift voor Filosofie* 65: 227–247.

__________. 2005. Anaximander of Miletus. In Patricia O'Grady, ed. *Meet the Philosophers of Ancient Greece,* 33–38. Aldershot: Ashgate.

__________. 2006. Die Angst zu fallen; Himmel und Erde in der antiken Kosmologie. *Prima Philosophia* 19: 23–43.

Cullen, Christopher. 1996. *Astronomy and Mathematics in Ancient China: the Zhou bi suan jin.* Cambridge: Cambridge University Press.

Daiber, Hans. 1980. *Aetius Arabus. Die Vorsokratiker in arabischer Überlieferung.* Wiesbaden: Franz Steiner Verlag GmbH.

Darsow, Wolfgang. 1954. Die Kore des Anaximandros. *Jahrbuch des deutschen archäologischen Instituts* 69: 101–117.

De Jong, J.J. 1994. *De wiskundige grondslagen van de Griekse en Romeinse tempelarchitectuur in theorie en praktijk tussen de 4e en 1e eeuw.* Leiden.

De Liahgre, Böhl and M.Th. Franz. 1952. *Het Gilgamesj-Epos. Nationaal heldendicht van Babylon.* Amsterdam: H.J. Paris.

Descartes, René. 1644. *Les Principes de la Philosophie.* Amsterdam: Elsevier.

__________. 1953. *Oevres et Lettres.* André Bridoux, ed., Paris: Éditions Gallimard.

De Vogel, Cornelia J. 1957. *Greek Philosophy, A Collection of Texts Selected and Supplied with Some Notes and Explanations,* Vol. I: *Thales to Plato.* Leiden: Brill.

Dicks, D.R. 1954. Ancient Astronomical Instruments. *The Journal of the British Astronomical Association* 64: 77–85.

__________. 1959. Thales. *The Classical Quarterly* 9: 294–309.

__________. 1966. Solstices, Equinoxes, and the Presocratics. *Journal of Hellenic Studies* 86: 26–42.

__________. 1970. *Early Greek Astronomy to Aristotle.* Ithaca, NY: Thames and Hudson.

Diels, Hermann. 1876. Chronologische Untersuchungen über Apollodors Chronica. *Rheinisches Museum* 31: 1–54.

__________. 1879. *Doxographi Graeci.* Berlin: Walter de Gruyter & Co.

__________. 1897. Über Anaximander's Kosmos. *Archiv für Geschichte der Philosophie* 10: 228–237.

__________. 1903. *Die Fragmente der Vorsokratiker.* Berlin: Weidmannsche Buchhandlung.

__________. 1910. *Die Fragmente der Vorsokratiker* (2nd edition). Berlin: Weidmannsche Buchhandlung.

__________. 1922. *Die Fragmente der Vorsokratiker* (4th edition). Berlin: Weidmannsche Buchhandlung.

__________. 1923a. Anaximandros von Milet. *Neue Jahrbücher für das klassische Altertum* 51: 65–75 (Also in *Kleine Schriften zur Geschichte der Philosophie*, 1–11).

__________. 1923b. *Kleine Schriften zur Geschichte der Philosophie.* Hildesheim (reprint 1969. Darmstadt: Wissenschaftliche Buchgesellschaft).

Diels, Hermann and Walther Kranz. 1954. *Die Fragmente der Vorsokratiker* (6th edition). Zürich/ Hildesheim: Weidmann.

Digges, Thomas. 1576. *A Perfit Description of the Caelestiall Orbes according to the mostaunciente doctrine of the Pythagoreans, latelye reuiued by Copernicus and by Geometricall Demonstrations approued.* Added to the second edition of: L. Digges, *A Prognostication euerlastinge.* London: Thomas Marsh. Repr. 1931 in pp. 78-95 of Francis R. Johnson and Sanford V. Larkey. Thomas Digges, the Copernican System, and the Idea of Infinity of the Universe in 1576. *The Huntington Library Bulletin* 5: 69–117.

Diodorus of Sicily. 2000. *Bibliotheca Historica.* Bd I of Charles H. Oldfather, ed. and transl., *Diodorus of Sicily in Ten Volumes.* Cambridge, MA: Harvard University Press.

Diogenes, Laertius. 1980. *Lives of Eminent Philosophers, in Two Volumes.* Robert D. Hicks, ed. and transl., Cambridge, MA: Harvard University Press, London: Heinemann.

Dreyer, John L.E. 1953. *A History of Astronomy from Thales to Kepler*. New York: Dover Publications, Inc. (Facs. reprint of: *History of the Planetary Systems from Thales to Kepler*. Cambridge: Cambridge University Press, 1905).

Dubs, Homer H. 1955. *The History of the Former Han Dynasty by Pan Ku*, Vol III. Baltimore: Waverley Press.

Duchesne-Guillemin, Jacques. 1966. D'Anaximandre à Empédocle: contacts Gréco-Iraniens. In A. Pagliano, ed., *Atti del convegno sul tema: la Persia a il mondo Greco-Romano (Roma, 11–14 aprile 1965)*, 423–431. Roma: Accademia Nazionale di Lincei.

Dümmler, Ferdinand. 1889. *Akademika, Beiträge zur Literaturgeschichte der sokratischen Schulen*. Gießen: Ricker (repr. 1987, Osnabrück: Biblio Verlag).

Dumont, Jean-Paul. 1988. *Les Présocratiques*. Paris: Gallimard.

Eastwood, Bruce S. 1992. Heraclides and Heliocentrism: Texts, Diagrams, and Interpretations. *Journal for the History of Astronomy* 23: 233–260.

Ebeling, Erich. 1931. *Tod und Leben nach der Vorstellung der Babylonier. I. Teil, Texte*. Berlin/ Leipzig: Walter de Gruyter & Co.

Ebert, Theodor, transl. and ed. 2004. *Platon. Werke. Übersetzung und Kommentar. I 4, Phaidon*. Göttingen: Vandenhoeck & Ruprecht.

Edwards, Iorwerth E.S. 1961. *The Pyramids of Egypt*. Harmondsworth: Penguin.

Eisler, Robert. 1910. *Weltenmantel und Himmelszelt*. München: Beck.

Eisler, Robert. 1949. The Polar Sighting Tube. *Archives internationales d'histoire des sciences* 28: 312–323.

Ekschmitt, Werner. 1969. Weltmodelle. Griechische Weltbilder von Thales bis Ptolemäus. Mainz am Rhein: Verlag Philipp von Zabern.

Emmer, Michele, ed. 1993. *The Visual Mind. Art and Mathematics*. Cambridge, MA: MIT Press.

Engmann, Joyce. 1991. Cosmic Justice in Anaximander. *Phronesis* 36(1): 1–25.

Eubulus. 1983. In Richard L. Hunter, ed., *The Fragments*. Cambridge: Cambridge Univerity Press.

Euclid. 1895. *Euclidis Optica, Opticorum Recensio Theonis, Catoptrica, cum Scholiis Antiquis*. Johan-Ludvig Heiberg, ed., Lipsiae: In Aedibus B.G. Teubneri.

__________. 1956. *The Thirteen Books of Euclid's Elements*. Thomas L. Heath, transl., New York: Dover Publications.

Euripides. 2008. Christopher Collard and Martin Cropp, ed. and transl., *Fragments*. Cambridge, MA: Harvard University Press.

Evans, Godfrey. 1970. The Astronomy of Heraclides Ponticus. *The Classical Quarterly* 20: 102–111.

Evans, James. 1998. *The History & Practice of Ancient Astronomy*. New York/Oxford: Oxford University Press.

Fehling, Detlev. 1985a. *Die sieben Weisen und die frühgriechische Chronologie*. Bern: Peter Lang.

__________. 1985b. Das Problem der Geschichte des griechischen Weltmodells vor Aristoteles. *Rheinisches Museum für Philologie* 128: 195–231.

__________. 1994. *Materie und Weltbau in der Zeit der frühen Vorsokratiker*. Innsbruck: Verlag des Instituts für Sprachwissenschaft der Universität Innsbruck.

Ferber, Rafael. 1986. Der Ursprung der Wissenschaft bei Anaximander von Milet. *Theologie und Philosophie* 61: 551–561. (Also in *Philosophia Naturalis* 24 (1987): 195–215).

Ferguson, Kitty. 1999. *Measuring the Universe. The Historical Quest to Quantify Space*. London: Headline.

Finkelberg, Aryeh. 1994. Plural World's in Anaximander. *American Journal of Philology* 115: 485–506.

Flammarion, Camille. 1888. L'Atmosphère: météorologie populaire. Paris: Librairie Hachette et Cie.

Fotheringham, John K. 1919. Cleoastratus. *Journal of Hellenic Studies* 39: 164–184.

Frank, Erich. 1923. *Plato und die sogenannten Pythagoreer. Ein Kapitel aus der Gechichte des griechischen Geistes*. Halle: Verlag von Max Niemeyer.

Fränkel, Hermann F. 1968. *Noten zu den Argonautika des Apollonios*. München: Beck.

Frankfort, Hans. 1933. *The Cenotaph of Seti I at Abydos. EES Excavation Memoir 39*, Vol. I: *Texts*, Vol. II: *Plates*. London: The Egypt Exploration Society.

Friedländer, Paul. 1914. Die Anfänge der Kugelgeographie. *Jahrbuch des kaiserlichen deutschen archäologischen Instituts* 29: 98–120.

Furley, David J. 1981. The Greek Theory of the Infinite Universe. *Journal of the History of Ideas* 42: 571–585.

__________. 1983. *Review of Dennis O'Brien, Theories of Weight in the Ancient World*, vol. I: *Democritus, Weight and Size. Oxford Studies in Ancient Philosophy* 1: 193–209 (repr. in *Cosmic Problems*, 91–102).

__________. 1987. *The Greek Cosmologists*, Vol. I: *The Formation of the Atomic Theory and Its Earliest Critics*. Cambridge: Cambridge University Press.

__________. 1989. *Cosmic Problems. Essays on Greek and Roman Philosophy of Nature*. Cambridge: Cambridge University Press.

Furley David J. and Reginald E. Allen, eds. 1970. *Studies in Presocratic Philosophy*, Vol. I: *The Beginnings of Philosophy*. New York and London: Humanities Press.

Gemelli Marciano, Laura M. 2007. *Die Vorsokratiker, Band I*. Düsseldorf: Artemis & Winkler.

Germain, Gabriel. 1954. *Homère et la mystique des nombres*. Paris.

Gershenson, Daniel E. and Daniel A. Greenberg. 1964. *Anaxagoras and the Birth of Physics*. New York: Blaisdell Publishing Company.

Gibbs, Sharon L. 1976. *Greek and Roman Sundials*. New Haven and London: Yale University Press.

__________. 1979. The First Scientific Instruments. In Kenneth Brecher and Michael Feirtag, eds., *Astronomy of the Ancients*, 39–59. Cambridge, MA and London: MIT Press.

Gilbert, Otto. 1967. *Die meteorologischen Theorien des griechischen Altertums*. Hildesheim: Georg Olms (orig. 1907, Leipzig: Teubner).

Gingerich, Owen. 2000. Plotting the Pyramids. *Nature* 408: 297–298.

Gobry, Ivan. 2000. *La cosmologie des Ioniens*. Paris: L'Harmattan.

Görgemanns, Herwig. 1970. *Untersuchungen zu Plutarchs Dialog De facie in orbe lunae*. Heidelberg: Winter.

Gottschalk, H.B. 1980. *Heraclides of Pontus*. Oxford: Clarendon Press.

Graham, Angus C., transl. 1990. *The Book of lieh-tzû: A Classic of the Tao*. New York: Columbia University Press.

Graham, Daniel W. 2006. *Explaining the Cosmos. The Ionian Tradition of Scientific Philosophy*. Princeton and Oxford: Princeton University Press.

__________ (transl. and ed.). 2010. *The Texts of Early Greek Philosophy. The Complete Fragments and Selected Testimonies of the Major Presocratics*. Cambridge: Cambridge University Press.

Gregory, Andrew. 1996. Astronomy and Observation in Plato's Republic. *Studies in History and Philosophy of Science* 27: 451–471.

__________. 2007. *Ancient Greek Cosmogony*. London: Duckworth.

__________. 2011. Anaximander's Zoogony. *Proceedings of the Third Biennial International Conference on Greek Studies*. Adelaide: Dept. of Languages-Modern Greek, Flinders University (forthcoming).

Gruben, Gottfried. 1963. Das archaische Didymaion. *Jahrbuch des deutschen archäologischen Instituts* 78: 78–177.

Guthrie, William K.C. 1962. *A History of Greek Philosophy, I. The Early Presocratics and the Pythagoreans*. Cambridge: Cambridge University Press.

__________. 1965. *A History of Greek Philosophy, II. The Presocratic Tradition from Parmenides to Democritus*. Cambridge: Cambridge University Press.

Haase, Frank. 2008. *Philosophie des Gnomon – Anaximander's Medientheorie*. München: Kopaed.

Hahm, David E. 1977. *The Origins of Stoic Cosmology*. Ohio: Ohio State University Press.

Hahn, Robert. 2001. *Anaximander and the Architects. The Contributions of Egyptian and Greek Architectural Technologies on the Origins of Greek Philosophy*. SUNY: Albany.

__________. 2003. Proportions and Numbers in Anaximander and Early Greek Thought. In: Dirk L. Couprie, Robert Hahn, and Gerard Naddaf, eds., *Anaximander in Context*, 70–163. SUNY: Albany.

__________. 2007. Heidegger, Anaximander, and the Greek Temple. *Wolkenkuckucksheim.* 12(1), *Heaven and Earth. Festschrift to Honor Karsten Harries*, http://www.tu-cottbus.de/theorienderarchitektur/wolke/eng/Subjects/071/Hahn/hahn.htm (no pages indicated).

__________. 2010. *Archaeology and the Origins of Philosophy*. SUNY: Albany.

Hall, J.J. 1969. πρηστῆρος αὐλος. *Journal of Hellenic Studies* 89: 57–59.

Hartner, Willy. 1969. Eclipse Periods and Thales' Prediction of a Solar Eclipse. *Centaurus* 14: 60–71.

Haslam Michael W., ed. and transl., 1986. *The Oxyrhynchus Papyri*, Vol. 53. London: The British Academy.

Havelock, Eric A. 1978. *The Greek Concept of Justice. From Its Shadow in Homer to Its Substance in Plato*. Cambridge: Cambridge University Press.

__________. 1983. The Linguistic Task of the Presocratics. In Kevin Robb, ed., *Language and Thought in Early Greek Philsophy*, 7–82. La Salle, IL: The Hegeler Institute.

Hawking, Stephen W. 1988. *A Brief History of Time: From the Big Bang to Black Holes*. New York: Bantam Books.

Heath, Thomas. 1913. *Aristarchus of Samos. The Ancient Copernicus*. Oxford: Clarendon Press.

Hegel, Georg Wilhelm Friedrich. 1986. *Dissertatio philosophica de orbitis planetarum = Philosophische Erörterung über die Planetenbahnen*. Latin text with German transl. by Wolfgang Neuser. Weinheim: Acta Humaniora VCH (orig. 1801, Habilitationsschrift Jena).

__________. 1816. *Vorlesungen über die Geschichte der Philosophie*.

Heidegger, Martin. 1957. *Der Satz vom Grund*. Tübingen: Max Niemeyer Verlag.

Heidel, William Arthur. 1921. Anaximander's Book, the Earliest Known Geographical Treatise. *Proceedings of the American Academy of Arts and Sciences* 56: 239–288.

__________. 1933. *The Heroic Age of Science*. Baltimore: Williams & Wilkins Company.

__________. 1937. *The Frame of the Ancient Greek Maps. With a Discussion of the Discovery of the Sphericity of the Earth*. New York: American Geographical Society.

Hendrix, John. 2004. *Platonic Architectonics. Platonic Philosophies and the Visual Arts*. New York: Peter Lang.

Herodotus. 1987–1997. *Historiae* in two vols. In: Haiim B. Rosén, ed., Leipzig: Teubner.

__________. 1998. *The Histories*. Robin Waterfield, transl., Oxford: Oxford University Press.

Herrmann, Albert. 1931. *Die Erdkarte der Urbibel*. Braunschweig: Kommisionsverlag von Georg Westermann.

Herrmann, Joachim. 1969. *Gesetze des Weltalls*. Stuttgart: Kosmos. Gesellschaft der Naturfreunde Franck'sche Verlagshandlung.

Herrmann, Karl F. 1849. *De philosophorum Ionicorum aetatibus*. Göttingen: Officina Academica Dieterichiana.

Hesiod. 1991. *Theogonie; Werke und Tage*. Albert Von Schirnding, ed. and transl., Darmstadt: Wissenschaftliche Buchgesellschaft.

__________. 1990. *Theogonia; Opera et Dies; Scrutum*. Friedrich Solmsen, ed., Oxonii: e Typographeo Clarendoniano.

Hippocrates. 1902. In Hugo Kuehlewein, ed., De articiculatione, Vol. II: *Hippocratis Opera quae feruntur omnia*. Lipsiae: Teubner.

Hippolytus. *Facta et dicta memorabilia*.

Holmberg, Uno. 1922–1923. *Der Baum des Lebens*. Helsingfors.

Hölscher, Uvo. 1970. Anaximander and the Beginnings of Greek Philosophy. In David J. Furley and Reginald E. Allen, eds., *Studies in Preoscratic Philosophy*, Vol. I: *The Beginnings of Philosophy*, 281–322. New York and London: Humanities Press.

Horowitz, Wayne. 1998. *Mesopotamian Cosmic Geography*. Winona Lake: Eisenbrauns.

Huffman, Carl A. 1993. *Philolaus of Croton, Pythagorean and Presocratic*. Cambridge: Cambridge University Press.

Isler, Martin. 1989. An Ancient Method of Finding and Extending Direction. *Journal of the American Research Center in Egypt* 26: 191–206.

__________. 1991a. The Merkhyt. *Varia Aegyptiaca* 7: 53–67.

__________. 1991b. The Gnomon in Egyptian Antiquity. *Journal of the American Research Center in Egypt* 28: 155–185.

Jacob, Christian. 1988. Inscrire la terre habitée sur une tablette. Réflections sur la fonction des cartes géographiques en Grèce ancienne. In Marcel Detienne, ed., *Les savoirs de l'écriture en Grèce ancienne*, 273–304. Lille: Presses Universitaires de Lille.

Jaki, Stanley L. 1972. *The Milky Way. An Elusive Road for Science*. New York: David & Charles.

Janssen, Tjitte Henk, ed. 2000. *Heel de wereld beschreven. Lucretius, Philo en anderen over kosmos, aarde en mens*. Amsterdam: Boom.

Jaspers, Karl. 1957. *Die grossen Philosophen. Erster Band. Aus dem Ursprung denkende Metaphysiker*. Munich: Piper.

Johnson, Francis R. and Sanford V. Larkey. 1931. Thomas Digges, the Copernican System, and the Idea of Infinity of the Universe in 1576. *The Huntington Library Bulletin* 5: 69–117.

Kahn, Charles H. 1970. On Early Greek Astronomy. *Journal of Hellenic Studies* 90: 99–116.

__________. 1994. Anaximander and the Origins of Greek Cosmology, Indianapolis: Hackett Publishing Company, Inc. (repr. of revised 2d ed.1985, Philadelphia PA: Centrum Philadelphia; first ed. 1960, New York: Columbia University Press).

__________. 2001. *Pythagoras and the Pythagoreans. A Brief History*. Indianapolis/Cambridge: Hackett Publishing Company, Inc.

Kambartel, Friedrich 1966. Der Satz vom zureichenden Grunde und das Begründungs-problem der Mechanik. Zu einer Bemerkung von Leibniz im 2. Schreiben an Clarke. *Zeitschrift für philosophische Forschung* 20: 457–470.

Kant, Immanuel. 1755. Allgemeine Naturgeschichte und Theorie des Himmels, oder Versuch von der Verfassung und dem mechanischen Ursprunge des ganzen Weltgebäudes nach Newtonischen Grundsätzen abgehandelt. In Wilhem Weischädel, ed., 1960. *Immanuel Kant. Werke in sechs Bänden*, Band I, 219–400. Wiesbaden: Insel-Verlag.

__________. 1763. Der einzig möglichen Beweisgrund zu einer Demonstration des Daseyns Gottes. In Wilhem Weischädel, ed., 1960. *Immanuel Kant. Werke in sechs Bänden*, Band I, 617–738. Wiesbaden: Insel-Verlag.

__________. 1781. *Kritik der reinen Vernunft*. In Wilhem Weischädel, ed., 1960. *Immanuel Kant. Werke in sechs Bänden*, Band II. Wiesbaden: Insel-Verlag.

__________. 1788. *Kritik der praktischen Vernunft*. In Wilhem Weischädel, ed., 1960. *Immanuel Kant. Werke in sechs Bänden*, Band IV, 103–302. Wiesbaden: Insel-Verlag.

Kauffmann, Hans. 1976. *Probleme griechischer Säulen*. Opladen: Westdeutscher Verlag.

Keith, Arthur B. 1917. Indian Mythology. In Louis H. Gray and George F. Moore, eds., Vol. VI: *The Mythology of All Races*. Boston: (reprint New York 1964).

Kelley D.H. and E.F. Milone. 2005. *Exploring Ancient Skies. An Encyclopedic Survey of Archaeoastronomy*. New York: Springer.

Kingsley, Peter. 1995. *Ancient Philosophy, Mystery, and Magic. Empedocles and Pythagorean Tradition*. Oxford: Oxford University Press.

Kirk, Geoffrey S., John E. Raven and Malcolm Schofield. 2009. *The Presocratic Philosophers*. Cambridge: Cambridge University Press (repr. of 2d ed. 1983).

Klepešta, Josef and Antonín Rükl. 1969. *Constellations*. London: Hamlyn.

Kobus, M.L. and J.J. Raimond Jr. 1936. *Cosmografie*. Groningen: J.B. Wolters.

Kočandrle, Radim. 2008. *Fysis Iónských Myslitelů. Rozprava nad Peripatetickou Dezinterpretací*. Praha: Pavel Mervart.

Krafft, Fritz. 1971a. *Geschichte der Naturwissenschaft I. Die Begründung einer Geschichte der Wissenschaft von der Natur durch die Griechen*. Freiburg: Verlag Rombach.

__________. 1971b. Anaximandros. In Kurt Fassmann, ed., *Die Grossen der Weltgeschichte*, 284–305. Zürich: Kindler.

Kratzert, Thomas. 1998. *Die Entdeckung des Raums. Vom hesiodischen "χαος" zur platonischen "χωρα"*. Amsterdam/Philadelphia: B.R. Grüner.

Kudlek, Manfred and Erich H. Mickler. 1971. *Solar and Lunar Eclipses of the Ancient near East from 3000 B.C. to 0 with Maps*. Kevelaer: Verlag Butzon & Bercker.

Lambert, Wilfred G. 1975. The Cosmology of Sumer and Babylon. In Carmen Blacker and Michael Loewe, eds., *Ancient Cosmologies*, 42–65. London: George Allen & Unwin Ltd.
Lebedev, Andrej V. 1986. Der geometrische Stil und die Kosmologie Anaximanders. In Nelli V. Motrošilova, ed., *Studien zur Geschichte der westlichen Philosophie. Elf Arbeiten jüngerer sowjetischer Autoren*, 31–54, transl. by Sigrun Bielfeldt. Frankfurt am Main: Suhrkamp Verlag.
Leclant, Jean, ed. 1980. *Le monde Égyptien. Les Pharaons.* Tome 3, *L'Égypte du crépuscule.* Paris: Gallimard.
Leibniz, Gottfried W. 1969. *Philosophical Papers and Letters.* Leroy E. Loemker, ed. and transl., Dordrecht: Reidel.
Leitz, Christian. 1995. *Altägyptische Sternuhren.* Leuven: Peeters.
Lewis, Michael J.T. 2001. *Surveying Instruments of Greece and Rome.* Cambridge: Cambridge University Press.
Livrea, Enrico, ed. 1973, see Apollonius Rhodius.
Lloyd, Geoffrey E.R. 1966. *Polarity and Analogy. Two Types of Argumentation in Early Greek Thought.* Cambridge: Cambridge University Press.
Lucretius (Titus Lucretius Carus). 1997. *De la nature = De rerum natura.* José Kany-Turpin, ed. and transl., Paris: Flammarion.
__________. 2001. *On the Nature of Things.* Martin F. Smith, transl., Indianapolis: Hackett.
Lull, José. 2005. El eclipse total de Sol del año 585 a.C. La sombra que envainó las espadas. *Boletín Huygens* 56, http://www.astrosafor.net/Huygens/2005/56/ Eclipse-585.htm.
__________. 2006. *La astronomía en el antiguo Egipto.* València: Universitat de València.
Mansfeld, Jaap. 1971. *The Pseudo-Hippocratic Tract Περὶ ἑβδομάδων ch. 1-11 and Greek Philosophy.* Assen: Van Gorcum.
__________. 1987. *Die Vorsokratiker.* Stuttgart: Philipp Reclam Jun.
__________. 1999. Sources. In Anthony A. Long, ed., *The Cambridge Companion to Early Greek Philosophy*, 22–44. Cambridge: Cambridge University Press.
Martin, Roland E. and Henri Stierlin. 1966. *Monde grec.* Fribourg: Office du livrre.
Martin, Th. Henri. 1864. Sur quelques prédictions d'éclipses mentionnées par des auteurs anciens. *Revue Archéologique* 9: 170–199.
__________. 1879. Mémoire sur les hypothèses astronomiques des plus anciens philosophes de la Grèce étrangers à la notion de la sphéricité de la Terre. *Mémoires de l'Institut National de France, Académie des Inscriptions et Belles-Lettres*, tôme XXIX, 2e partie: 29–252 (reprint New York 1976).
Mayall, Robert N. and Margaret W. Mayall. 1938. *Sundials.* Cambridge: Sky Publishing Corporation.
McKirahan, Richard D. 2001. Anaximander's Infinite Worlds. In Anthony Preus, ed., *Before Plato. Essays in Ancient Greek Philosophy VI*, 49–65. SUNY: Albany.
Menander. *Fragments.*
Morrison, John S. 1955. Parmenides and Er. *The Journal of Hellenic Studies* 75: 59–68.
__________. 1959. The Shape of the Earth in Plato's Phaedo. *Phronesis* 4: 101–119.
Mosshammer, Alden A. 1981. Thales' Eclipse. *Transactions of the American Philological Association* 111: 145–155.
Mueller, Ian. transl. 2005. *Simplicius on Aristotle's "On the Heavens" 2.10–14.* Ithaca, NY: Cornell University Press.
Mugler, Charles. 1953. *Deux thèmes de la cosmkologie grecque: devenir cyclique et pluralité des mondes.* Paris: Klincksiek.
Murray, John. 1895. *Report of the Scientific Results of the Voyage of H.M.S. Challenger During the Years 1872–76.* London: Eyre & Spottiswoode.
Naddaf, Gerard. 1998. On the Origin of Anaximander's Cosmological Model. *Journal of the History of Ideas* 59: 1–28.
__________. 2001. Anaximander's Measurements Revisited. In Anthony Preus, ed. *Before Plato. Essays in Ancient Greek Philosophy VI*, 5–21. SUNY: Albany.

__________. 2003. Anthropogony and Politogony in Anaximander of Miletus. In Dirk L. Couprie, Robert Hahn, and Gerard Naddaf, eds. *Anaximander in Context*, 7–69. SUNY: Albany.

__________. 2005. *The Greek Concept of Nature*. SUNY: Albany.

NASA. *Five Millennium Catalog of Solar Eclipses*, http://eclipse.gsfc.nasa.gov/SEcat5/SEcatalog.html.

Needham, Joseph. 1959. *Science and Civilisation in China*, Vol. 3: *Mathematics and the Sciences of the Heavens and the Earth*. Cambridge: Cambridge University Press.

Neugebauer, Otto. 1952. *The Exact Sciences in Antiquity*. Princeton: Princeton University Press.

__________. 1975. *A History of Ancient Mathematical Astronomy*, 3 vols. Berlin, Heidelberg & New York: Springer.

__________. 1983. *Astronomy and History. Selected Essays*. New York: Springer.

__________. 1980. On the Orientation of Pyramids, *Centaurus* 24: 1–3.

Neugebauer, Otto and Richard A. Parker. 1960–1969. *Egyptian Astronomical Texts I–III*. Providence, RI: Brown University Press.

Neugebauer, Paul V. 1922. *Tafeln zur astronomischen Chronologie III, Hilfstafeln zur Berechnung von Himmelserscheinungen*. Leipzig: Hinrichs'sche Buchhandlung.

Neuhaeuser, Iosephus. 1883. *Dissertatio de Anaximandri Milesius sive vetustissima quaedam rerum universitatis conceptio restituta*. Bonnae: Max Cohen et Filius.

Nietzsche, Friedrich. 1973a. *Die Philosophie im tragischen Zeitalter der Griechen* (1873). In Giorgio Colli and Mazzino Montinari, eds., *Kritische Gesamtausgabe* (KGW) III.2, 393–366. Berlin/New York: Walter de Gruyter.

__________. 1973b. *Die fröhliche Wissenschaft* (1882) In: Giorgio Colli and Mazzino Montinari, eds., *Kritische Gesamtausgabe* (KGW) V.2, 11-336. Berlin/New York: Walter de Gruyter.

__________. 1974. *Notes of Summer 1885*. In: Giorgio Colli and Mazzino Montinari, eds. *Kritische Gesamtausgabe* (KGW) VII.3. Berlin/New York: Walter de Gruyter.

North, John. 2008. *Cosmos. An Illustrated History of Astronomy and Cosmology*. Chicago & London: The University of Chicago Press.

O'Brien, Dennis. 1967. Anaximander's Measurements. *The Classical Quarterly* 17: 423–432.

__________. 1968. Derived Light and Eclipses in the Fifth Century. *Journal of Hellenic Studies* 88: 114–127.

__________. 1981. *Theories of Weight in the Ancient World. Four Essays on Democritus, Plato and Aristotle*. Vol. I: *Democritus, Weight and Size. An Essay in the Reconstruction of Early Greek Philosophy*. Paris/Leiden: Les Belles Lettres/Brill.

Østby, Erik. 1992. Der dorische Tempel von Pherai. *Opuscula Atheniensia* 19: 85–113.

O'Grady, Patricia. 2002. *Thales of Miletus. The Beginnings of Western Science and Philosophy*. Aldershot: Ashgate.

__________. Thales. In J. Fieser and B. Dowden, eds., *Internet Encyclopedia of Philosophy*, http://utm.edu/research/iep.

Olympiodorus. 1900. In Guilelmus Stüve, ed., *Aristotelis Meteora Commentaria* Berolini: Reimer.

O'Neil, William M. 1986. *Early Astronomy from Babylonia to Copernicus*. Sydney: Sydney University Press.

Panchenko, Dmitri. 1994a. Thales's Prediction of a Solar Eclipse. *Journal for the History of Astronomy* 25: 275–288.

__________. 1994b. Ὅμοιος and Ὁμοιότης in Thales and Anaximander. *Hyperboreus* 1: 28–55.

__________. 1997. Anaxagoras' Argument Against the Sphericity of the Earth. *Hyperboreus* 3: 175–178.

__________. 1999. Who Found the Zodiac? *Antike Naturwissenschaft und ihre Rezeption* 9: 33–44.

__________. 2000. Anaximander, Hecataeus and Dionysius on the Mycenean Script. *Hyperboreus* 5: 418–420.

__________. 2002a. Eudemus Fr. 145 Wehrli and the Ancient Theories of Lunar Light. In István Bodnár & William W. Fortenbaugh, eds., *Eudemus of Rhodes*, 323–336. New Brunswick and London: Transaction Publishers.

__________. 2002b. The City of the Branchidae and the Question of Greek Contribution to the Intellectual History of India and China. *Hyperboreus* 8: 244–255.

__________. 2008. Parmenides, the Nile, and the Circumnavigation of Africa by the Phoenicians. In J.M. Caudan Morón, F.J. González Ponce, A.L. Chávez Reino, eds., *Realidad e literatura en la visión grecorromano de África. Homenaje al Prof. Jehan Desanges*, 189–193. Sevilla.

Pannekoek, Anton. 1961. *A History of Astronomy*. New York: Dover Publications.

Parker, Richard A. 1974. Ancient Egyptian Astronomy. In Frank R. Hodson, ed., *The Place of Astronomy in the Ancient World*, 51–66. London: Oxford University Press for the British Academy.

Parpola, Simo. 1993. *Letters from Assyrian and Babylonian Scholars*. Helsinki: Helsinki University Press.

Pascal, Blaise. 1962. *Pensées*. Présentées par Jean Guitton. Paris: Éditions Gallimard (Le Livre de Poche).

Pecker, Jean-Claude. 2001. *Understanding the Heavens. Thirty Centuries of Astronomical Ideas from Ancient Thinking to Modern Cosmology*. Berlin/Heidelberg/New York: Springer.

Pédech, Paul. 1976. *La géographie des Grecs*. Paris: PUF.

Pedersen, Olaf and Mogens Pihl. 1974. *Early Physics and Astronomy*. London/New York: American Elsevier.

Philo, Judaeus Alexandrinus. 1941. *On the Eternity of the World*. In: F.H. Colson, ed. and transl. *Philo*, vol. IX, 184–291. London: William Heinemann Ltd., Cambridge, MA: Harvard University Press.

Piankoff, Alexandre and Natacha Rabova-Valentino. 1957. *Mythological Papyri. In Two Parts*. New York: Pantheon Books.

Pindar. 1997. *The Odes and Selected Fragments*, Vol II: *Olympian Odes, Pythian Odes*. William H. Race, ed. and transl., Cambridge, MA: Harvard University Press.

Plass, Paul. 1972. A Note on *presteros aulos*. *Journal of Hellenic Studies* 92: 179–180.

Plato. *Apology of Socrates*.

Plato. 1926. *Cratylus*. Harold N. Fowler, ed. and transl., London: Heinemann, New York: Putnam.

Plato. 1926. *Laws*. Robert G. Bury, ed. and transl., London: Heinemann, New York: Putnam.

Plato. 1927. *Epinomis*. Walter R.M. Lamb, ed. and transl., London: Heinemann, New York: Putnam.

Plato. 1927. *The lovers*. Walter R.M. Lamb, ed. and transl., London: Heinemann, New York: Putnam.

Plato. 1966. *Timaeus*. R.G. Bury, ed. and transl., London: Heinemann Ltd, Cambridge, MA: Harvard University Press.

Plato. 1982. *Phaedo*. Harold N. Fowler, ed. and transl., Cambridge, MA: Harvard University Press, London: Heinemann Ltd.

Plato. 1982. *Phaedrus*. Harold N. Fowler, ed. and transl., Cambridge, MA: Harvard University Press, London: Heinemann Ltd.

Plato. 1999. *Republic*, 2 vols. Paul Shorey, ed. and transl., Cambridge, MA: Harvard University Press.

Pliny (Plinius Secundus, C.). 1938. *Natural History I = Praefatio, Libri I, II*. Harris Rackham, ed. and transl., London: Heinemann, Cambridge, MA: Harvard University Press.

__________. 1942. *Natural History II = Libri III-VII*. Harris Rackham, ed. and transl., London: Heinemann, Cambridge, MA: Harvard University Press.

__________. 1950. *Natural History V = Libri XVII-XIX*. Harris Rackham, ed. and transl., London: Heinemann, Cambridge, MA: Harvard University Press.

__________. 1962. *Natural History X = Libri XXXVI-XXXVII*. D. E. Eichholz, ed. and transl., London: Heinemann, Cambridge, MA: Harvard University Press.

Plutarch. 1935. *Le "peri to prosopou" de Plutarque*. (*On the Face in the Moon*). P. Raingeard, ed. and transl., Paris: Les Belles Lettres.

__________. 1971. *Vies de Plutarque, Tome VI: Pyrrhos – Marius; Lysandre – Sylla*. Robert Flacelière and Émile Chambray, ed. and transl., Paris: Les Belles Lettres.

__________. 1978. *Vies de Plutarque, Tome XIV: Dion – Brutus*. Robert Flacelière and Émile Chambray, ed. and transl., Paris: Les Belles Lettres.

__________. 2002. (Plutarco). *La generazione dell' anima nel Timeo* (*On the Generation of the Soul in the Timaeus*). In: Franco Ferrari and Laura Baldi, eds. and transl., Napoli: M. d'Auria Editore.

Porter, Bertha, Ethel Burney, and Rosalind L.B. Moss. 1972. *Topographical Bibliography of Ancient Egyp Hieroglyphic Texts, Reliefs, and Paintings. II. Theban Temples*. Oxford: Clarendon Press.

Ptolemy. 1952. *Almagest*. transl. R. Catesby Taliaferro. Chicago, IL: Encyclopedia Brittanica Inc.

Rawlins, D. and K. Pickering. 2001. Ancient Chronology: The Orientation of the Pyramids. *Nature* 412: 699.

Redshift 5.1. 2005 (CD-ROM). London: Focus Multimedia.

Rescher, Nicholas. 1958. Cosmic Evolution in Anaximander. *Studium Generale* 11: 718–731. [Reprinted (with small modifications) in 1982, *Essays in Philosophical Analysis*, 3–32. Washington, DC: University Press of America (repr. of 1969. Pittsburgh: The University of Pittsburg Press)].

Reuther, Oscar. 1957. *Der Heratempel von Samos*. Berlin: Gebr. Mann.

Riedel, M. 1987. ϛρχη und απειρον. Über das Grundwort Anaximanders. *Archiv für Geschichte der Philosophie* 69: 1–17.

Robinson, James M. 1968. *An Introduction to Early Greek Philosophy*. Boston: Houghton Mifflin Company.

__________. 1971. Anaximander and the Problem of the Earth's Immobility. In John Anton and George L. Kustas, eds., *Essays in Ancient Greek Philosophy*, 111–118. SUNY: Albany.

Rochberg-Halton, Francesca. 1983. Stellar Distances in Early Babylonian Astronomy: A New Perspective on the Hilprecht Text (HS 229). *Journal of Near Eastern Studies* 42: 209–217.

Roller, Duane W. 1989. Columns in Stone: Anaximander's Conception of the World. *L'Antiquité Classique* 58: 185–189.

Romant, Bernard. 1978. *Life in Egypt in Ancient Times*. Genève: Minerva.

Ronan, Colin A. 1971. *Discovering the Universe. A History of Astronomy*. New York: Basic Books, London: Heinemann Educational Books.

Röper, Gottlieb. 1852. Emendationsversuche zu Hippolyti philosophumena (2). *Philologus* 7: 606–637.

Rosenmeyer, Thomas G. 1956. Phaedo 111 c4 ff. *The Classical Quarterly* 6: 193–197.

Rossmann, Fritz, ed. 1986. *Nikolaus Kopernikus. Erster Entwurf seines Weltsystems, sowie eine Auseinandersetzung Johannes Keplers mit Aristotle über die Bewegung der Erde*. Darmstadt: Wissenschaftliche Buchgesellschaft (repr. of 1948, München: Rinn).

Rovelli, Carlo. 2009a. Anaximander's Legacy. Collapse. *Philosophical Research and Development* 5: 50–71.

__________. 2009b. *Sospesa sul nulla. Cos'è la scienza? La grande eredità di Anassimandro di Mileto e il dibattito contemporaneo tra pensiero scientifico e religione*, http://www.cpt.univ-mrs.fr%7Erovelli/anassimandro14.pdf.

__________. 2009c. Anaximandre de Milet ou la naissance de la pensée scientifque. Paris: Dunod.

Runia, David T. 2008. The Sources for Presocratic Philosophy. In Patricia Curd and Daniel W. Graham, eds., *The Oxford Handbook of Presocratic Philosophy*, 27–54. Oxford: Oxford University Press.

Ruska, J. 1970. *Turba Philosophorum. Ein Beitrag zur Geschichte der Alchemie*. Berlin: Springer Verlag (repr. of 1931).

Sachs, Abraham J. 1988. *Astronomical Diaries and Related Texts from Babylonia*, compl. and ed. by Hermann Hunger, vol. I, *Diaries From 652 B.C. to 262 B.C.* Vienna.

Sarton, George. 1959. *A History of Science. Ancient Science through the Golden Age of Greece*. Cambridge, MA: Harvard University Press/London: Oxford University Press.

Sartorius, Maximilian A. 1883. Die Entwicklung der Astronomie bei den Griechen bis Anaxagoras und Empedokles, in besonderem Anschluss and Theophrast. Erste Hälfte. *Zeitschrift für Philosophie und philosophische Kritik* 82: 197–231. Also 1883, Halle: Heinemannsche Buchdruckerei.

Sayili, Aydin. 1953. The "Observation Well". *Ankara Universitesi Dil ve Tarih-Cografya Fakultesi Dergisi* 11: 149–155. Also available on the Internet, http://www.muslimheritage.com/uploads/The_Observation_Wel.pdf.

Schaber, Wilfried. 1982. *Die archaischen Tempel der Artemis von Ephesos.* Waldsassen-Bayern: Stiftland Verlag.

Schäfer, Heinrich. 1928. *Ägyptische Kunst und Heutige Kunst und Weltgebäude der Alten Ägypter, Zwei Aufsätze.* Berlin/Leipzig: Vwerlag von Walter de Gruyter & Co.

Schmitz, Hermann. 1988. *Anaximander und die Anfänge der griechischen Philosophie.* Bonn: Bouvier.

Schneider, Peter. 1996. Neue Funde vom archaischen Apollontempel in Didyma. *Säule und Gebalk: Diskussionen zur archäologischen Bauforschung* 6: 78–83.

Schulz, Regine and Matthias Seidel. 1997. *Ägypten. Die Welt der Pharaonen.* Köln: Könemann.

Seneca. *Naturales Quaestiones.*

Sider, David. 1973. Anaxagoras on the Size of the Sun. *Classical Philology* 86: 128–129.

Simplicius. 1894. Johan-Ludvig Heiberg, ed., *Simplicii In Aristotelis De caelo Commentaria.* Berolini: Reimer.

__________. 2005. Ian Mueller, transl., *On Aristotle's on theHeavens 2.10-14.* Ithaca, NY: Cornell University Press.

Simplicius. *In Aristotelis Physicae Commentaria.*

Sloley, R.W. 1924. Ancient Clepsydrae. *Ancient Egypt* 1924: 43–50.

__________. 1931. Primitive Methods of Measuring Time with Special Reference to Egypt. *Journal of Egyptian Archaeology* 17: 166–178.

Spence, Kate. 2000. Ancient Egyptian Chronology and the Astronomical Orientation of Pyramids. *Nature* 408: 320–324.

__________. 2001. Ancient Chronology: Astronomical Orientation of the Pyramids. *Nature* 412: 699–700.

Staudacher, Willibald. 1968. *Die Trennung von Himmel und Erde. Ein vorgriechischer Schöpfungsmythos bei Hesiod und den Orphikern.* Darmstadt: Wissenschaftliche Buchgesellschaft (reprint of 1942, Inaugural Dissertation Tübingen).

Steele, John M. 2000. Eclipse Prediction in Mesopotamia. *Archive for the History of Exact Sciences* 54: 421–454.

__________. 2008. *A Brief Introduction to Astronomy in the Middle East.* London/San Francisco/Beirut: SAQI.

Steele, John M. and F.Richard Stephenson. 1997. Lunar Eclipse Times Predicted by the Babylonians. *Journal for the History of Astronomy* 28: 119–131.

Stephenson, Bruce, Marvin Bolt and Anna F. Friedman. 2000. *The Universe Unveiled. Instruments and Images through History.* Cambridge: Cambridge University Press.

Stephenson, Francis R. 1997. *Historical Eclipses and Earth's Rotation.* Cambridge/New York: Cambridge University Press.

Stephenson, Francis R. and Louay J. Fatoohi. 1997. Thales' Prediction of a Solar Eclipse. *Journal for the History of Astronomy* 28: 279–282.

Sticker, Bernard, ed. and transl., 1967. *Bau und Bildung des Weltalls, Kosmologische Vorstellungen in Dokumenten aus zwei Jahrtausenden.* Freiburg/Basel/Wien: Verlag Herder.

Stobaeus, Joannes. 1974. Anthologium. In Curtius Wachsmuth, ed., *Libri duo priores qui inscribi solent eclogae physicae et ethicae*, Vol. 1-1. Berolini: Weidmannsche Verlagsbuchhandlung (repr. of Berolini 1884).

Strabo. 1923. *Geographica.* ed. H.L. Jones.

Szabó, Árpád and Erkka Maula. 1982. *Enklima. Untersuchungen zur frühgeschichte der griechischen Astronomie, Geographie und der Sehenentafeln.* Athens: Academy of Athens.

Tannery, Paul. 1887. *Pour l'histoire de la science hellène: de Thalès à Empédocle*. Paris: Félix Alcan (repr. 1930, Paris: Gauthier-Villars et Cie and 1987, New York: Garland Publishing).

Tredennick, Hugh. 1955. *Plato. The Last Days of Socrates. The Apology, Crito, Phaedo*. Harmondsworth: Penguin.

___________. 1897. Pseudonymes antiques. *Revue des études grecques* 10: 127–137.

Tarán, Leonardo. 1965. Parmenides: A Text with Transl., Comm. and Critical Essays. Princeton, NJ: Princeton University Press.

Teichmüller, Gustav. 1874. *Studien zur Geschichte der Begriffe*. Berlin: Weidmann'sche Buchhandlung (repr. Hildesheim: Georg Olms Verlagsbuchhandlung 1966).

Theognis. 1979. In T. Hudson-Williams, ed., *The Elegies of Theognis and Other Elegies in the Theognidean Sylloge*. New York: Arno Press.

Thurston, Hugh. 1994. *Early Astronomy*. New York: Springer Verlag.

Tredennick, Hugh. 1955. *Plato. The Last Days of Socrates. The Apology, Crito, Phaedo*. Harmondsworth: Penguin.

Treue, W., ed. 1986. *Achse, Rad, und Wagen. Fünftausend Jahre Kultur- und Technikgeschichte*. Göttingen: Vandenhoeck und Ruprecht.

Turner, A.J. 1984. *The Time Museum*, Vol. I: *Time Measuring Instruments. Part 2: Water-Clocks, Sand-Glasses, Fire-Clocks*. Rockford: Johnson Press.

Tzuong-Tsieng Moh. *A Note on the Ancient Chinese Astrological Measurements*, http://omni.cc.purdue.edu/~wtv/article.html.

Unger, Georg F. 1884. Die Zeitverhältnisse des Anaxagoras und Empedokles. *Philologus* 4(suppl.): 511–550.

Upgren, Arthur. 2002. *The Turtle and the Stars. Observations of an Earthbound Astronomer*. New York: Owl Books.

Van der Waerden, Bartel L. 1965. *Die Anfänge der Astronomie. Erwachende Wissenschaft II*. Groningen: P. Noordhoff Ltd.

___________. 1969. *Science Awakening I*. Groningen: Wolters Nordhoff.

___________. 1974. *Science Awakening II: The Birth of Astronomy*. Leyden: Noordhoff International Publishing/New York: Oxford University Press.

Van Wolde, Ellen J. 2009. Why the Verb ברא Does Not Mean 'to Create' in Genesis 1.1-2.4a. *Journal for the Study of the Old Testament* 34: 3–23.

Verdenius, Willem J. 1958. Notes on Plato's Phaedo. *Mnemosyne* 11: 192–243.

Vian, Francis. 1981. *Apollonius de Rhodes: "Argonautiques", Tome III, Chant IV*. Paris: Les Belles Lettres.

Vitruvius Pollio, Marcus. 1975. In Laura Cherubini, ed., *De architectone libri X*. Pisa: Giardini.

___________. 1999. *Ten Books on Architecture*. D. Rowland, transl., Cambridge: Cambridge University Press.

Von Bomhard, Anne-Sophie. 1999, *Der ägyptische Kalender. Ein Werk für die Ewigkeit*. London: Periplus.

Von Fritz, Kurt. 1971. *Grundprobleme der Geschichte der antiken Wissenschaft*. Berlin and New York: Walter de Gruyter.

Wasserstein, A. 1955. Thales' Determination of the Diameters of the Sun and Moon. *The Journal of Hellenic Studies* 75: 114–116.

Weeks, Kent R. 2005. *The Treasures of Luxor and the Valley of the Kings*. Italy: White Star Spa.

Wehrli, Fritz. 1969. *Die Schule des Aristoteles. Texte und Kommentar. Heft VII: Herakleides Pontikos*. Basel/Stuttgart: Schwabe & Co. Verlag.

Wells, Ronald A. 1992. The Mythology of Nut and the Birth of Ra. *Studien zur altägyptischen Kultur* 19: 305–321.

___________. 1997. Astronomy in Egypt. In Christopher Walker, ed., *Astronomy Before the Telescope*, 28–40. London: British Museum Press.

Wenskus, Otto. 1990. *Astronomische Zeitangaben von Homer bis Theophrast*. Stuttgart: Franz Steiner Verlag.

West, Martin L. 1971. *Early Greek Philosophy and the Orient*. Oxford: Clarendon Press.

West, Stephanie. 1982. Some Watchers of the Skies. *The Classical Quarterly* 32: 121–123.

White, Stephen A. 2002. Thales and the Stars. In Victor Caston and Daniel W. Graham, eds., *Presocratic Philosophy. Essays in Honour of Alexander Mourelatos*, 3–18. Aldershot: Ashgate.

__________. 2008. Milesian Measures: Time, Space, and Matter. In Patricia Curd and Daniel W. Graham, eds., *The Oxford Handbook of Presocratic Philosophy*, 89–133. Oxford: Oxford University Press.

Wöhrle, Georg. 1993. *Anaximenes aus Milet. Die Fragmente zu seiner Lehre*. Stuttgart: Franz Steiner Verlag.

__________. 1995. Wer entdeckte die Quelle des Mondlichts? *Hermes* 123: 244–247.

Workman, Allen. 1953. La terminologie sculpturale dans la philosophie présocratique. *Actes du XIème Congrès International de Philosophie, Bruxelles 20–26 Août 1953*, Vol. XII: *Histoire de la philosophie: méthodologie Antiquité et Moyen Âge*: 45–50. Amsterdam: North-Holland Publishing Company/Louvain, Éditions E. Nauwelaerts.

Wright, M.R. 1995. *Cosmology in Antiquity*. London and New York: Routledge.

Yalouris, Nicolas. 1972. Das Akroter des Heraions in Olympia. *Mitteilungen des Deutschen archäologischen Instituts (Athenische Abteilung)* 87: 85–98.

Zeller, Edward and Wilhelm A. Nestle. 1920. *Die Philosophie der Griechen in ihrer geschivhtliche Entwicklung. 1. Teil, 2. Hälfte: Allgemeine Einleitung, Vorsokratische Philosophie* (6th edition). Leipzig: Reisland.

Name Index

Index of Classical Authors

('DK' refers to: Hermann Diels and Walther Kranz, *Die Fragmente der Vorsokratiker*. Zürich/ Hildesheim, Weidmann (6th edition), 1954. Under the names of the authors are mainly those texts mentioned that are not in DK. When appropriate these items are added to the list of the Presocratic philosopher in question as well)

MIX
Papier aus verantwortungsvollen Quellen
Paper from responsible sources
FSC® C105338

If you have any concerns about our products,
you can contact us on
ProductSafety@springernature.com

In case Publisher is established outside the EU,
the EU authorized representative is:
Springer Nature Customer Service Center GmbH
Europaplatz 3, 69115 Heidelberg, Germany

Printed by Libri Plureos GmbH
in Hamburg, Germany